Masterclass in Neuroendocrinology is published in collaboration with the International Neuroendocrine Federation (INF) https://www.inf-neuroendocrinology.org/. This series aims to illustrate and promote the use of the latest and most impactful technologies in basic and clinical research in the exciting field of neuroendocrinology. Neuroendocrinology is defined as the study of the control of endocrine function by the brain and the actions of hormones on the brain.

The series is made up of volumes that highlight different topics in neuroendocrinology that are of current interest and encompass the study of both normal and abnormal function, and the developmental origins of disease. They include the study of neural networks in the brain that regulate and form neuroendocrine systems, and also include the study of behaviours and mental states that are influenced or regulated by hormones. They also necessarily include the study of peripheral physiological systems that are regulated by neuroendocrine mechanisms. Some recent examples include: Developmental Neuroendocrinology, Neuroendocrine Clocks and Calendars, and Neuroanatomy of Neuroendocrine Systems.

Each book

- is edited by leading experts in the field
- is written by a team of internationally respected researchers
- is focused on an area of neuroendocrinology research
- demonstrates the present state of knowledge and understanding in the field of neuroendocrinology and highlights the translation of such knowledge to improve human health and well-being
- includes assessments of different experimental approaches both *in vivo* and *in vitro*, and of how the resulting data are interpreted.

The Series is aimed at a wide audience, as neuroendocrinology integrates neuroscience and endocrinology. It is intended for established scientists, clinicians and early-career researchers.

Founding Series Co-Editors: William E. Armstrong and John A. Russell

The series is indexed in Scopus.

Paula J. Brunton • David R. Grattan
Editors

Neuroendocrine Regulation of Mammalian Pregnancy and Lactation

Springer

Editors
Paula J. Brunton
Centre for Discovery Brain Sciences
University of Edinburgh
Edinburgh, UK

David R. Grattan
Centre for Neuroendocrinology, Department of Anatomy
University of Otago
Dunedin, New Zealand

ISSN 2662-2068 ISSN 2662-2076 (electronic)
Masterclass in Neuroendocrinology
ISBN 978-3-031-51137-0 ISBN 978-3-031-51138-7 (eBook)
https://doi.org/10.1007/978-3-031-51138-7

This Springer imprint is published by the registered company Springer Nature Switzerland AG
The registered company address is: Gewerbestrasse 11, 6330 Cham, Switzerland

Paper in this product is recyclable

Series Preface

This series began as a joint venture between the International Neuroendocrine Federation and Wiley-Blackwell, and now is continuing with Springer Nature as publisher for the federation. The broad aim of the series is to provide established researchers, trainees, and students with authoritative, up-to-date accounts of the present state of knowledge, and prospects for the future, across a range of topics in the flourishing field of neuroendocrinology. The series is aimed at a wide audience as neuroendocrinology integrates the disciplines of neuroscience and endocrinology. We define neuroendocrinology as the study of how endocrine function is controlled by the brain and the actions of hormones on the brain. It encompasses the study of normal and abnormal function, and the developmental origins of disease. It includes investigation of the neural networks in the brain that regulate and form neuroendocrine systems, and also the study of behaviours and mental states that are influenced or regulated by hormones. In addition, neuroendocrinology encompasses the understanding and study of peripheral physiological systems that are regulated by neuroendocrine mechanisms. While neuroendocrinology embraces many issues of concern to human health and well-being, research in reductionist animal models is usually required to fully understand these issues.

Contemporary research in neuroendocrinology involves a wide range of techniques and technologies, from subcellular and systems to the whole-organism level. A particular aim of the series is to provide expert advice and discussion about experimental or technical protocols in neuroendocrinology research, and to further advance the field by giving information and advice about novel techniques, technologies, and interdisciplinary approaches.

To achieve our aims, each book focuses on a particular theme in neuroendocrinology. For each book we recruit editors, who are leaders in their field, to engage an international team of experts to contribute chapters in their individual areas of expertise. The mission of each contributor is to provide an update of current knowledge and recent discoveries, and to discuss new approaches, "gold-standard" protocols, translational possibilities, and future prospects. Authors are asked to write for a broad audience, to use references selectively, and to consider the use of video clips and explanatory text boxes; each chapter is peer-reviewed and has a glossary.

The masterclass series is open-ended; books in the series published to date are:

- *Neurophysiology of Neuroendocrine Neurons* (2014, ed. WE Armstrong & JG Tasker)
- *Neuroendocrinology of Stress* (2015, ed. JA Russell & MJ Shipston)
- *Molecular Neuroendocrinology: From Genome to Physiology* (2016, ed. D Murphy & H Gainer)
- *Computational Neuroendocrinology* (2016, ed. DJ MacGregor & G Leng)
- *Neuroendocrinology of Appetite* (2016, ed. SL Dickson & JG Mercer)
- *The GnRH Neuron and its Control* (2018, ed. AE Herbison & TM Plant)
- *Model Animals in Neuroendocrinology* (2019, ed. M Ludwig & G Levkowitz)

The first books of the series published by Springer Nature are:

- *Neurosecretion: Secretory Mechanisms* (2020, ed. JR Lemos & G Dayanithi)
- *Developmental Neuroendocrinology* (2020, ed. S Wray & S Blackshaw)
- *Neuroendocrine Clocks and Calendars* (2020, ed. FJP Ebling & HD Piggins)
- *Glial-Neuronal Signaling in Neuroendocrine Systems* (2021, ed. JG Tasker, JS Bains, & JA Chowen)
- *Neuroanatomy of Neuroendocrine Systems* (2022, ed. V Grinevich & A Dobolyi)
- *Neuroendocrine-Immune System Interactions* (2023, ed. JP Konsman & TM Reyes)
- *Cardiovascular Neuroendocrinology* (2023, ed, T Cunningham & G Yosten)

Under development are *Neuroendocrinology of Behavior and Emotions* (ed. H. K. Caldwell & H. E. Albers) and *Evolutionary and Comparative Neuroendocrinology* (ed. V Grinevich & R Oliveira).

Feedback and suggestions are welcome.

International Neuroendocrine Federation—https://www.inf-neuroendocrinology.org/
Series Editors

Edinburgh, UK — Mike Ludwig
Dunedin, New Zealand — Rebecca Campbell

Volume Preface

Pregnancy and lactation are remarkable physiological states where complex interactions between the nervous and endocrine systems lead to dramatic changes in a female's physiology. Several complex and interlinked adaptations emerge, driven primarily through the actions of hormones that serve to ensure a successful pregnancy outcome. The hormonal profile of mammalian pregnancy is distinctly different from the non-pregnant state. Some of these hormones (e.g. estrogen and progesterone) are familiar to female mammals, though the pattern of secretion changes and they are present at much greater levels than in the non-pregnant state, whereas others (e.g. human chorionic gonadotropin and placental lactogen) are seen for the first time during a first pregnancy.

Recently, there have been major advances in our understanding of the molecular mechanisms involved in regulating the neuroendocrine networks that govern mammalian pregnancy, parturition, lactation, and maternal behaviour. New technologies have enabled mapping of the neuronal circuits that control these maternal functions, imaging of their activity, and functional manipulation using genetic tools to establish how they operate. These adaptations are essential for pregnancy maintenance, optimal fetal growth and development, timely parturition, and the nourishment and care of the offspring postpartum. The neuroendocrine system plays a pivotal role in exquisitely coordinating these changes, and this is the focus of this book.

We begin with maternal recognition of pregnancy and the role of the placenta as an endocrine organ. In order for the hormone-induced changes in maternal physiology to be initiated, the blastocyst must signal its presence to the mother, in a process termed the maternal recognition of pregnancy (Chap. 1). Maintaining progesterone secretion by the corpus luteum is crucial to sustain pregnancy with species differences in the hormones providing luteotrophic support until the placenta develops. As pregnancy advances, the placenta becomes a prominent source of peptide and steroid hormones, which as well as supporting the pregnancy also act to prime the maternal brain in preparation for motherhood (Chap. 2).

Next follows a description of the adaptations in maternal neuroendocrine systems that are necessary for normal pregnancy, parturition, lactation, and maternal behaviour. There is a physiological shift in the brain regulatory networks controlling appetite and metabolism (Chap. 3) geared towards promoting increased food intake and energy storage. Increased energy intake in pregnancy not only supplies the growing uterus and placenta(e) but also supports the increased metabolic demands of the mother and her developing fetus(es). In addition, the accumulation of surplus energy stores as fat facilitates milk production during lactation. Such adaptive changes, where the body is not responding to immediate requirements but rather is changing in preparation for future demands, are ideally suited to neuroendocrine control mechanisms. Hormonal signals can be used to predict the changes necessary in maternal physiology that will be necessary later to ensure a successful outcome. The hypothalamic–pituitary–adrenal (HPA) axis also undergoes adaptations to manage the increased metabolic demands of pregnancy and lactation and maternal stress responses are restrained to minimise exposure of the offspring to excessive levels of potentially damaging glucocorticoids (Chap. 4).

The transition from pregnancy to lactation and motherhood is another critical step facilitated by neuroendocrine mechanisms. The culmination of pregnancy and initiation of labour differs between species but in most mammals is triggered by a change in the balance of hormones produced by the mother, placenta, and fetus, in particular corticotropin-releasing hormone and glucocorticoids (Chap. 5). In late pregnancy, adaptations in the control of the neurohypophysial oxytocin system allow large stores of oxytocin to accumulate in the posterior pituitary gland to meet the demands of parturition and lactation. An increase in circulating estrogen sensitizes the uterus and mammary gland to oxytocin at the end of pregnancy. Hence, once labour is initiated, oxytocin augments myometrial contractions, promoting fetal expulsion (Chaps. 5 and 6), while postpartum, oxytocin plays an essential role in the milk ejection reflex in response to infant suckling (Chap. 6). Simultaneously, the neuroendocrine control of prolactin secretion shifts to permit a state of hyperprolactinemia during lactation, stimulating milk production to feed the offspring (Chap. 7). Following birth, new mothers display a dramatic change in behaviour. They care for their young, feed them, and protect them from environmental threats. The expression of these essential components of maternal behaviour is a consequence of hormonal priming of the neural circuits in late pregnancy in preparation for motherhood (Chap. 8).

Finally, the implications of these central adaptations for cognitive function and postpartum mood disorders are considered. Pregnancy effects on cognition are well recognised, with colloquial terms such as "mommy brain" or "pregnancy brain" commonly used in many societies. These terms often have a negative connotation, with mothers associating pregnancy with absent-mindedness or forgetfulness, but research actually shows improved performance in many cognitive tasks in pregnancy. The changes are perhaps better considered as natural, adaptive changes to the maternal brain that are prioritising a new infant-directed focus, and are necessary for expression of species-appropriate maternal behaviours (Chap. 9). Adaptations in maternal brain function in pregnancy and lactation, however, are evidently not without cost. Mental illness during pregnancy and the postpartum period is a growing public health concern and a societal and economic burden. The dramatic changes in neuroendocrine function during pregnancy and in particular the withdrawal of hormones postpartum have been proposed to contribute to perinatal mood disorders (Chap. 10).

The regulatory systems that adapt to support pregnancy covered in this volume are not an exhaustive list. Indeed, there are multiple other systems where neuroendocrine mechanisms are also known to play a critical role (Napso et al. 2018), such as altered cardiovascular control and osmoregulation which supports increased blood flow to the uterus and placenta ensuring a sufficient supply of oxygen and nutrients (Brunton et al. 2008; Russell and Brunton 2014). There is also evidence that pregnancy hormones influence adaptations in the regulation of other systems, including the sleep–wake cycle (Harrington et al. 2022; Irvine et al. 2023), thermoregulation (Grattan and Ladyman 2020), respiration (Bayliss and Millhorn 1992; Lomauro et al. 2019), the skeletal system (Winter et al. 2020), immune system (Motomura et al. 2023), and the gastrointestinal system and microbiome (Astbury et al. 2015; Yeo et al. 2022). Perhaps, once more research enables a fuller understanding of the neuroendocrine mechanisms involved in each of these processes, another volume will be justified!

This book is written for students, early career researchers, educators, and established researchers—anyone who is interested in understanding how the hormonal changes of pregnancy impact on the function of the maternal brain. The contributing authors are internationally recognised experts in the field. Together, the chapters in this book provide context for understanding the neuroendocrine adaptations in mammalian pregnancy and lactation. Research examples utilising both classic and cutting-edge approaches are employed to illustrate recent advances in the field, and areas where further research is needed are highlighted.

Understanding the neuroendocrine adaptations that occur during pregnancy and lactation is important. This is one of the most profound physiological challenges a women will face in her lifetime, and despite advances in medicine, there remain unacceptably high levels of complications. Comprehensive knowledge of the mechanisms that ensure optimal health of both the mother and her offspring can provide insights into potential therapeutic targets for pregnancy-related complications and disorders involving neuroendocrine dysregulation. As research in this field continues to advance, deeper insights into the complexities of these adaptations will pave the way for improved maternal and neonatal care.

Edinburgh, United Kingdom Paula J. Brunton
Dunedin, New Zealand David R. Grattan

References

Astbury S, Mostyn A, Symonds ME, Bell RC (2015) Nutrient availability, the microbiome, and intestinal transport during pregnancy. Appl Physiol Nutr Metab 40:1100–1106

Bayliss DA, Millhorn DE (1992) Central neural mechanisms of progesterone action: application to the respiratory system. J Appl Physiol 73:393–404

Brunton PJ, Arunachalam S, Russell JA (2008) Control of neurohypophysial hormone secretion, blood osmolality and volume in pregnancy. J Physiol Pharmacol 59(Suppl 8):27–45

Grattan DR, Ladyman SR (2020) Neurophysiological and cognitive changes in pregnancy. Handb Clin Neurol 171:25–55

Harrington YA, Parisi JM, Duan D, Rojo-Wissar DM, Holingue C, Spira AP (2022) Sex hormones, sleep, and memory: interrelationships across the adult female lifespan. Front Aging Neurosci 14:800278

Irvine A, Gaffney MI, Haughee EK, Horton MA, Morris HC, Harris KC, Corbin JE, Merrill C, Perlis ML Been LE (2023) Elevated estradiol during a hormone simulated pseudopregnancy decreases sleep and increases hypothalamic activation in female Syrian hamsters. J Neuroendocrinol e13278

Lomauro A, Aliverti A, Frykholm P, Alberico D, Persico N, Boschetti G, Debellis M, Briganti F, Nosotti M, Righi I (1985) Adaptation of lung, chest wall, and respiratory muscles during pregnancy: preparing for birth. J Appl Physiol 127:1640–1650

Motomura K, Miller D, Galaz J, Liu TN, Romero R, Gomez-Lopez N (2023) The effects of progesterone on immune cellular function at the maternal-fetal interface and in maternal circulation. J Steroid Biochem Mol Biol 229:106254

Napso T, Yong HEJ, Lopez-Tello J, Sferruzzi-Perri AN (2018) The role of placental hormones in mediating maternal adaptations to support pregnancy and lactation. Front Physiol 9:1091

Russell JA, Brunton PJ (2014) Maternal brain adaptations in pregnancy. In: Plant TM, Zeleznik AJ (eds) Knobil and Neill's physiology of reproduction (4th edn). Elsevier, San Diego, CA

Winter EM, Ireland A, Butterfield NC, Haffner-Luntzer M, Horcajada MN, Veldhuis-Vlug AG, Oei L, Colaianni G, Bonnet N (2020) Pregnancy and lactation, a challenge for the skeleton. Endocr Connect 9:R143–R157

Yeo E, Brubaker PL, Sloboda DM (2022) The intestine and the microbiota in maternal glucose homeostasis during pregnancy. J Endocrinol 253:R1–R19

Contents

About the Editors

Paula J. Brunton is currently Senior Lecturer at the Centre for Discovery Brain Sciences, University of Edinburgh, UK. She received her bachelor's degree in Physiology in 1998 and her PhD in Neuroendocrinology in 2002, both from the University of Edinburgh. Her expertise lies in the area of stress neurobiology, neuroendocrinology, and behaviour. Her key research themes are focused on understanding the impact of maternal stress exposure during pregnancy on the mother, the pregnancy, her offspring, and subsequent generations, with a particular emphasis on unearthing the underlying central mechanisms involved and how the effects can be prevented or reversed. Her research has revealed the neuroendocrine mechanisms underlying altered stress responsivity in pregnancy and has also demonstrated that maternal exposure to stress during pregnancy is linked with adverse consequences for the offspring, some of which persist to the next generation. She has published over 60 peer-reviewed research articles, invited reviews, and book chapters, in the fields of neuroendocrinology and neuroscience. She has been Senior Editor for the *Journal of Neuroendocrinology* since 2017 and was elected to the Board of Trustees of the British Society for Neuroendocrinology in 2020. She organised the 4th International Parental Brain Conference in 2010 and continues to serve on the scientific advisory committee for this meeting. She is also actively involved in delivering taught undergraduate degree programmes in Biomedical Sciences at both the University of Edinburgh and Zhejiang University, China, on topics within the fields of endocrinology and neuroendocrinology.

David R. Grattan is Professor in the Department of Anatomy and was Director of the Centre for Neuroendocrinology at the University of Otago (2018–2023). He obtained a PhD in Physiology in 1991 from Victoria University of Wellington (New Zealand) and then held a post-doctoral fellowship at the University of Maryland School of Medicine (Baltimore, MD, USA), before returning to New Zealand to take up a Lecturer position in the Department of Anatomy at the University of Otago in 1995. He was promoted to full Professor in 2009, and served as the Head of this Department from 2011 to 2014. He has published over 150 research articles, invited reviews, and book chapters, in the fields of neuroendocrinology, endocrinology, and neuroscience, with a particular focus on the pituitary hormone prolactin and its role in the maternal brain during pregnancy and lactation. From 2009 to 2014, he was the

Editor-in-Chief of the *Journal of Neuroendocrinology*, and is currently Associate Editor for *Endocrinology*. He also chaired the organising committee for the 8th International Congress of Neuroendocrinology in Sydney, Australia, 2014. Dave has a strong interest in promoting the field of neuroendocrinology, the fascinating nexus between endocrinology and neuroscience.

Maternal Recognition of Pregnancy 1

David R. Grattan and Sharon R. Ladyman

Abstract

Hormone-induced adaptations occur in many aspects of maternal physiology during gestation. Before these changes can be initiated, the newly formed blastocyst must signal its presence to the mother. This process is known as the maternal recognition of pregnancy and occurs in all mammalian species, but there are important species differences in how this process is achieved. Because progesterone is essential to support uterine function in pregnancy, a critical step is to maintain the function of the corpus luteum, formed from the freshly ovulated follicle, to sustain progesterone secretion. Humans and other primates have an extended luteal phase during their menstrual cycle, and post ovulatory progesterone secretion continues for a period beyond the time of implantation. Hence, these species use a placenta-derived hormone, human chorionic gonadotropin, to act as a luteotrophic hormone. Commonly used laboratory rodent species, however, have a short luteal phase during their oestrous cycle, and progesterone secretion lasts for only a few hours after ovulation. These species use a mating-induced signal to support corpus luteum function to initiate pregnancy. A unique neuroendocrine reflex initiated by vagino-cervical stimulation during mating induces a pattern of twice-daily prolactin surges that provide luteotrophic support for the first part of pregnancy, maintaining progesterone secretion until implantation can occur. After formation of the placenta, placental lactogens take over as the primary luteotrophic hormones. This chapter reviews the initial pro-

D. R. Grattan (✉) · S. R. Ladyman
Centre for Neuroendocrinology, University of Otago, Dunedin, New Zealand

Department of Anatomy, School of Biomedical Sciences, University of Otago, Dunedin, New Zealand
e-mail: dave.grattan@otago.ac.nz

P. J. Brunton, D. R. Grattan (eds.), *Neuroendocrine Regulation of Mammalian Pregnancy and Lactation*, Masterclass in Neuroendocrinology 15,
https://doi.org/10.1007/978-3-031-51138-7_1

cesses whereby the establishment of pregnancy is signalled to the maternal endocrine system, sustaining progesterone secretion and enabling subsequent initiation of hormone-induced adaptive changes in maternal physiology and behaviour. In the interests of limiting the scope of the chapter, we have focused on humans and rodents, as the major biomedical research species. We have not attempted to describe the processes in the blastocyst and uterus leading to implantation, which are clearly important for pregnancy but do not have a significant neuroendocrine involvement. Similarly, we have not covered immune interactions between the conceptus and the mother.

Keywords

Human chorionic gonadotropin · Progesterone · Corpus luteum · Prolactin · Placenta · Placental lactogen · Pseudopregnancy

1.1 Introduction

The hormone-induced adaptations that occur in many aspects of maternal physiology during **gestation** are hallmark characteristics of the neuroendocrine response to pregnancy. The extended secretion of progesterone ("pro-gestation hormone") is essential for the initiation and maintenance of **pregnancy**. Therefore, before any other changes can be initiated, the developing **blastocyst** must signal its presence to the mother and maintain the corpus luteum in the ovary to prevent the loss of progesterone secretion that would normally occur at the end of each female reproductive cycle. This process is known as the "*maternal recognition of pregnancy*" and occurs in all mammalian species (Short 1969; Bazer 2015). There are important species differences, however, in how this process is achieved, due to marked differences in the normal duration of corpus luteum function and in mechanisms of **placentation**. In this chapter, we focus on two major examples: humans, as an example of a species with an extended **luteal phase** during its menstrual cycle; and the commonly used laboratory rodent species, which have a short luteal phase during their **oestrous cycle**, and therefore have evolved quite different mechanisms for the maternal recognition of pregnancy.

1.2 Critical Role of Progesterone for Pregnancy

For the duration of pregnancy, the developing **embryo**/foetus is exclusively dependent on the mother for its growth, nutrition, and protection. Indeed, many of the adaptations to maternal physiology described in this book can be viewed as the foetus subverting its mother's physiology to support its own development. Thus, maternal metabolism and glucose homeostasis are altered to provide nutrients to the foetus (see Chap. 3), and maternal immune function is altered to prevent rejection of the implanting embryo and to change inflammatory responses that might impair foetal development. Uterine and mammary tissues undergo marked changes to prepare for the future requirements of the foetus/**baby**. Cardiovascular and respiratory systems similarly change to optimize delivery of nutrients to the developing foetus. Moreover, both pre- and post-partum behaviour is changed to encourage protection and nurturing of offspring, with nesting behaviour initiated during pregnancy, and complex maternal behaviours such as pup retrieval, nursing and maternal aggression initiated after birth to complement the process of lactation (see Chap. 8). All of these adaptations are driven by the hormonal changes that are associated with pregnancy. But the first and arguably most important hormonal change is the extended secretion of progesterone that starts in the luteal stage of the reproductive cycle and is then maintained for the duration of pregnancy. Progesterone is absolutely required for pregnancy.

The initial source of progesterone during pregnancy is from the corpus luteum that develops from the recently ovulated ovarian follicle, and then in some species, such as humans, the **placenta** of the developing embryo contributes essential progesterone in the latter parts of gestation. Progesterone is a steroid hormone, synthesized from cholesterol typically via a Δ^4-ketosteroid pathway (involving conversion to pregnenolone and then progesterone). It is secreted into the blood, where it circulates predominantly bound to steroid-binding proteins and albumin. It has a relatively short half-life in the body (around 5 min) and is predominantly metabolized in the liver by the enzymes 5α-reductase and 3α-hydroxysteroid dehydrogenase, with major metabolites including allopregnanolone and pregnanolone (Taraborrelli 2015). However, biologically significant metabolism also occurs in target tissues, including the brain, where functionally significant amounts of allopregnanolone can be produced from progesterone (Sundstrom-Poromaa et al. 2020). Interestingly, liver metabolism of progesterone means that oral progesterone absorbed via the hepatic portal vein is relatively ineffective at raising serum progesterone levels, an important consideration in the use of progesterone as a therapeutic or contraceptive compound. This has resulted in the development of alternative formulations (e.g., micronized progesterone, (Fitzpatrick and Good 1999)) or the use of synthetic progestins in oral treatments that might have different biological properties compared with natural progesterone (Pletzer et al. 2023). Progesterone acts via two related isoforms of the progesterone receptor, the full-length PRb and the truncated version, PRa (DeMayo and Lydon 2020), and by membrane receptors (progesterone receptor membrane component, PGRMC1 and 2) and membrane-associated kinases. In addition, its metabolites, such as allopregnanolone, can have important neuronal actions by positive allosteric modulation of $GABA_A$ receptors on neurons.

In the initial stages of pregnancy, progesterone has a critical role in preparing the uterine endometrium for **implantation** of the blastocyst (see Box 1.1). It promotes the transition of the endometrium from the oestradiol-dependent proliferative phase prior to the time of ovulation, in which re-epithelialization, stromal proliferation, and angiogenesis occur, to the secretory phase, in which the endometrium transforms into a form suitable for implantation, driven by increased levels of progesterone from the post-ovulatory corpus luteum. It is also critical for the **decidualization** response, where fibroblastic stromal cells underlying the endometrium undergo localized proliferation and transformation to become endothelioid cells, enabling trophoblastic cells of the blastocyst to invade and establish the early placenta. Early progesterone receptor (PR)-knockout studies showed that classical PR-mediated nuclear signalling in the uterus was necessary for decidualization and implantation (Lydon et al. 1995; Mulac-Jericevic et al. 2000). Once pregnancy is established, progesterone promotes immunotolerance, and stimulates the growth and vascularity of the uterus to support foetal and placental growth. It also suppresses myometrial contractility and helps maintain cervical integrity, promoting a suitable uterine environment for appropriate foetal growth and development. Loss of progesterone at any time during pregnancy will cause termination of the pregnancy, and hence the progesterone receptor antagonist RU486 (mifepristone) has been used as a drug to induce abortion. Mechanisms to maintain progesterone secretion are a common feature of diverse, species-specific processes to enable maternal recognition of pregnancy.

Box 1.1: Miscarriage

Georgeanna Seegar Jones (1912–2005), the director of the Johns Hopkins' Laboratory of Reproductive Physiology in the mid-1900s, first described luteal-phase deficiency as a primary cause of infertility and pregnancy loss due to inadequate endometrium preparation and support (Jones 1949). She was credited with being the first person to use progesterone to treat women with a history of miscarriages (Damewood and Rock 2005; Di Renzo et al. 2020). It is estimated that natural conception in humans is only 30% per cycle (Zinaman et al. 1996), with early pregnancy loss (prior to clinical detection) common due to implantation failure, and up to 50–60% of all conceptions failing to advance beyond 20 weeks of gestation (Norwitz et al. 2001). Miscarriage, or spontaneous loss of clinically established pregnancy, is also common, affecting up to 15% of pregnancies (Devall and Coomarasamy 2020). A proportion of these pregnancy losses is attributed to a non-responsive endometrium caused by a lack of progesterone action (Zhang et al. 2013). The corpus luteum is supported by luteinizing hormone driven by pulsatile secretion of gonadotropin-releasing hormone (GnRH) (Mesen and Young 2015), and there is some evidence that luteal phase deficiency may result from impaired neuroendocrine support of the corpus luteum leading to impaired progesterone production (Soules et al. 1989). Classic experiments in rhesus monkeys demonstrated a key role for ongoing luteotrophic support by pulsatile GnRH and LH throughout the luteal phase (Hutchison and Zeleznik 1984). A remarkable recent study found that around a third of women of European descent express a variant of the progesterone receptor that originates from the Neanderthal genome (Zeberg et al. 2020). This is around ten times more prevalent than other Neanderthal gene variants found in modern humans, suggesting it provides a favourable effect on fertility. The study found that women with this receptor variant produced more progesterone receptors in their endometrial cells, tended to have fewer miscarriages, and more live births (Zeberg et al. 2020). The role of progesterone in supporting early pregnancy is unequivocal (Mesen and Young 2015), but 70 years after Seegar Jones' discovery, it remains debated as to whether supplementation of progesterone is an effective treatment for miscarriage (Devall and Coomarasamy 2020), perhaps because an impaired response to progesterone may also contribute to luteal phase deficiency. More research is required to understand this critical process.

1.2.1 Suppression of Concurrent Pregnancy

One key function that occurs during early pregnancy in most mammals is the suppression of fertility to prevent concurrent pregnancies. While there are a few mammalian species that have evolved a superfetation strategy (sequential fertilization of oocytes from multiple ovarian cycles), including European Badgers, Brown Hares, and American Mink (Roellig et al. 2010; Roellig et al. 2011), in most species there is no further ovulation once pregnancy is initiated. This focuses maternal resources towards supporting the growing foetus and towards preparation for lactation (see Chap. 3). It is highly likely that the high progesterone levels characteristic of early pregnancy will suppress ovulatory activity, as progesterone is known to profoundly inhibit activity of the kisspeptin neurons required for normal follicular development (McQuillan et al. 2019). Interestingly, it seems that waves of follicular activity continue to occur during pregnancy (Kirillova et al. 2021), presumably representing the gonadotropin-independent phases of follicular development. In the presence of high progesterone, however, gonadotropin secretion is low, and so follicles are not selected and stimulated through the gonadotropin-sensitive growth phases required to develop into large antral follicles and subsequent ovulation. In rodents, the mating-induced prolactin surges may also contribute to the suppression of pulsatile LH secretion and consequent fertility. Prolactin is known to have an inhibitory action on LH secretion, acting through direct inhibition of hypothalamic kisspeptin neurons (Brown et al. 2014, 2019; Hackwell et al. 2023).

1.3 Establishment of Human Pregnancy

1.3.1 Human Chorionic Gonadotropin

In humans (and other primates), a spontaneous luteal phase occurs after ovulation, with progesterone secretion from the newly developed corpus luteum peaking approximately 7–8 days after the preovulatory luteinizing hormone (LH) surge and subsequent ovulation. This peak coincides with the 2-day endometrial window when a blastocyst can attach to the endometrial surface. In the absence of fertilization, luteal support from the LH surge is lost and **luteolysis** will occur a few days later, leading to a fall in progesterone and the onset of menstruation. If fertilization has occurred, the blastocyst will implant in this time-window approximately 7–9 days after ovulation. Human chorionic gonadotropin (hCG) is synthesized in the **syncytiotrophoblast** of the implanting blastocyst as early as 6–7 days after fertilization, and will immediately pass into the maternal circulation as soon as the blastocyst embeds in the uterine endothelium. Therefore, hCG is present in the maternal circulation as soon as implantation occurs, and prior to the normal time of luteolysis in the menstrual cycle (Gridelet et al. 2020). hCG binds to LH receptors in the corpus luteum to support luteal function and prevent luteolysis, and therefore provides the key signal enabling maternal recognition that pregnancy has been initiated (see Fig. 1.1). Administration of exogenous hCG will prolong luteal function and

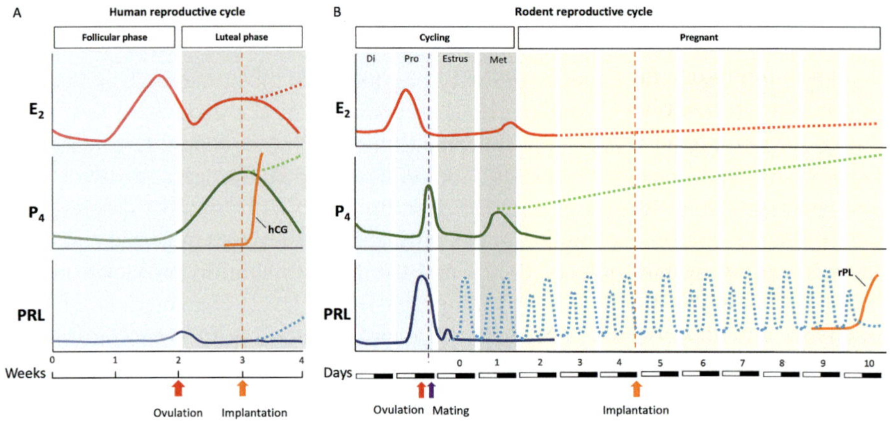

Fig. 1.1 Hormone changes in early pregnancy. Diagrammatic representation of hormonal changes in early pregnancy comparing (**a**) human and (**b**) rodent (modified from Phillipps et al. 2020). Note the similarities in the pattern of oestradiol (E_2; red line), progesterone (P_4; green line), and prolactin (blue line) during the follicular phase (preceding ovulation; blue shaded region; Di: dioestrus, Pro: **proestrus**) and early luteal phase (grey shaded region; Oestrus and Met: **metestrus**). The solid lines represent hormone levels in the reproductive cycle, with the dotted lines representing the changes in secretion that accompany pregnancy. In humans, implantation of the blastocyst leads to rapid production of human chorionic gonadotropin (orange), signalling the presence of the embryo to the maternal system and sustaining progesterone production. In rodents, however, the luteal phase is very short (<2 days), meaning that progesterone secretion would not last until the time of implantation (orange dotted line) on day 4–5 of pregnancy. Hence, mating induces a pattern of twice-daily prolactin surges that provides luteotrophic support to maintain progesterone secretion from the corpus luteum and sustain pregnancy (pale yellow-shaded regions represent the extended luteal phase in the presence of luteotrophic support from prolactin). Typically, rodents show a prolonged period of dioestrus-like vaginal smears during pregnancy, dominated by leukocytes with increasing levels of mucous present. After implantation and formation of the placenta (around days 8–10 gestation), placental lactogen (rPL) secretion becomes the primary luteotrophic support, and results in termination of the prolactin surges by negative feedback to the hypothalamus and pituitary gland

increase progesterone secretion in non-pregnant women. Conversely, passive immunization against hCG in marmosets neutralized chorionic gonadotropin function and prevented rescue of the corpus luteum, resulting in the prevention of pregnancy. Similar data have been achieved with immunization against hCG in women (Talwar 1999). These data suggest that hCG is both necessary and sufficient for the prevention of luteal regression in primates.

hCG was conceptually discovered almost 100 years ago, with the demonstration by Ascheim and Zondek that the blood and urine of pregnant women contained a "gonad-stimulating substance" (Lunenfeld 2004; d'Hauterive et al. 2022). It is a glycoprotein hormone composed of two subunits, closely related to the pituitary gonadotropins. It has a relatively long half-life in blood (24–36 h) and about 20% is excreted in urine. This characteristic of hCG has been capitalized upon to establish simple pregnancy testing in humans (see Box 1.2). Apart from its critical action to support the corpus luteum and stimulate progesterone secretion in early pregnancy, hCG may exert other actions through the LH receptors. In particular, hCG binds to its receptor in the decidua, promoting the secretion of factors such as prolactin and leukaemia inhibitory factor and decreasing secretion of interleukin-6, factors that influence embryo implantation (d'Hauterive et al. 2022). It also supports angiogenesis within the uterine endothelium (Gridelet et al. 2020). Interestingly, LH receptors have also been found in the brain (Ryu et al. 2022), leading to speculation that hCG could influence neuronal activity in the mother during pregnancy. For example, it has been suggested that elevated hCG acting on LH receptors in the brainstem may explain the nausea and vomiting that occurs in some women during normal pregnancy (Cole 2010). However, despite the extremely high levels present in the blood, there is little evidence that hCG crosses the blood-brain barrier and has a significant "neuroendocrine" function in the maternal brain.

Box 1.2: Pregnancy Test

The most common self-administered pregnancy test currently used is a lateral flow test for human chorionic gonadotropin (hCG) in urine. The image shows an example of a positive pregnancy test. hCG is released from the developing conceptus in high amounts very soon after implantation (see also Fig. 1.1), and 20% of hCG is excreted in urine. Thus, hCG can be detected in urine as early as 8 days after conception (even a few days before an expected menstrual period) (d'Hauterive et al. 2022). Because of the convention that human pregnancy is dated from the time of the last menstruation, a positive test at the time of a missed menstrual period will be counted as 4 weeks pregnant (with weeks 1 and 2 actually occurring prior to ovulation!).

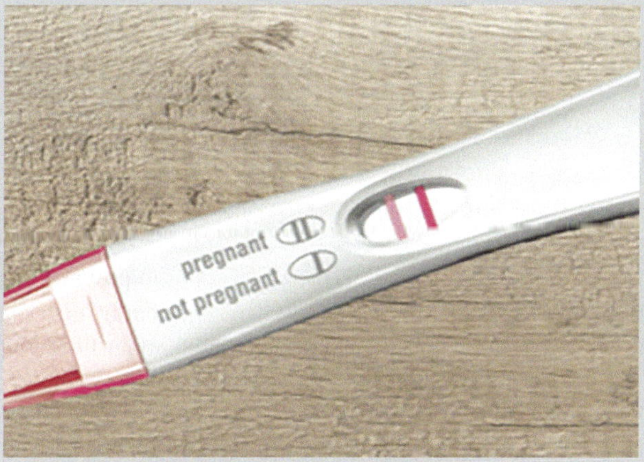

The history of using hCG as a pregnancy test is fascinating, with multiple different bioassays developed. Prior to the 1920s there was no definitive test to determine pregnancy, except waiting for the onset of morning sickness, breast tenderness, weight gain, or even the first movements of the foetus to confirm any external examination that was undertaken. There is recorded

(continued)

Box 1.2 (continued)

evidence that ancient Egyptians had a "pee-based" test whereby the curious woman would urinate on barley and wheat plants, then growth of the plants would indicate a pregnancy, and which plant grew would indicate the sex of the baby. While there is no evidence of the latter being correct, a study to recreate this test has shown that it is about 70% accurate in positively revealing a pregnancy with the high oestrogen levels in the urine thought to underlie the growth in the plants (Ghalioungui et al. 1963). In the 1920s, Berlin-based gynaecologists Selmar Aschheim and Bernhard Zondek demonstrated that urine from pregnant women contained a substance that promoted development and growth of the ovary (Finkel 1931). This study involved twice-daily injections of urine from a pregnant woman into pre-pubertal female mice. This led to the development of the so-called "A-Z test", in which the urine from a woman could be tested to confirm pregnancy as little as 2 weeks after implantation. This test was time-consuming, however, as it needed to be done in a specialized laboratory and required twice-daily injections into at least five mice that were all killed at the end of the test for dissection and analysis of the uterus and ovary. Not long after they developed the "A-Z test", Ascheim and Zondek's research ended abruptly when these men, both Jewish, were forced to flee Nazi Germany. By then, however, American doctor Maurice Friedman had adapted the test in the United States using rabbits, leading to the colloquial phrase "the rabbit died" to indicate a pregnancy (in actual fact though, the rabbit would die whatever the outcome of the test) (Kelley 2010). Subsequently, in the 1930s, British scientist Lancelot Hogben, while working in South Africa, injected a local type of frog, the *Xenopus laevis*, with urine from a pregnant woman and noticed that this rapidly triggered the appearance of a clutch of eggs on the frog's back. The fact that frogs did not need to be killed to get the results and the reduced time to get the results (less than 24 h after urine administration) made the frog a superior animal bioassay, and it became the popular choice for pregnancy testing. *Xenopus laevis* were exported around the world for this purpose. In 1947, it was demonstrated that urine from pregnant women was able to induce the detachment of spermatozoa from the Sertoli cells in the testis of toads, similar to the action of toad gonadotrophins from the pituitary, leading to yet another animal-based bioassay. In this test, the presence or absence of spermatozoa in the toad urine just hours after the injection of urine from the suspected pregnant women was assessed, with the presence of spermatozoa indicating a positive test result (Galli Manini 1947). Remarkably, while these animal-based tests that relied on the presence of hCG in the urine for the positive test began to be used from 1927, the hormone itself was not actually isolated until the 1950s (Kelley 2010). The advent of antibodies to detect hCG and the testing process refined over time to an easy "pee on a stick" protocol, means a laboratory and animal-based bioassay approaches are no longer required. The history of pregnancy tests in the 1900s, however, illustrates how scientific knowledge can be used and applied even if all of the details are not yet fully understood.

The production of a chorionic gonadotropin that is required to sustain progesterone secretion during pregnancy is largely restricted to primates. Horses have an equivalent chorionic gonadotropin, but in many other species with a long luteal phase, such as sheep, cows, and pigs, such a hormone is not produced. Corpus luteum function is controlled by a balance between luteotrophic factors, such as LH from the maternal pituitary gland, and luteolytic factors that promote degradation of the corpus luteum (Bazer 2015). In these species, pituitary luteal support from LH seems to be sufficient to sustain the corpus luteum indefinitely, and termination of the reproductive cycle is achieved through a luteolytic signal (prostaglandin F2α) from the uterus. Following implantation, pregnancy-specific factors in the implantation site (trophoblast interferon) suppress local prostaglandin production and prevent the induction of luteolysis (Arosh et al. 2016). Functionally, this provides that same outcome, maternal recognition of pregnancy and maintenance of luteal progesterone secretion.

1.3.2 Other Placental Hormones

In women, the corpus luteum remains the primary source of progesterone to maintain pregnancy for around the first 6–9 weeks of gestation, after which the placenta becomes the primary source of progesterone for the remainder of the pregnancy (Costa 2016). The placenta is also a primary source of oestrogens in pregnancy, together with the somatomammotropins, a class of protein hormones related to growth hormone (human growth hormone variant; hGH-V) and prolactin (human placental lactogen, hPL; also known as chorionic somatomammotropin). Hence, by the end of the first **trimester**, the human conceptus is largely self-sufficient in its endocrine requirements, and, indeed, is signalling through both steroid and peptide hormones to influence maternal physiology (see Chap. 2). In many other species, the placenta does not produce sufficient levels of steroid hormones, and there remains a dependence on ovarian luteal progesterone to maintain the pregnancy. In those species (e.g., rodents, pigs, cows), removal of either the pituitary gland or ovaries will lead to a loss of progesterone in the blood and termination of the pregnancy.

1.4 Maternal Recognition of Pregnancy in Species with a Short Luteal Phase

1.4.1 Mating-Induced Signals for Maternal Recognition of Pregnancy

The primary biomedical research animals are rodents such as rats and mice, and in many ways, they provide excellent models for the study of pregnancy biology. However, it is important to be aware of key differences and take these into consideration in experimental design. Unlike humans, who experience a luteal phase of the menstrual cycle lasting 7–10 days, rodents have a short oestrous cycle with a particularly truncated luteal phase (metestrus) that only lasts a few hours (Freeman 1994). This means there is insufficient time for implantation to occur and maternal recognition of pregnancy to be established before the normal time of luteolysis and loss of progesterone secretion (see Fig. 1.1). Because of their size and relative ease of collecting blood samples, most work to date has characterized hormone levels in the rat (Freeman 1994). A recent study monitored hormone levels throughout the oestrous cycle in mice (Wall et al. 2023), showing some similarities with the rat and some specific differences, most notably an earlier peak in oestradiol levels. Nevertheless, the very short luteal phase was confirmed in that study. In the absence of a placental signal, rodents have evolved to use mating as a surrogate marker of pregnancy. Mating induces a pattern of twice-daily surges of prolactin secretion from the female pituitary gland (Gunnet and Freeman 1983), and this prolactin is the predominant luteotrophic hormone maintaining progesterone secretion during the first week of gestation (prompting some people to use the alternative name of "luteotropin" or "luteotrophic hormone" for prolactin (White 1949)). From around day 7–8 of gestation, placental lactogen becomes detectable in the maternal blood, with both rats and mice having two separate placental lactogens (PL) produced by the placental trophoblast cells: PL-1, peaking around mid-pregnancy, and PL-2, increasing from mid-pregnancy to peak around day 14, then remaining high until parturition (Voogt et al. 1982; Tonkowicz and Voogt 1983) (see Fig. 4.3 in Chap. 4).

In addition to placental lactogens, the decidualized stromal cells of the uterus also produce prolactin and other factors belonging to the prolactin family of hormones (e.g., Prl8a2, and Prlr3c1), and these are thought to contribute to supporting the implanting blastocyst (Prigent-Tessier et al. 1999). For example, decidual Prl8a2 is induced by Notch signalling in the uterus, and this appears to contribute to sustaining corpus luteum function in mice through the mid-pregnancy period when luteotrophic support is transitioning from pituitary prolactin to placental lactogen (Bao et al. 2021). This prolactin variant, however, does not act through the prolactin receptor (Rasmussen et al. 1996), and so its luteotrophic action must be independent of that of pituitary prolactin. Decidual prolactins are insufficient to sustain luteal function in the absence of pituitary prolactin, and thus, may have other functions. This is supported by the observation that progesterone replacement does not completely prevent pregnancy loss in prolactin- or prolactin receptor-knockout mice (Binart et al. 2000; Vomachka et al. 2000), suggesting that prolactin may act to support embryo survival distinct from its luteotrophic action. It is likely that decidual prolactin or prolactin-like hormones may mediate this function. For example, decidual prolactin has been shown to suppress the expression of several factors that contribute to foetal loss, including inflammatory cytokines (Bao et al. 2007; Alam et al. 2015). In women, an absence of decidual prolactin has been associated with first trimester miscarriage and with upregulation of these inflammatory cytokines (Garzia et al. 2013).

Interestingly, the rise in placental lactogen from mid-pregnancy in rodents leads to the inhibition of the mating-induced prolactin surges, as these placental lactogens bind to the same receptors as pituitary prolactin and therefore activate the dopamine-mediated negative feedback system to block prolactin secretion (see Chap. 7). Thus, placental lactogen production effectively establishes a condition of hyperprolactinemia during pregnancy, bypassing the maternal regulatory mechanisms that would normally suppress prolactin secretion.

1.4.2 A "Unique Neuroendocrine Response"

The pattern of prolactin secretion following mating has been best characterized in rats, and was aptly described as a "unique neuroendocrine response" (Gunnet and Freeman 1983). What makes this particularly unusual is that it is induced by a single stimulus, mating, but then manifests as a **circadian** pattern of prolactin secretion characterized by twice-daily surges of prolactin that repeat over many days, apparently invoking some sort of memory of the mating stimulus. There have been multiple studies evaluating what it is about the mating stimulus that establishes and sustains such a long-lasting pattern of secretion. It is initiated by the vagino-cervical stimulation that occurs during coitus, and can be replicated artificially by mechanical or electrical stimulation of the cervix. Thus, it does not appear to require specific behavioural or pheromonal interactions with the males. Nevertheless, the quality of the signal can be altered by specific female sexual behaviours, and is most robust if the female can control the pacing of interactions with males and, therefore, the timing of the stimulation that is received (Frye and Erskine 1990; Erskine 1995).

Female reproductive behaviour will normally only occur in a specific period of behavioural oestrus driven by the presence of sufficiently high levels of oestrogen and progesterone. These ovarian steroids are produced by the developing preovulatory ovarian follicle and peak at a time associated with ovulation. Thus, females will only engage in mating at a time that coincides with ovulation, maximizing the opportunity for fertile pregnancy to result from each mating. The mating-induced luteotrophic response is therefore also optimally timed to occur when a post ovulatory corpus luteum is forming, providing maternal recognition of pregnancy by rescuing and maintaining ovarian progesterone production. The same ovarian steroid hormone signal that drives ovulation and reproductive behaviour primes the neuroendocrine circuits to establish the mating-induced prolactin surges (Gunnet and Freeman 1983). Indeed, even in the absence of mating, elevated levels of oestrogen and progesterone can induce prolactin surges and cause **pseudopregnancy**-like changes in the oestrous cycle (de Greef and Zeilmaker 1979). Similarly, stress may induce prolactin secretion, and under the appropriate conditions, can result in a self-sustaining pseudopregnancy. These observations provide some insight into the mechanisms driving the prolactin surges (see below).

Box 1.3: Pregnancy Block, or the "Bruce Effect"

Because maternal recognition of pregnancy in rodents is regulated by a neuroendocrine pathway, it is possible for exogenous factors such as pheromones or stress to impact on the successful maintenance of the pregnancy. The "Bruce effect", named after the person who first described it (Bruce 1959), is a condition of pheromone-induced pregnancy failure when a pregnant female mouse is exposed to an unfamiliar male mouse (usually of a different strain to the male that mated with her to induce the pregnancy). The key mechanism is a neuroendocrine pathway, mediated by a pheromone-induced increase in dopamine production from the **tuberoinfundibular dopamine (TIDA) neurons** (Rosser et al. 1989), blocking the prolactin surges of early pregnancy and thereby removing luteotrophic support. As mice are completely dependent on luteal progesterone to support the pregnancy, this results in abortion. The pheromone involved is exocrine gland-secreting peptide 1 (ESP1), which is produced in tear fluid of adult male mice and is detected by the specific receptor V2Rp5 in the vomeronasal sensory neurons of the female. Pregnancy is terminated due to a failure of luteal function, and this can be mimicked by exposure to exogenous ESP1 (Hattori et al. 2017). From an evolutionary perspective, it is assumed that the Bruce effect may have a reproductive advantage of avoiding infanticide by unmated males, with the new dominant male contributing to parenting of his own pups (Elwood and Kennedy 1990).

The fact that vagino-cervical stimulation in rats caused oestrous cycles to cease was first described over 80 years ago by Greep and Hisaw, who describe how this induced a period of pseudopregnancy characterized by persistent dioestrus-like vaginal smears (Greep and Hisaw 1938). With the development of radioimmunoassays enabling measurement of pituitary hormones in blood in the 1970s, Smith, Freeman, and colleagues provided comprehensive descriptions of the patterns of hormones associated with these changes, particularly identifying the role of elevated prolactin in sustaining progesterone secretion after mating (Smith et al. 1975). They described the twice-daily surges of prolactin, referring to them as a diurnal surge (peaking in the afternoon towards the end of the light period) and a nocturnal surge (peaking late in the dark period) (Gunnet and Freeman 1983). Female sex behaviour establishes sufficient vagino-cervical stimulation to maximize the signal initiating these surges, with a threshold level of stimulation establishing an "all or none" neuroendocrine response, the magnitude of which is not altered by additional stimulation (Erskine 1995). Prolactin surges are more likely to be generated by multiple intromissions (>5), particularly if those are spaced apart. If allowed to control interactions with the male, the female will establish these conditions to optimize the luteotrophic response. The **lordosis** response, part of normal female reproductive behaviour, is required to enable intromissions, but is insufficient by itself to stimulate the prolactin surges (Erskine 1995).

Interestingly, because there is no requirement for a placental signal, pregnancy-like hormonal changes can occur in the absence of fertilization, such as after artificial stimulation or mating with a vasectomized or infertile male. This condition, known as *pseudopregnancy*, is characterized in rats by identical patterns of twice daily prolactin surges from the female pituitary gland and sustained progesterone secretion from the corpus luteum, as normally seen in pregnancy (Gunnet and Freeman 1983). In the absence of placental lactogens from a fertile pregnancy, this pattern of secretion will persist for even longer than that seen in pregnancy (where the initiation of placental lactogen secretion terminates the maternal prolactin surges by negative feedback). The eventual termination appears to be driven by luteolytic factors from the uterus that inhibit progesterone secretion and block prolactin release from the pituitary gland (Freeman 1979; Gorospe et al. 1981). While not as well-characterized as in rats, mouse pregnancy is also dependent on mating-induced prolactin. Some studies have observed twice-daily surges of prolactin, as seen in rats (Larsen and Grattan 2010), but others report only a single daily surge during the light phase (Yang et al. 2009). These differences may be due to strain or housing conditions, or differential timing of the prolactin surges.

1.4.3 Neuroendocrine Regulation of the Mating-Induced Prolactin Surges

The circadian pattern of prolactin secretion induced by mating involves interactions between the hypothalamus and pituitary lactotrophs. As mentioned above, the same pattern can be triggered by exogenous steroid hormones (de Greef and Zeilmaker 1979), but will also occur in ovariectomized rats, demonstrating that the ovarian steroids are not essential for expression of these surges. Similarly, this pattern of prolactin surges can be induced by central injections of prolactin itself (Helena et al. 2009), or a prolactin-releasing factor such as oxytocin (Kennett and McKee 2012). These observations facilitated a mathematical model that predicted such a pattern could essentially be generated by the normal negative feedback pathways that control prolactin secretion (Bertram et al. 2006; Egli et al. 2006). In this model, a rise in prolactin induced by mating (or any other stimulus) will, after a delay, trigger a compensatory activation of the tuberoinfundibular dopamine (TIDA) neurons that inhibit prolactin secretion (see Chap. 7). Elevated dopamine will subsequently suppress prolactin secretion, but then the falling prolactin will lead to a reduction in TIDA neuronal activity and a fall in dopamine release, thereby allowing a rebound rise in prolactin secretion. The self-sustaining rise and fall in prolactin secretion seems to be facilitated and entrained to the circadian cycle by the involvement of an endogenous rhythm that stimulates prolactin secretion at particular times of day (Arey et al. 1989). While normally masked by the inhibitory tone coming from the TIDA neurons, when unmasked by a reduction in TIDA neuronal activity, this will promote prolactin secretion at particular phases of the circadian cycle.

In the case of mating, it is proposed that a mating-induced release of oxytocin acts as a prolactin-releasing factor at the pituitary gland to provide the initial stimulus to prolactin secretion (Kennett and McKee 2012). This is because lactotrophs express the oxytocin receptor (Breton et al. 1995), and administration of an oxytocin receptor antagonist at the time of cervical stimulation will block the induction of the subsequent prolactin surges (McKee et al. 2007). The circadian element involved in controlling the mating-induced prolactin surges appears to involve vasoactive intestinal peptide (VIP) neurons in the suprachiasmatic nucleus, which innervate the TIDA neurons and inhibit their activity during the dark phase to entrain the timing of the reduction in dopamine and the nocturnal surge of prolactin (Poletini et al. 2010). Noradrenergic inputs to the ventromedial hypothalamus also appear to be critical in transmitting this mating response (Northrop et al. 2006; Northrop and Erskine 2008), perhaps acting on the oxytocin receptor-containing neurons in this region, as pharmacological blockade of alpha1 adrenergic receptors in this region will block the induction of prolactin surges. Opioid-induced suppression of dopamine may also be required, as blockade of kappa or mu opioid receptors will also block the prolactin surges (Andrews and Grattan 2002, 2003). After initiation, reciprocal feedback between prolactin and dopamine is then thought to sustain the prolactin surges over multiple days until it is termination by chronic, placental lactogen-induced activation of the TIDA neurons.

1.4.4 Other Rapid Responses Associated with the Maternal Recognition of Pregnancy

A number of additional physiological changes occur rapidly in early pregnancy and may contribute to (or be dependent on) the maternal recognition of pregnancy. These may be induced independently by mating, or changed in response to the mating-induced changes in hormones.

1. Oxytocin release within the hypothalamic paraventricular nucleus is elevated during mating, and even more so if the oestrous female is able to control sexual interactions (i.e., under **paced mating** conditions) (Nyuyki et al. 2011). This might be part of the pathway whereby mating induces the prolactin surges (Northrop and Erskine 2008), as it is also known that these surges are optimally induced under paced mating conditions (Erskine 1995).
2. Within 14 h of mating in mice, core body temperature in females rises, arguably one of the earliest detectable signals that pregnancy is underway (Smarr et al. 2016) (see Fig. 1.2). This elevation results in daytime temperatures being significantly higher, manifesting as a flattening of the normal circadian variation between the dark (active) and light (inactive) periods. It seems likely that this reflects the thermogenic actions of progesterone, which will be rising in response to the rescue of the corpus luteum. Changes in core body temperature can be used to stage the different phases of the menstrual cycle, with significantly elevated temperatures observed in the luteal phase.
3. There is also a reduction in locomotor activity seen very early in pregnancy, although this effect is reportedly much less consistent than the change in core

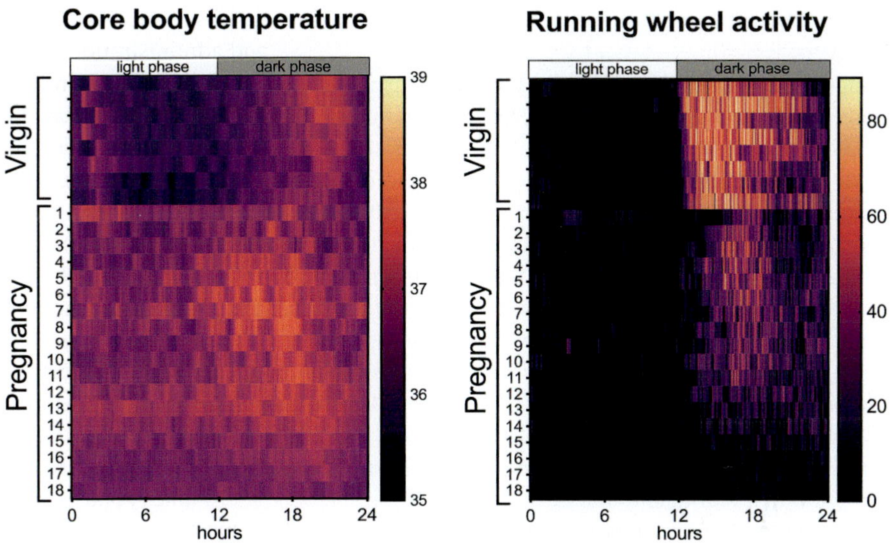

Fig. 1.2 Body temperature and activity in early pregnancy. Heat maps showing rapid changes in core body temperature (left) and running wheel activity (right) in the mouse in early pregnancy, starting immediately after mating (our unpublished data, based on the observations from Kreigsfeld's group (Smarr et al. 2016)). Data are generated from conscious, freely behaving animals using remote recording from implanted telemetry devices. The data are pooled from ten animals (core temperature) and eight animals (running wheel) recorded continuously before and during pregnancy. The heat maps show the normal circadian variation in core body temperature and running wheel activity seen before pregnancy, with temperature and activity lower in the light phase (when mice sleep), and higher in the dark phase (when mice are awake). These variables are rapidly altered on day 1 of pregnancy in the mouse (pregnancy day is given on the y axis). *Left panel*: Core body temperature (colour scale, as indicated on right of each figure, represents °C) immediately rises, and higher temperatures are then sustained, particularly in the light phase, such that the circadian pattern is largely lost. Temperatures remain high in pregnancy, until falling slightly in the last few days of pregnancy. *Right panel*: Physical activity (colour scale represents metres travelled on the running wheel per 5 min) is also rapidly reduced on day 1 of pregnancy, and this reduced activity is maintained throughout the pregnancy. These changes occur prior to implantation, and so must be driven by hormonal changes associated with mating (evidence of the maternal recognition of pregnancy)

body temperature (Smarr et al. 2016). Voluntary running wheel activity is also dramatically reduced very early in pregnancy (Fig. 1.2), an effect that seems to be driven by elevated prolactin, acting in the preoptic region of the brain (Ladyman et al. 2021).

All of these changes occur very early in pregnancy, prior to the implantation of the embryo, and are therefore driven by the hormonal changes associate with mating. Thus, mating-related signals provide a surrogate indicator for the maternal recognition that pregnancy is possible.

1.5 Conclusion

Maternal recognition of pregnancy is necessary to initiate the appropriate adaptations to maintain progesterone secretion. In humans and other animals with a long luteal phase, this typically involves the early production of a chorionic gonadotropin from the implanted conceptus to signal to the ovary and maintain corpus luteum function. In animals such as rodents, with a spontaneous luteal phase that is too short to last until implantation and initial formation of the placenta, the mating stimulus provides a surrogate for the maternal recognition of pregnancy and initiates a number of hormonal changes. The most prominent of these is the mating-induced release of prolactin, which serves the same purpose as chorionic gonadotropin, rescuing the corpus luteum and maintaining progesterone secretion. This initial recognition of pregnancy is followed by further elaboration of placental endocrine functions, and the characteristic of the feto-placental takeover of the maternal endocrine system (in whole or in part) to enable patterns of hormone secretion that facilitate foetal development. These placental hormones contribute to the wide range of pregnancy adaptations described in other chapters of this book.

Perspectives

The "maternal recognition of pregnancy" is a critical process, and in the context of topics covered in this book, it initiates the neuroendocrine adaptations in the mother during pregnancy. However, there is large species variation in the mechanism of maternal recognition of pregnancy, making investigation complicated. Critical health problems, such as miscarriage, need better understanding, but it is important to determine suitable biomedical models for investigation. Future research must focus on what mechanisms are unique to the model in question, and that are common across mammals and therefore might directly inform translational outcomes. For research into the neuroendocrine adaptations to pregnancy, the specific mechanism underlying initial maternal recognition of pregnancy in the different experimental models might not be important, because the consequent hormone-induced changes in function might be similar. It is important that researchers understand these processes, however, so that appropriate conclusions and translational inferences can be made.

Key References (See the Reference List for Citation Details)

Bazer (2015). Provides an overview of the origins of the concept of "maternal recognition of pregnancy", and a thorough description of the range of species differences in this process.

Bertram et al. (2006). Provides the theoretical understanding of how a pattern of twice-daily prolactin surges can be generated from a single stimulus (mating). This formed the basis of much experimental work to elucidate the pathways involved.

Erskine (1995) and Gunnet and Freeman (1983). Comprehensive reviews of the mechanisms controlling the secretion of prolactin following mating. This is an essential luteotrophic signal in rodents to support early pregnancy.

Smith et al. (1975). The first complete characterization of the patterns of hormone secretion associated with the maternal recognition of pregnancy in rodents.

References

Alam SM, Konno T, Soares MJ (2015) Identification of target genes for a prolactin family paralog in mouse decidua. Reproduction 149(6):625–632

Andrews ZB, Grattan DR (2002) Opioid control of prolactin secretion in late pregnant rats is mediated by tuberoinfundibular dopamine neurons. Neurosci Lett 328(1):60–64

Andrews ZB, Grattan DR (2003) Opioid receptor subtypes involved in the regulation of prolactin secretion during pregnancy and lactation. J Neuroendocrinol 15(3):227–236

Arey BJ, Averill RL, Freeman ME (1989) A sex-specific endogenous stimulatory rhythm regulating prolactin secretion. Endocrinology 124(1):119–123

Arosh JA, Banu SK, McCracken JA (2016) Novel concepts on the role of prostaglandins on luteal maintenance and maternal recognition and establishment of pregnancy in ruminants. J Dairy Sci 99(7):5926–5940

Bao L, Tessier C, Prigent-Tessier A, Li F, Buzzio OL, Callegari EA, Horseman ND, Gibori G (2007) Decidual prolactin silences the expression of genes detrimental to pregnancy. Endocrinology 148(5):2326–2334

Bao H, Sun Y, Yang N, Deng N, Ni Z, Tang Y, Li G, Du L, Wang YL, Chen D, Wang H, Kong S (2021) Uterine Notch2 facilitates pregnancy recognition and corpus luteum maintenance via upregulating decidual Prl8a2. PLoS Genet 17(8):e1009786

Bazer FW (2015) History of maternal recognition of pregnancy. Adv Anat Embryol Cell Biol 216:5–25

Bertram R, Egli M, Toporikova N, Freeman ME (2006) A mathematical model for the mating-induced prolactin rhythm of female rats. Am J Physiol Endocrinol Metab 290(3):E573–E582

Binart N, Helloco C, Ormandy CJ, Barra J, Clement-Lacroix P, Baran N, Kelly PA (2000) Rescue of preimplantatory egg development and embryo implantation in prolactin receptor-deficient mice after progesterone administration. Endocrinology 141(7):2691–2697

Breton C, Pechoux C, Morel G, Zingg HH (1995) Oxytocin receptor messenger ribonucleic acid: characterization, regulation, and cellular localization in the rat pituitary gland. Endocrinology 136(7):2928–2936

Brown RS, Herbison AE, Grattan DR (2014) Prolactin regulation of kisspeptin neurones in the mouse brain and its role in the lactation-induced suppression of kisspeptin expression. J Neuroendocrinol 26(12):898–908

Brown RSE, Khant Aung Z, Phillipps HR, Barad Z, Lein HJ, Boehm U, Szawka RE, Grattan DR (2019) Acute suppression of LH secretion by prolactin in female mice is mediated by kisspeptin neurons in the arcuate nucleus. Endocrinology 160(5):1323–1332

Bruce HM (1959) An exteroceptive block to pregnancy in the mouse. Nature 184:105

Cole LA (2010) Biological functions of hCG and hCG-related molecules. Reprod Biol Endocrinol 8:102

Costa MA (2016) The endocrine function of human placenta: an overview. Reprod Biomed Online 32(1):14–43

Damewood MD, Rock JA (2005) In memoriam: Georgeanna Seegar Jones, M.D.: her legacy lives on. Fertil Steril 84(2):541–542

de Greef WJ, Zeilmaker GH (1979) Serum prolactin concentrations during hormonally induced pseudopregnancy in the rat. Endocrinology 105(1):195–199

DeMayo FJ, Lydon JP (2020) 90 years of progesterone: New insights into progesterone receptor signaling in the endometrium required for embryo implantation. J Mol Endocrinol 65(1):T1–T14
Devall AJ, Coomarasamy A (2020) Sporadic pregnancy loss and recurrent miscarriage. Best Pract Res Clin Obstet Gynaecol 69:30–39
d'Hauterive SP, Close R, Gridelet V, Mawet M, Nisolle M, Geenen V (2022) Human chorionic gonadotropin and early embryogenesis: review. Int J Mol Sci 23(3):1380–1395
Di Renzo GC, Tosto V, Tsibizova V (2020) Progesterone: History, facts, and artifacts. Best Pract Res Clin Obstet Gynaecol 69:2–12
Egli M, Bertram R, Toporikova N, Sellix MT, Blanco W, Freeman ME (2006) Prolactin secretory rhythm of mated rats induced by a single injection of oxytocin. Am J Physiol Endocrinol Metab 290(3):E566–E572
Elwood RW, Kennedy HF (1990) The relationship between infanticide and pregnancy block in mice. Behav Neural Biol 53(2):277–283
Erskine MS (1995) Prolactin release after mating and genitosensory stimulation in females. Endocr Rev 16(4):508–528
Finkel H (1931) The diagnosis of pregnancy by the Aschheim-Zondek test. N Engl J Med 204:203–209
Fitzpatrick LA, Good A (1999) Micronized progesterone: clinical indications and comparison with current treatments. Fertil Steril 72(3):389–397
Freeman ME (1979) A direct effect of the uterus on the surges of prolactin induced by cervical stimulation in the rat. Endocrinology 105(2):387–390
Freeman ME (1994) The neuroendocrine control of the ovarian cycle of the rat. In: Knobil E, Neill JD (eds) The physiology of reproduction, vol 2. Raven, New York, p 613
Frye CA, Erskine MS (1990) Influence of time of mating and paced copulation on induction of pseudopregnancy in cyclic female rats. J Reprod Fertil 90(2):375–385
Galli Manini C (1947) Pregnancy test using the male toad. J Clin Endocrinol Metab 7(9):653–658
Garzia E, Clauser R, Persani L, Borgato S, Bulfamante G, Avagliano L, Quadrelli F, Marconi AM (2013) Prolactin and proinflammatory cytokine expression at the fetomaternal interface in first trimester miscarriage. Fertil Steril 100(1):108–115 e101–102
Ghalioungui P, Khalil S, Ammar AR (1963) On an ancient Egyptian method of diagnosing pregnancy and determining foetal sex. Med Hist 7(3):241–246
Gorospe WC, Mick C, Freeman ME (1981) The obligatory role of the uterus for termination of prolactin surges initiated by cervical stimulation. Endocrinology 109(1):1–4
Greep RO, Hisaw FL (1938) Pseudopregnancies from electrical stimulation of the cervix in the diestrum. Proc Soc Exp Biol Med 39(2):359–360
Gridelet V, Perrier d'Hauterive S, Polese B, Foidart JM, Nisolle M, Geenen V (2020) Human chorionic gonadotrophin: new pleiotropic functions for an "old" hormone during pregnancy. Front Immunol 11:343
Gunnet JW, Freeman ME (1983) The mating-induced release of prolactin: a unique neuroendocrine response. Endocr Rev 4(1):44–61
Hackwell ECR, Ladyman SR, Brown RSE, Grattan DR (2023) Mechanisms of lactation-induced infertility in female mice. Endocrinology 164(5):bqad049
Hattori T, Osakada T, Masaoka T, Ooyama R, Horio N, Mogi K, Nagasawa M, Haga-Yamanaka S, Touhara K, Kikusui T (2017) Exocrine gland-secreting peptide 1 is a key chemosensory signal responsible for the Bruce effect in mice. Curr Biol 27(20):3197–3201 e3193
Helena CV, McKee DT, Bertram R, Walker AM, Freeman ME (2009) The rhythmic secretion of mating-induced prolactin secretion is controlled by prolactin acting centrally. Endocrinology 150(7):3245–3251
Hutchison JS, Zeleznik AJ (1984) The rhesus monkey corpus luteum is dependent on pituitary gonadotropin secretion throughout the luteal phase of the menstrual cycle. Endocrinology 115(5):1780–1786
Jones GE (1949) Some newer aspects of the management of infertility. J Am Med Assoc 141(16):1123–1129

Kelley K (2010) The Aschheim-Zondek test for pregnancy. Embryo Project Encyclopedia from http://embryo.asu.edu/handle/10776/2078

Kennett JE, McKee DT (2012) Oxytocin: an emerging regulator of prolactin secretion in the female rat. J Neuroendocrinol 24(3):403–412

Kirillova A, Martazanova B, Mishieva N, Semenova M (2021) Follicular waves in ontogenesis and female fertility. Biosystems 210:104558

Ladyman SR, Carter KM, Gillett ML, Aung ZK, Grattan DR (2021) A reduction in voluntary physical activity in early pregnancy in mice is mediated by prolactin. Elife 10:e62260

Larsen CM, Grattan DR (2010) Prolactin-induced mitogenesis in the subventricular zone of the maternal brain during early pregnancy is essential for normal postpartum behavioral responses in the mother. Endocrinology 151(8):3805–3814

Lunenfeld B (2004) Historical perspectives in gonadotrophin therapy. Hum Reprod Update 10(6):453–467

Lydon JP, DeMayo FJ, Funk CR, Mani SK, Hughes AR, Montgomery CA Jr, Shyamala G, Conneely OM, O'Malley BW (1995) Mice lacking progesterone receptor exhibit pleiotropic reproductive abnormalities. Genes Dev 9(18):2266–2278

McKee DT, Poletini MO, Bertram R, Freeman ME (2007) Oxytocin action at the lactotroph is required for prolactin surges in cervically stimulated ovariectomized rats. Endocrinology 148(10):4649–4657

McQuillan HJ, Han SY, Cheong I, Herbison AE (2019) GnRH pulse generator activity across the estrous cycle of female mice. Endocrinology 160(6):1480–1491

Mesen TB, Young SL (2015) Progesterone and the luteal phase: a requisite to reproduction. Obstet Gynecol Clin North Am 42(1):135–151

Mulac-Jericevic B, Mullinax RA, DeMayo FJ, Lydon JP, Conneely OM (2000) Subgroup of reproductive functions of progesterone mediated by progesterone receptor-B isoform. Science 289(5485):1751–1754

Northrop LE, Erskine MS (2008) Selective oxytocin receptor activation in the ventrolateral portion of the ventromedial hypothalamus is required for mating-induced pseudopregnancy in the female rat. Endocrinology 149(2):836–842

Northrop LE, Shadrach JL, Erskine MS (2006) Noradrenergic innervation of the ventromedial hypothalamus is involved in mating-induced pseudopregnancy in the female rat. J Neuroendocrinol 18(8):577–583

Norwitz ER, Schust DJ, Fisher SJ (2001) Implantation and the survival of early pregnancy. N Engl J Med 345(19):1400–1408

Nyuyki KD, Waldherr M, Baeuml S, Neumann ID (2011) Yes, I am ready now: differential effects of paced versus unpaced mating on anxiety and central oxytocin release in female rats. PLoS One 6(8):e23599

Phillipps HR, Yip SH, Grattan DR (2020) Patterns of prolactin secretion. Mol Cell Endocrinol 502:110679

Pletzer B, Winkler-Crepaz K, Maria Hillerer K (2023) Progesterone and contraceptive progestin actions on the brain: A systematic review of animal studies and comparison to human neuroimaging studies. Front Neuroendocrinol 69:101060

Poletini MO, Kennett JE, McKee DT, Freeman ME (2010) Central clock regulates the cervically stimulated prolactin surges by modulation of dopamine and vasoactive intestinal polypeptide release in ovariectomized rats. Neuroendocrinology 91(2):179–188

Prigent-Tessier A, Tessier C, Hirosawa-Takamori M, Boyer C, Ferguson-Gottschall S, Gibori G (1999) Rat decidual prolactin. Identification, molecular cloning, and characterization. J Biol Chem 274(53):37982–37989

Rasmussen CA, Hashizume K, Orwig KE, Xu L, Soares MJ (1996) Decidual prolactin-related protein: heterologous expression and characterization. Endocrinology 137(12):5558–5566

Roellig K, Goeritz F, Fickel J, Hermes R, Hofer H, Hildebrandt TB (2010) Superconception in mammalian pregnancy can be detected and increases reproductive output per breeding season. Nat Commun 1(6):78

Roellig K, Menzies BR, Hildebrandt TB, Goeritz F (2011) The concept of superfetation: a critical review on a 'myth' in mammalian reproduction. Biol Rev Camb Philos Soc 86(1):77–95
Rosser AE, Remfry CJ, Keverne EB (1989) Restricted exposure of mice to primer pheromones coincident with prolactin surges blocks pregnancy by changing hypothalamic dopamine release. J Reprod Fertil 87(2):553–559
Ryu V, Gumerova A, Korkmaz F, Kang SS, Katsel P, Miyashita S, Kannangara H, Cullen L, Chan P, Kuo T, Padilla A, Sultana F, Wizman SA, Kramskiy N, Zaidi S, Kim SM, New MI, Rosen CJ, Goosens KA, Frolinger T, Haroutunian V, Ye K, Lizneva D, Davies TF, Yuen T, Zaidi M (2022) Brain atlas for glycoprotein hormone receptors at single-transcript level. Elife 11
Short R (1969) Implantation and the maternal recognition of pregnancy. In: Ciba Foundation symposium—foetal autonomy 2
Smarr BL, Zucker I, Kriegsfeld LJ (2016) Detection of successful and unsuccessful pregnancies in mice within hours of pairing through frequency analysis of high temporal resolution core body temperature data. PLoS One 11(7):e0160127
Smith MS, Freeman ME, Neill JD (1975) The control of progesterone secretion during the estrous cycle and early pseudopregnancy in the rat: prolactin, gonadotropin and steroid levels associated with rescue of the corpus luteum of pseudopregnancy. Endocrinology 96(1):219–226
Soules MR, Clifton DK, Cohen NL, Bremner WJ, Steiner RA (1989) Luteal phase deficiency: abnormal gonadotropin and progesterone secretion patterns. J Clin Endocrinol Metab 69(4):813–820
Sundstrom-Poromaa I, Comasco E, Sumner R, Luders E (2020) Progesterone—friend or foe? Front Neuroendocrinol 59:100856
Talwar GP (1999) Vaccines and passive immunological approaches for the control of fertility and hormone-dependent cancers. Immunol Rev 171:173–192
Taraborrelli S (2015) Physiology, production and action of progesterone. Acta Obstet Gynecol Scand 94(Suppl 161):8–16
Tonkowicz PA, Voogt JL (1983) Termination of prolactin surges with development of placental lactogen secretion in the pregnant rat. Endocrinology 113(4):1314–1318
Vomachka AJ, Pratt SL, Lockefeer JA, Horseman ND (2000) Prolactin gene-disruption arrests mammary gland development and retards T-antigen-induced tumor growth. Oncogene 19(8):1077–1084
Voogt J, Robertson M, Friesen H (1982) Inverse relationship of prolactin and rat placental lactogen during pregnancy. Biol Reprod 26(5):800–805
Wall EG, Desai R, Khant Aung Z, Yeo SH, Grattan DR, Handelsman DJ, Herbison AE (2023) Unexpected plasma gonadal steroid and prolactin levels across the mouse estrous cycle. Endocrinology 164(6):bqad070
White A (1949) The chemistry and physiology of adenohypophyseal luteotropin (prolactin). In: Harris RS, Thimann KV (eds) Vitamins & hormones, vol 7. Academic, New York, pp 253–292
Yang JJ, Larsen CM, Grattan DR, Erskine MS (2009) Mating-induced neuroendocrine responses during pseudopregnancy in the female mouse. J Neuroendocrinol 21(1):30–39
Zeberg H, Kelso J, Paabo S (2020) The neandertal progesterone receptor. Mol Biol Evol 37(9):2655–2660
Zhang S, Lin H, Kong S, Wang S, Wang H, Wang H, Armant DR (2013) Physiological and molecular determinants of embryo implantation. Mol Aspects Med 34(5):939–980
Zinaman MJ, Clegg ED, Brown CC, O'Connor J, Selevan SG (1996) Estimates of human fertility and pregnancy loss. Fertil Steril 65(3):503–509

The Placenta as a Neuroendocrine Organ

2

Rosalind M. John

Abstract

Pregnancy is a unique situation, in which two (or more) individuals are in intimate contact, with development of the foetus supported by a specialised structure called the placenta. The placenta transports nutrients and waste, and provides a physical barrier between the maternal and foetal circulation. Importantly for this chapter, the placenta functions as an endocrine organ, synthesising hormones that flood the maternal circulation, inducing and maintaining the critical adaptations required for a successful pregnancy. While mammals have adopted different strategies to synthesise hormones in sufficiently high concentrations, and there are considerable variations in the nature of placental hormone gene families, the adaptations driven by placental hormones are conserved at a fundamental level. These adaptations include changes in behaviour of the mother to ensure that offspring receive appropriate care and nutrition. Experimental studies in rodents have identified placental hormones that contribute to the priming of this maternal behaviour. Studies in humans suggest that aberrant expression of placental hormones may contribute to the higher incidence of mood disorders associated with pregnancy. Hormones synthesised by, or dependent on, the placenta may also act directly on the foetal brain influencing neurodevelopment. Through this bi-directional signalling of hormones, the placenta has potential to influence both maternal and offspring behaviour. For these reasons, studies on placental endocrine lineage development are fundamentally important for our understanding of brain health.

R. M. John (✉)
Cardiff School of Biosciences, Cardiff University, Wales, UK
e-mail: JohnRM@cardiff.ac.uk

P. J. Brunton, D. R. Grattan (eds.), *Neuroendocrine Regulation of Mammalian Pregnancy and Lactation*, Masterclass in Neuroendocrinology 15,
https://doi.org/10.1007/978-3-031-51138-7_2

Keywords

Placental hormones · Placental lactogen · Maternal caregiving · Neurodevelopment

2.1 Introduction

During pregnancy, the mammalian mother-to-be undergoes substantial metabolic and behavioural adaptations required to support foetal growth and early postnatal nurturing, driven predominantly by signals from the **placenta** (Costa 2016; John and Hemberger 2012; Napso et al. 2018; Perez-Garcia et al. 2018; Smiley et al. 2019). The placenta is a transient organ that develops in pregnancy to support the growth of the foetus. Although all Eutherian mammals possess highly complex placentas, there are considerable variations in placentas between species, with differences in shape, size, arrangement with respect to the maternal and foetal circulation, their invasiveness and the length of gestation supported (Carter 2012, 2018; Carter and Enders 2004; Cross 2006; Georgiades et al. 2002). Despite these diversities, placentas perform three common and essential functions: the transport of nutrients and waste, the provision of a physical barrier between the maternal and foetal circulation, and the production of placental hormones. Through these functions, mammals are able support growth in utero for long periods and give birth to live young. In some cases, these young are able to walk within minutes of being born while others, like rodents and humans, are born in a relatively immature state, remaining heavily dependent on parental care, which can last for several years. Consequently, a key component of successful reproduction in mammals is the induction of maternal caregiving behaviour. While maternal caregiving is not unique to mammals, and indeed parenting behaviour is not unique to females, data from rodents demonstrate that the mammalian mother is prepared for her role in caregiving before she gives birth. This preparation or "priming" takes place across pregnancy, in part mediated by hormones produced from, or dependent on, the foetally derived placenta. In addition to priming the maternal brain, there are genes expressed by the placenta with the potential to directly influence foetal brain development. Placental hormones are therefore important not only for the metabolic adaptations that occur in pregnancy, which are the traditional focus of reproductive endocrinology studies, but also for the discipline of reproductive neuroendocrinology.

2.2 Maternal Behaviour

Rodents have often been used to study the intimate and intricate relationship between the mother, the foetus, and the placenta, and the relationship between the mother and her offspring after birth providing considerable insight into the remarkable process of mammalian pregnancy. Rodents do not provide the perfect model for human pregnancy as they give birth to multiple young which are less well developed than human infants, and there are differences in placental structure and placental hormones between rodents and humans (Carter 2020). However, a number of elegant and informative experiments in rodents demonstrate key principles related to maternal behaviour (see Chap. 8). First and foremost, studies in rodents have proven experimentally that the maternal brain is primed during pregnancy by hormones so that the new mother responds immediately to the presence of her offspring, changes that persist through subsequent pregnancies as "maternal memory" (Bridges and Scanlan 2005; Gallagher et al. 2019; Moltz and Robbins 1965; Nephew and Bridges 2008; Scanlan et al. 2006; Seegal and Denenberg 1974). During pregnancy changes in maternal behaviour are already evident, including decreased physical activity and increased feeding, directed by hormonal changes (Ladyman et al. 2020). These changes are necessary to support the increased metabolic demands imposed by the foetus and placenta, and to lay down extra stores of fat to support lactation after birth. As their pups are born, rodent mothers respond by ingesting the placenta, foetal membranes and fluids, and engage in licking, suckling, and retrieving behaviours, alongside changes in maternal aggression and anxiety levels (Champagne et al. 2007; Champagne et al. 2003; Franks et al. 2011; Lonstein 2007). Maternal-like behaviours such as retrieving pups, licking, and nest building can be observed in virgin rodents. For examples, virgin female rats manifest maternal behaviour after 5–7 days of contact with foster pups (Rosenblatt 1967) and virgin female mice within 2 days (Noirot 1969). However, in addition to this longer latency to express maternal-like behaviours, virgin females are slower to retrieve and crouch over pups, treat them more roughly, and spend less time crouching over pups when compared to postpartum females (Stolzenberg and Rissman 2011). Therefore, while virgin female rodents, and to some extent males, can display aspects of parenting behaviour, it is the postpartum mother who displays the highest levels of maternal responsivity immediately on exposure to her pups, indicative of prenatal programming. In addition, while all rodents build nests, pregnant rodents build larger and more complex ones (Latham and Mason 2004; Weber and Olsson 2008).

Maternal care is important not just for short-term survival of offspring, but also for their lifelong health, with neglected offspring susceptible to metabolic and behavioural disorders (Bredy et al. 2003; Cameron et al. 2008; Champagne et al. 2003; Curley and Champagne 2016; Meaney 2001; Szyf et al. 2005; Weaver et al. 2004). Considerable research has identified the key neural circuitry and neuroendocrine pathways that centrally mediate maternal caregiving with key brain regions including the amygdala, hippocampus, nucleus accumbens, olfactory bulb, bed nucleus of the terminalis, ventral tegmental area and periaqueductal grey, and the medial preotic area of the hypothalamus (Dobolyi et al. 2014; Dulac et al. 2014; Bridges 2015; Grattan 2012; Kohl et al. 2018; Kohl and Dulac 2018; Larsen and Grattan 2012; Numan and Young 2016; Smiley et al. 2019; Stolzenberg and Champagne 2016), as eloquently described in Chap. 8. In addition to changes in the existing neuronal populations within these regions, changes in neurogenesis have been reported contributing to neuroplasticity (Bridges 2015; Furuta and Bridges 2005; Kim et al. 2010; Larsen and Grattan 2010; Shingo et al. 2003). Evidence that factors in the blood might play a role in driving changes in the maternal brain to instruct maternal caregiving came initially from parabiotic transfusion of blood from late pregnant rats into virgin female recipients, which resulted in a rapid expression of maternal care in the virgins (Terkel and Rosenblatt 1972). The placenta was more formally demonstrated to be a source of factors priming maternal caregiving in experiments where wild type female mice were exposed to placenta with either underdevelopment or overdevelopment of the endocrine-synthesising cells, and these mothers displayed decreased or increased pup focused behaviour, respectively (Creeth et al. 2018).

2.3 The Placenta

In order to appreciate the role of placental hormones in neurodevelopmental processes, it is important to describe the development of the mammalian placenta and the endocrine lineages which synthesise placental hormones. The majority of mammals, including humans and mice, belong to a group called the eutherians, which are often referred to as placental mammals. However, it should be noted that both distinct placental structures and maternal care are present in marsupials (Guernsey et al. 2017) and even the young of the egg-laying monotremes possess the capacity to extract maternal nutrients while inside the body of the mother, via their leathery shell (Carter 2018). All mammalian embryos initially obtain nutrients through extraembryonic membranes which in some species form a choriovitelline or yolk sac placenta which is in contact with maternal tissue but does not invade (Freyer and Renfree 2008; Ross and Boroviak 2020). While there is expression of some peptide hormones at this stage, including prolactin-like hormones in marsupials (Menzies et al. 2011) and chorionic somatomammotropin hormone in humans (Cindrova-Davies et al. 2017), the major source of hormones in a eutherian pregnancy is the chorioallantoic placenta, which is a specialised structure that can be invasive. The chorioallantoic placenta, which is also present in marsupials with longer gestations, allows a more significant bidirectional exchange of nutrients, wastes, and gases between mother and foetus, supporting substantial growth of the foetus. In addition, specialised lineages form that act as factories for the synthesis and secretion of peptide hormones, and in the case of the primate placenta, also significant quantities of steroid hormones.

Box 2.1: Developmental Timepoints

For mouse studies the first day of visible plug is referred to as embryonic day (E) 0.5. Older publications can refer to the first day of pregnancy as E1 but E0.5 is deemed the more accurate term because mating takes place at night. When discussing events predominantly concerning the mothers, the day of pregnancy can be referred to gestational day (G) with first day of visible plug referred as G1. This means that E14.5 for the foetus/placenta is G15 for the dam. Theiler stages (TS) is another tool that is used to define mouse development and this is done by a set of morphological criteria rather than the time since mating (Theiler 1989). This staging scheme is useful because the developmental rate of individual embryos can differ, resulting in different morphological characteristics. For most mice strains the length of gestation is 19–21 days and from birth perinatal events are recorded from postnatal day (P)1. For rats, the first day of observable sperm is referred to as gestational day (GD) 1 with the average length of gestation being 21–23 days.

2.4 Placental Development and Placental Endocrine Lineages in Rodents

The chorioallantoic placenta forms from a specialised lineage called the trophectoderm, which is the first lineage specified in mammals (Woods et al. 2018). In mice, embryonic day (E) 0.5 is the first day on which mating is confirmed, and the length of gestation varies between 19–21 days, depending on the strain (Box 2.1). **Trophectoderm** lineage specification takes place between E2.75 and E3.25 at the 16–32 cell stage. Cells on the outside of the preimplantation embryo take on a trophectoderm identity consolidated through an epigenetic mechanism distinguishing **trophoblast** from embryonic cells (Ng et al. 2008). Between E3.25 and E3.75, the murine blastocyst forms, comprising a layer of trophectoderm surrounding an inner cell mass, which will develop into the embryo proper, and a fluid filled blastocoelic cavity (Fig. 2.1a). The trophectoderm lying opposite the inner cell mass is referred to as mural trophectoderm and these cells terminally differentiate to form primary parietal trophoblast giant cells (TGCs). The trophectoderm lying directly over the inner cell mass is referred to as polar trophectoderm. These cells attach to the lining of the uterus. Trophectoderm proliferation, differentiation, and invasion of the uterine wall then rapidly results in multiple specialised trophoblast cell types (Woods et al. 2018). Initially, the diploid trophectoderm cells form extra-embryonic ectoderm, an epithelial cell type with stem cell capacity. The region of extra-embryonic ectoderm furthest from the developing embryo then differentiates to become the ectoplacental cone, while the surrounding cells continue to differentiate into secondary parietal TGCs (Simmons et al. 2008). The region of extra-embryonic ectoderm nearest to the embryo subsequently differentiates, at the time of gastrulation, to form chorionic ectoderm. Alongside development of these trophoblast lineages, extra-embryonic mesoderm arising from cells of the inner cell mass differentiates to form the foetal placental vasculature. At around E8.5, the allantois (progenitor tissue of the umbilical cord) grows out from the embryo to make contact with the chorion (composed of extra-embryonic ectoderm derived from trophectoderm and extra-embryonic mesoderm derived from proximal epiblast) with completion of chorioallantoic fusion by E9.5 (Fig. 2.1b), which is approaching halfway through a mouse pregnancy. After fusion, further differentiation of cells within the ectoplacental cone and chorion occurs to form the mature chorioallantoic placenta.

The mature placenta is organised into two main foetally derived compartments, named the junctional zone and the labyrinthine zone, and the maternally derived decidua formed by modification of the uterine endometrium at the site where the conceptus implanted (Fig. 2.1c) (Hemberger et al. 2020; John and Hemberger 2012; Woods et al. 2018). In some older publications, the junctional zone can be labelled simply as the spongiotrophoblast zone, but this is misleading as there are two major cell types in this region, only one of which is the spongiotrophoblast. As the mature placenta forms, the secondary *parietal* TGCs come to lie as a single discontinuous layer between the junctional zone and the decidua. These secondary *parietal* TGCs can be referred to in older publications simply as *parietal* TGCs or just TGCs, but this ignores the fact that primary *parietal* TGCs exist earlier in development, and

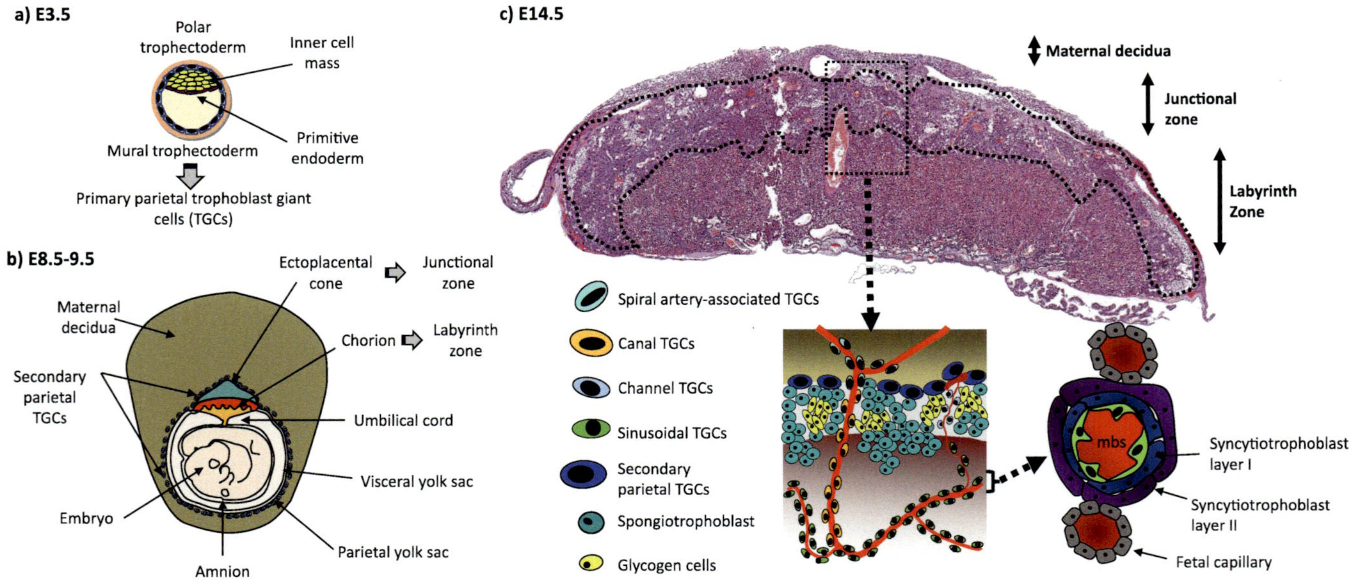

Fig. 2.1 Development of the mouse placenta. (**a**) Embryonic day (E) 3.5 blastocyst stage preimplantation embryo. (**b**) Transitional stage between choriovitelline placenta and chorioallantoic placenta. Progenitors located in the ectoplacental cone differentiate to form the spongiotrophoblast and glycogen cell lineages of the junctional zone, and most of the trophoblast giant cell lineages (TGCs). Progenitors in the chorion differentiate to form the two layer syncytiotrophoblast and, potentially, also the sinusoidal TGCs. (**c**) Mature mouse placenta. Maternal blood enters from maternal spiral arteries lined by spiral artery TGCs, which have replaced the maternal endothelial cells. Maternal spiral arteries transition to blood-filled canals lined by canal TGC that pass through the junctional zone into the labyrinth. Towards the bottom of the labyrinth, sinusoidal TGCs envelope the maternal blood spaces. These are in turn enveloped by syncytiotrophoblast layers I and II, which are in direct contact with the endothelium of the foetal blood vessels. Deoxygenated maternal blood filters out of the labyrinth through small channels in the junctional zone that are lined by Channel TGC to form blood pools surrounded by very large secondary parietal TGCs lying just beneath the maternal decidua. Placental hormone-loaded maternal blood is then returned to the maternal circulation. Figure adapted from John and Hemberger (2012), Rai and Cross (2014), and Woods et al. (2018)

that different classes of TGCs are found later in placental development (Simmons et al. 2008). As a consequence of the differences in terminology used to describe certain regions or cell types in the rodent placenta, some findings in older publications are challenging to interpret.

2.4.1 The Junctional Zone

The junctional zone is derived exclusively from progenitor cells located within the ectoplacental cone (Figs. 2.1b and 2.2) (Hu and Cross 2011) and is composed of two lineages, the spongiotrophoblast and the glycogen cell lineages (Simmons et al. 2008). The spongiotrophoblast lineage continually increases in cell number from E9.5 until E16.5, when there are over six million cells present, and then cell division essentially ceases (Coan et al. 2006). The glycogen cell lineage, named after the stores of glycogen that accumulate over gestation, expands slightly later than the spongiotrophoblast, reaching similar maximal numbers but with a significant proportion of these cells migrating into the maternal decidua (Coan et al. 2006).

2.4.2 The Labyrinth Zone

The labyrinth zone primarily functions in the transport of nutrients, gases, and waste. Within this region the foetal endothelium encloses the foetal circulation and is surrounded by a trilaminar layer of trophoblast-derived cells, consisting of two multinucleated syncytiotrophoblast layers that function in nutrient transport (I and II; previously called layers II and II) and a single mononuclear layer of sinusoidal trophoblast giant cells (previously called trophoblast layer I) in direct contact with maternal blood (Figs. 2.1c and 2.2). These trophoblast elements are derived principally from progenitor cells located within the chorion region, although this may not be the case for the sinusoidal TGCs. As three trophoblast layers separate the maternal and foetal circulation, the mouse placenta is referred to as haemotrichorial (Hemberger et al. 2020; John and Hemberger 2012; Woods et al. 2018).

2.4.3 Trophoblast Giant Cells

TGC lineages are defined by their giant nuclei, which result from a process called endoreduplication, in which rounds of DNA synthesis are not coupled with mitoses, doubling in DNA content for each round of DNA replication. This high degree of polyploidy is thought to enable large-scale protein synthesis. In addition to the secondary *parietal* TGCs, lying between the junctional zone and decidua, and the sinusoidal TGCs, located in the labyrinth, there are four additional and distinct TGC lineages (Hu and Cross 2011; Rai and Cross 2014; Simmons et al. 2007) (Figs. 2.1c and 2.2). The spiral artery TGCs line the maternal blood system on entry to the

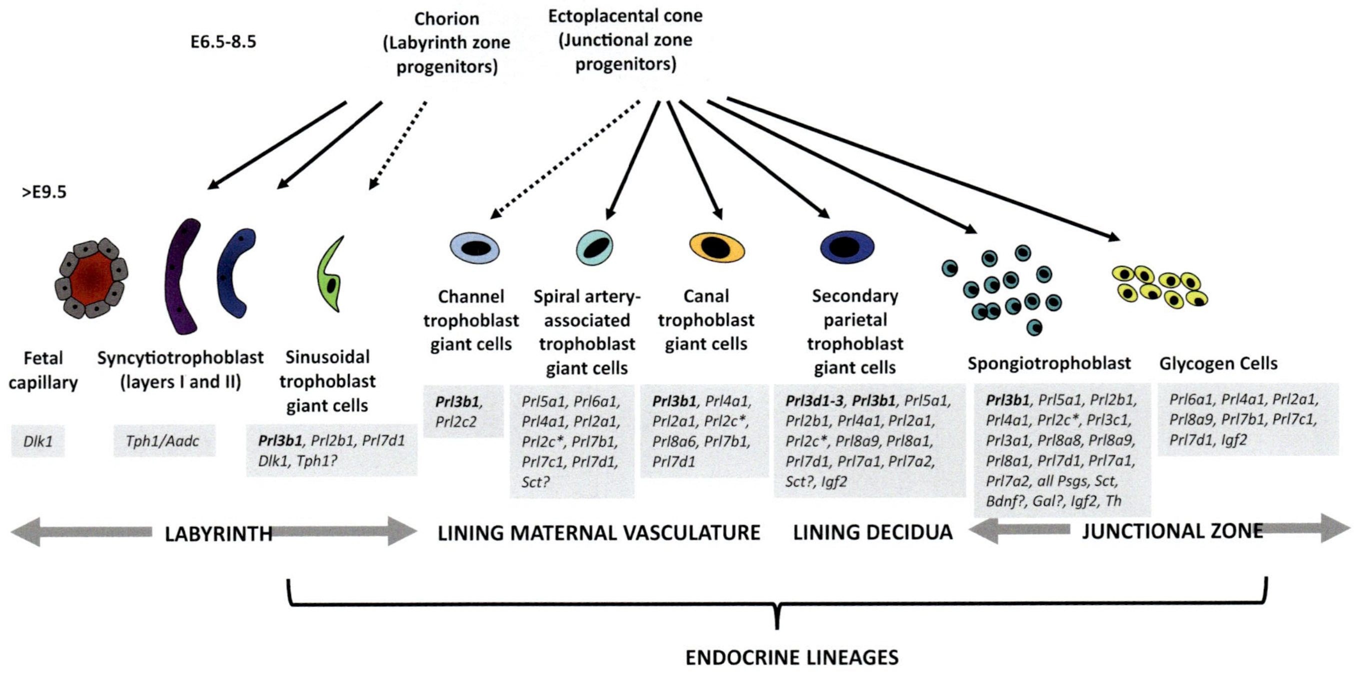

Fig. 2.2 Placental lineages in mice that express placental hormones. In situ hybridisation and immunohistochemistry studies have uncovered expression of prolactin family members in several lineages in the mature mouse placenta. Critically for studies of maternal behaviour, *Prl3d1-3* (*PL-1*) is detectable in the primary and secondary parietal TGCs from as early as E6.5 until approximately E9.5 and *Prl3b1* (aka *PL-II*) is detectable initially within cells located within the ectoplacental between E8.5 and E10.5, and then at increasingly higher levels by secondary parietal TGCs from E9.5 to term, and by the spongiotrophoblast, and the sinusoidal-, canal-, and channel TGC lineages from E11.5 to term (Simmons et al. 2008; Gasperowicz et al. 2013; Tunster et al. 2016a, b). Additional hormones and enzymes with potential neuroendocrine function of which expression has been localised to specific lineages are also indicated, but it must be noted that expression levels are very low relative to *Prl3b1*. ? Indicates where localisation is not entirely certain. Figure adapted from Creeth and John (2020)

placenta and are involved in remodelling the uterine wall by intercalating and replacing the maternal endothelial cells of the spiral arteries. The canal TGCs line the maternal blood canals in the junctional zone and labyrinth and at the base of the labyrinth these are replaced by the sinusoidal TGCs that line the maternal blood canals in the labyrinth. Finally, the channel TGCs line the maternal blood spaces located just beneath the decidua where maternal blood leaves the placenta (Gasperowicz et al. 2013; Rai and Cross 2014; Simmons et al. 2007, 2008). Thus, all the TGC types identified to date are in direct contact with either maternal blood or, as in the case of the *parietal* TGCs, both maternal blood and maternal decidua.

Box 2.2: RNAscope-Aided Recognition of Placental Endocrine Lineages

All three placental regions can be distinguished histologically in haematoxylin- and eosin-stained sections in which cell and nuclear shapes are preserved (Fig. 2.1c), but a confident identification of distinct types can be aided by an additional process of cell identification either by the more traditional in situ hybridisation (ISH) of RNA probes for marker genes or by newer technologies such as RNAscope Fluorescent Multiplex assays. RNAscope involves multiplex nucleic acid ISH assays that allow the detection and visualisation of gene expression within intact cells using multiple fluorescent probes (Tunster et al. 2018). *RNAscope* detection relies on signal amplification and suppression of background noise, and is a major improvement over traditional RNA ISH methods. Samples can be formalin-fixed paraffin-embedded tissues, fixed frozen (low expressors) or fresh frozen (if highly auto fluorescent), all prepared under RNAse-free conditions. Usually two probes, for markers of specific regions or cell types, can be combined with a probe for one gene of interest and 4′,6-diamidino-2-phenylindole (DAPI) nuclear staining to localise gene expression, as we did for the imprinted gene *Peg3* (Fig. 2.3) (Tunster et al. 2018). Volumes of probes must be optimised and will depend on the expression level of the specific markers. For example, *Prl3b1* and *Pcdh12* are useful markers to distinguish spongiotrophoblasts, TGCs, and glycogen cells (Table 2.1). However, *Prl3b1* is expressed at exceedingly high levels whereas *Pcdh12* is expressed at relatively low levels in the mature mouse placenta. This can be addressed by diluting *Prl3b1* 50-fold compared to *Pcdh12* (Tunster et al. 2018). Similarly, the fluorophores are diluted based on the signal intensity determined during optimisation. As the placenta is blood-filled, there will be autofluorescence that can interfere with the detection of weaker signals and it is therefore advisable to use a different channel for the more lowly expressed probes. The choice of probe dilution and channel can be further guided using RNA sequencing or similar techniques to gain a measure of relative expression levels for the timepoint under study.

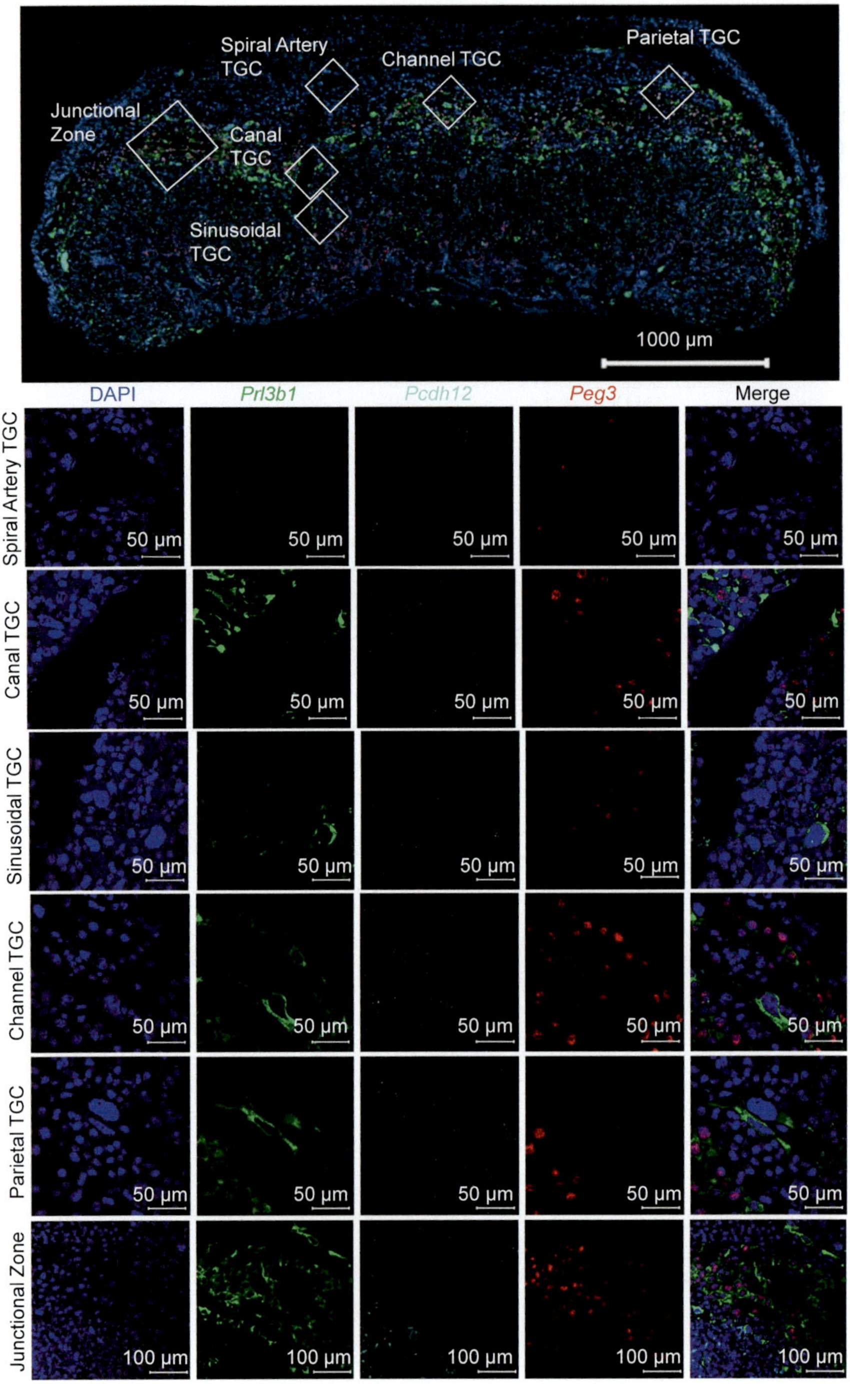

Fig. 2.3 RNAscope-led identification of the placental endocrine lineages and cellular localisation of an imprinted gene (Box 2.2). *Prl3b1* (*Pl2*; green) identifies spongiotrophoblast and all the TGC lineages except spiral artery TGCs (Simmons et al. 2008). *Pcdh12* identifies glycogen cells (Bouillot et al. 2006). In situ amplification of *Peg3* (red) reveals high expression in spongiotrophoblast, with lower levels of expression in glycogen cells within junctional zone and those migrating into decidua, and in all TGC lineages except spiral artery TGCs. Nuclei labelled with 4′,6-diamidino-2-phenylindole (DAPI; blue). Adapted from Tunster et al. (2018)

Table 2.1 Cell types and lineage markers for the mouse placenta

Cell type	Function/character	Markers
Primary parietal trophoblast giant cells	Cells with polyploid nuclei that differentiate from the outer layer of trophodermal cells of the blastocyst lying opposite to the ICM	***Prl3d***
Ectoplacental cone	Composed of Tpbpa +ve and −ve cells. Gives rise to *parietal* TGCs, channel TGCs, canal TGCs, spiral artery TGCs, spongiotrophoblast, and glycogen cells	(Subset of cells): ***Tpbpa***, ***Pcdh12***, *Phlda2*, *Ascl2*, *Blimp*
Extraembryonic ectoderm (ExE)	Becomes chorion and is a source of sinusoidal TGCs	*Eomes*
Chorion	Derived from ExE, differentiates to form the two types of syncytiotrophoblast cells (layers I and II) through cell–cell fusion	*Gcm1*
Foetal endothelium	Endothelial cells lining the lumen of foetal blood	***Flk1***
Syncytiotrophoblast layer I	Multinucleated cellular layer located between sinusoidal TGCs and syncytiotrophoblast layer II	***Syna***, *Ly6e*
Syncytiotrophoblast layer II	Multinucleated cellular layer located between the foetal-derived blood vessels and syncytiotrophoblast layer I	***Synb***, *Gcm1*, *Cebpa*
Trophoblast giant cells (TGCs)	Large, polyploid nuclei with large cytoplasm and extensive Golgi and endoplasmic reticulum, undergo several rounds of DNA replication without intervening mitosis and cell division (endoreduplication)	***Prl2c*** (expressed in all of the subtypes except sinusoidal TGCs early in gestation)
Sinusoidal TGC	Mononuclear TGCs in contact with maternal blood spaces in labyrinth, lining syncytiotrophoblast layer, becomes "perforated" at E12.5	***Prl3b1***, ***Prl2b1***, ***Ctsq***
Canal TGCs	Large pipe-like structures delivering maternal blood from spiral arteries to the bottom of the labyrinth	***Prl3b1***, ***Prl2c***, ***Prl7b1*** (Blimp)
Channel TGCs	Line the small channels collecting deoxygenated blood out of labyrinth	***Prl3d***, ***Prl2c***, *Ctsq*
Secondary parietal TGCs	Polyploid cells. Lie between junctional zone and decidua and surround large lakes (lacunae) of deoxygenated maternal blood leading into the uterine veins	***Prl3b1***
Spiral artery TGCs	Invade into the spiral arteries displacing maternal endothelial cells	***Prl2c***, ***Prl7b1*** (*Blimp*, *Rgs5*, *Lnc7*)
Junctional zone	Region lying between decidua and labyrinth, analogous to basal plate in human placenta and sometimes referred to as such, or as just spongiotrophoblast	***Tpbpa***, ***Tpbpb***, ***Flt1***
Labyrinth	Highly branched, large surface area for nutrient and gas exchange	No single marker

(continued)

Table 2.1 (continued)

Cell type	Function/character	Markers
Spongiotrophoblast	Major endocrine lineage of the mature mouse placenta	***Prl8a8*** and ***Prl7a2*** *(only)*, ***Psgs***, ***Prl2c*** (also expresses *Prl3b1*)
Glycogen cells	Glycogen cells form a layer around luminal cells of the maternal blood spaces in Junctional zone. Accumulate glycogen and migrate into the maternal uterine wall from E12.5 until term	***Pcdh12*** and ***Gjb3*** (only), ***Prl7b1***, (*Gbe1*, *Ugp2*, *Igf2*, *Blimp*) and PAS staining

2.4.4 Rodent Placental Endocrine Lineages

Lineages expressing placental hormones in mice are the spongiotrophoblast, glycogen cells, and the six TGCs types (Hemberger et al. 2020; Simmons et al. 2008; Woods et al. 2018) (Fig. 2.2). Of these, the spongiotrophoblast is the major endocrine lineage by cell number, with the highest total number of cells (Coan et al. 2006; Creeth and John 2020; Hemberger et al. 2020; Napso et al. 2018). However, the large nuclei of the TGCs resulting from endoreduplication and further over-replication of specific regions encoding placental hormones are likely to enable large-scale protein synthesis (Hannibal and Baker 2016). Hormones synthesised by TGCs probably enter the maternal circulation through direct contact with the maternal vasculature/blood spaces, while hormones synthesised by parietal TGCs and spongiotrophoblasts enter the maternal circulation through the large pools (lacunae) of blood that accumulate beneath the decidua and exit the placenta via the uterine veins (Rai and Cross 2014). Both the spongiotrophoblast and the TGCs are therefore major sources of placental hormones in mice and other rodents.

2.5 Key Hormones of Pregnancy Implicated in the Instruction of Maternal Behaviours in Rodents

Polypeptide hormones and steroid hormones, either expressed by the placenta or dependent on the placenta for their synthesis, are key to driving adaptations in the mother during pregnancy. During pregnancy the human placenta expresses over 100 peptides with predicted hormonal function (Liu et al. 2018) while 41 hormone-encoding genes with significant levels of placental expression have been reported in mice (Knox et al. 2011). In addition, the placenta is involved, either directly (primates) or indirectly (rodents) in the synthesis of steroid hormones, and has several genes with potentially neuromodulator function. The functions of these signals from the placenta in driving the metabolic and immunological changes, and the induction of mammary gland development in the mother required for a successful pregnancy, have been extensively reviewed (Costa 2016; John and Hemberger 2012;

Napso et al. 2018; Perez-Garcia et al. 2018). The focus here will be on those placentally derived hormones with potential neuroendocrine functions. Both targeted studies and genome-wide transcriptomic studies have identified a number of genes expressed in the rodent placenta with the potential to influence the development or function of the brain. Some of these hormones are expressed at exceedingly high levels, suggesting important roles in the mother, whereas other potential neuromodulators, either directly expressed by the placenta or with components of their synthesis pathways expressed in the placenta, are expressed at very low levels (Creeth et al. 2019) (Fig. 2.4), suggesting that their site of action is likely to be localised to the placenta or potentially the foetal brain, rather than the maternal brain.

2.6 The Growth Hormone/Prolactin Gene Family

Considerable evidence from rodent studies highlight the function of pituitary-synthesised prolactin and placentally synthesised "placental lactogen" as key peptide hormones involved in the priming of the maternal brain in preparation for maternal caregiving (Bridges 2015; Creeth et al. 2019; Grattan 2012). These peptide hormones, originally identified through their milk-stimulating activities (Chap. 7), belong to an extensive and closely related prolactin/growth hormone gene family that includes the pituitary-expressed growth hormone. All family members arose through gene duplication events from a single common ancestral peptide hormone which first duplicated to generate prolactin and growth hormone (Miller and Eberhardt 1983) (Fig. 2.5a). In rodent lineages, multiple duplications of the ancestral *prolactin* gene subsequently occurred to generate an extensive family of prolactin-related genes primarily expressed in the placenta. Mice possess 22 genes referred to collectively as *prolactin family members*, which are located together on mouse chromosome 13 (Simmons et al. 2008) with *Gh* located on chromosome 11 (Fig. 2.5b) and rats have at least 25 family members located together on rat chromosome 17 (Alam et al. 2006). Like their ligands, the prolactin and growth hormone receptors also share a common ancestor (Ocampo Daza and Larhammar 2018).

Despite being related to prolactin, only prolactin family 3, subfamily d, member 1 (Prl3d1; previously named PL-I) and prolactin family 3, subfamily b, member 1 (Prl3b1; previously named PL-II) have been shown to bind the prolactin receptor and therefore function as cognate placental lactogens (Soares et al. 2007). *Prl3d2* and *Prl3d3* are two genes closely related to *Prl3d1* and predicted to bind the receptor (Newbern and Freemark 2011; Soares et al. 2007) (Fig. 2.5b). The three *Prl3d* genes are challenging to distinguish by most molecular approaches and are often collectively referred to as *Prl3d*. The remaining prolactin family members function in a variety of processes influencing angiogenesis and modulating the maternal immune response (Soares 2004). Findings after targeted deletion of *Prl4a1* (Ain et al. 2004), *Prl7b1* (Bu et al. 2016), and *Prl7d1* (Zhang et al. 2019) also suggest a role for these peptide hormones in maintaining pregnancy under conditions of stress.

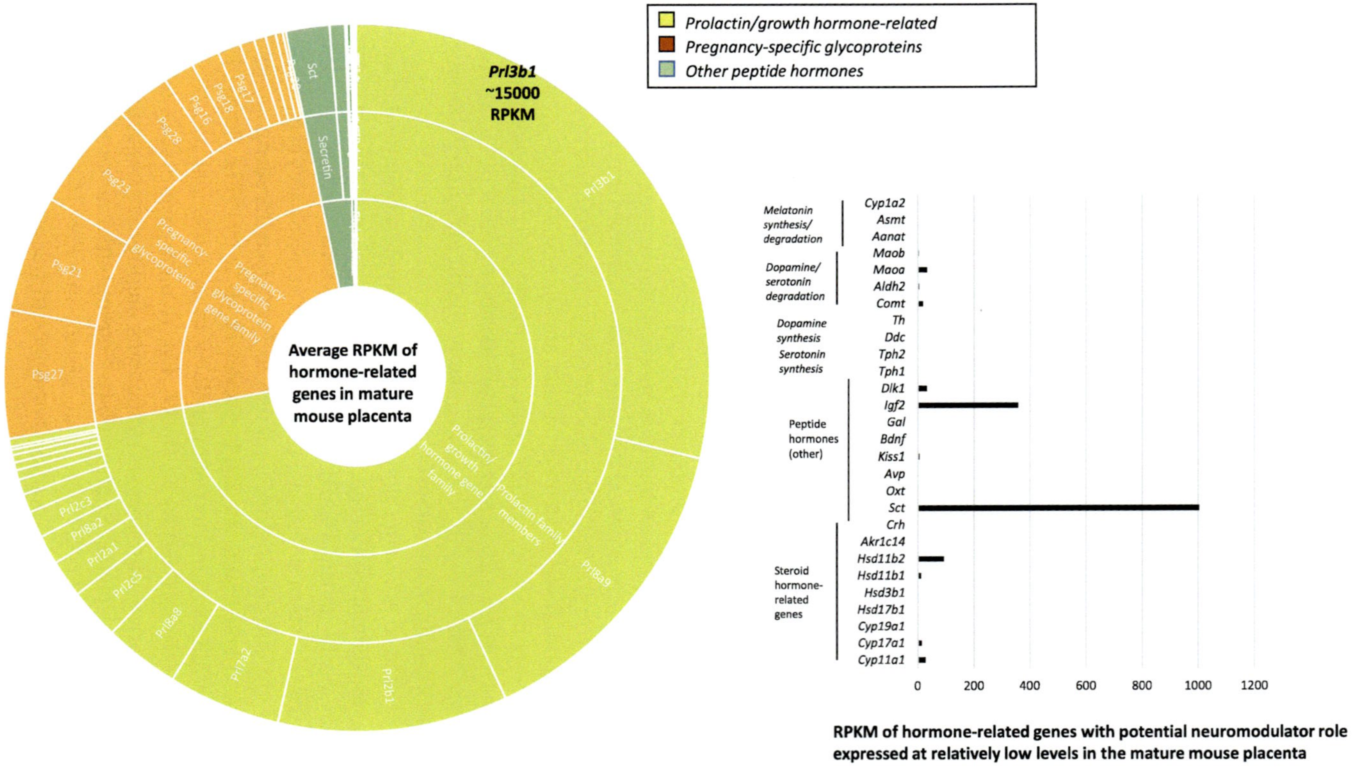

Fig. 2.4 Global hormone transcriptional profiling of the mature mouse placenta. Pie chart showing relative expression levels of key peptide hormone and hormone-related genes with potential neuroendocrine roles in the mouse placenta, determined from total RNA sequencing of whole E14.5 C57BL/6 mouse placenta performed to a depth of 50–60 Mb reads per sample; counts averaged from eight placental samples, four male and four female. The outermost circle of the pie chart provides individual gene names while inner circle indicates gene family. *Prl3b1*, which encodes mouse **placental lactogen**, is the most highly expressed peptide hormone, at approaching 15000 RPKM in this dataset. Genes expressed at considerably lower levels in comparison to *Prl3b1* are also shown in the bar graph. RPKM: reads per kilobase of transcript, per million mapped reads

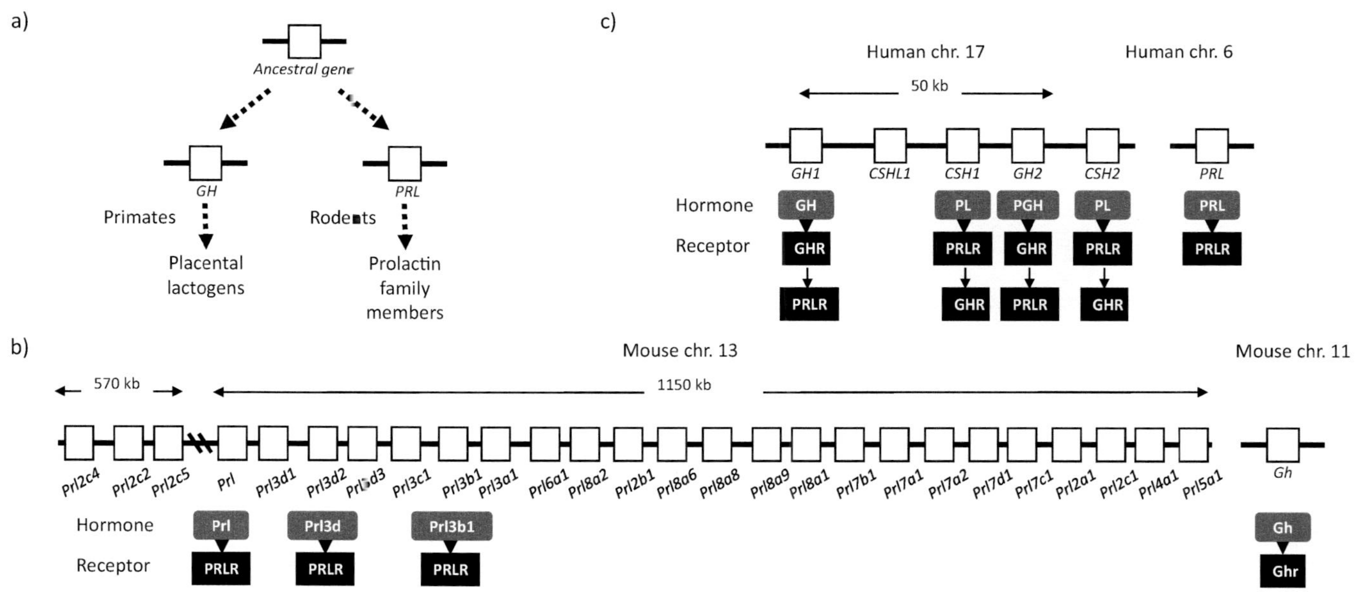

Fig. 2.5 The rodent and human prolactin/growth hormone gene family. (**a**) *GH* and *PRL* are derived from a common ancestral gene. In rodents *PRL* underwent multiple duplications to produce placenta-specific prolactin family members. In primates, placental lactogens and growth hormones arose through duplication of *GH*. (**b**) Twenty-two peptide hormones related to prolactin/growth hormone are expressed in the mouse placenta. (**c**) Four peptide hormones related to prolactin/growth hormone are expressed in the human placenta with overlapping binding affinity for the prolactin and growth hormone receptors. Figure adapted from Simmons et al. (2008). GH, growth hormone; PRL, prolactin; CSH, chorionic somatomammotropin hormone; GHR, growth hormone receptor; PRLR, prolactin receptor; Prl3d1, prolactin family 3 subfamily d, member 1; Prl3b1, prolactin family 3, subfamily b, member 1

In mice, expression of *Prl3d1–3* is detectable in the primary and secondary *parietal* TGCs from as early as E6.5 until approximately E9.5 when expression is maximal, and when the mature chorioallantoic placenta forms (Simmons et al. 2008). Studies using quantitative PCR have reported expression of all three *Prl3d* variants in the junctional zone, near the point where maternal blood exits the placenta, and labyrinth later in gestation (Rawn et al. 2015), but Northern blot data suggest that the levels are relatively low compared to expression at the earlier stage (Simmons et al. 2008). Beyond E9.5, *Prl3b1* encodes the major source of placental lactogen. *Prl3b1* expression is detectable initially within cells located within the ectoplacental cone between E8.5 and E10.5 and then at increasingly higher levels by *parietal* TGCs from E9.5 to term, and by the spongiotrophoblast, and sinusoidal-, canal-, and channel-TGC lineages from E11.5 to term (Simmons et al. 2008) (Fig. 2.2). It is key to note that early studies on the mouse placenta primarily refer generically to TGCs as a source of placental lactogen, or sometimes more specifically the *parietal* TGCs. However, as is evident from our RNAscope analysis of the mid-term placenta, the majority of signal for *Prl3b1* mRNA is localised to the spongiotrophoblast within the junctional zone (Fig. 2.3). When global hormone profiling is performed by RNA sequencing of transcripts from whole murine placenta at E14.5 of pregnancy, *Prl3b1* one of the most highly expressed transcripts, representing 1.75–1.9% of the total reads per sample for a C57BL/6 (BL6) and *129S2/*SvHsd (129) placenta, respectively (Fig. 2.4). The lineages synthesising *Prl3b1* in the mature mouse placenta are all closely associated with maternal blood on its journey through the placenta and back into the maternal circulation. Spiral artery TGCs are the only TGC lineage that does not express *Prl3b1* (Simmons et al. 2008) and these are located at the point where maternal blood enters into the placenta (Fig. 2.1c). It should be noted that the maternally derived decidua is a source of prolactin (Prigent-Tessier et al. 1999) which functions locally to maintain pregnancy (Bao et al. 2007). This localised function is consistent with *prolactin* expression in whole E14.5 placenta being 50,000 times lower than *Prl3b1*.

The first experimental evidence that prolactin-like hormones were involved in the stimulation of maternal care came when prolactin injection into virgin female rats was shown to stimulate pup retrieval behaviour (Riddle et al. 1935). From the 1970s onwards, a number of studies further implicated prolactin, and then the placentally derived placental lactogens, as key to this behavioural adaptation through applying hormonal exposures involving injections of progesterone, oestradiol, and prolactin into ovariectomised (removal of endogenous steroid hormonal signals) female rats with subsequent regimes including placental lactogens (Bridges et al. 1985, 1990, 1996, 1997; Bridges and Freemark 1995; Bridges and Ronsheim 1990; Moltz et al. 1970). Some experimenters used recombinant human placental lactogen while others used rat placental lactogen I and II, homologous to mouse Prl3d and Prl3b1. Given that these peptides all signal via the same maternal receptor, the infusion approach does not distinguish the relative importance of prolactin versus placental lactogen in the priming of maternal caregiving behaviour. In vivo in the pregnant rodent, the maternal brain is first exposed to prolactin when secretion from the pituitary is stimulated by the physical act of mating (Gunnet and Freeman 1983), followed by surges in the afternoon and night until mid-gestation (see Chap. 1), which have been observed in both mouse (Barkley et al. 1978) and rat pregnancy (Freeman and Neill 1972) but not in all mammals, as reviewed (Bridges 2015; Soares 2004). From mid-gestation, prolactin secretion is suppressed by placental lactogen encoded by *Prl3d1–3* until around E9.5 and by *Prl3b1* from then until near term (Soares et al. 2007). Prolactin is repressed until the night before parturition, when secretion restarts in response to progesterone withdrawal (Bridges 2015; Soares 2004). As a consequence of these dual sites of peptide synthesis, there is sequential and increasing exposure of the maternal brain first to locally secreted prolactin and then to the placentally derived placental lactogen during pregnancy. Evidence that both pituitary prolactin and placental lactogen contribute to the priming of maternal behaviour is suggested by two findings. First, experimentally induced low levels of prolactin in pregnancy, induced by the dopamine antagonist bromocriptine, result in increased postpartum anxiety and decreased pup retrieval in rats (Larsen and Grattan 2012), demonstrating the requirement for pituitary prolactin. Second, mice with a disrupted prolactin gene still exhibit some maternal behaviour (Horseman et al. 1997) demonstrating the requirement for other hormones with activity mirroring prolactin's. These data, along with the temporal synthesis through gestation, support the sequential action of prolactin and then placental lactogen on the maternal brain in priming maternal behaviour. Once pups are born, pituitary-derived prolactin is chronically elevated during lactation, when maternal behaviours are expressed. It is important to note that maternal motivation to interact with pups is highest after birth and requires prolactin receptor activity, with poor motivation observed in mid- or late-pregnant mice (Swart et al. 2021). This could mean that placental hormones are unable to induce maternal behaviour alone, that is, the process of delivery or a secondary stimulus from the pups is required to activate the primed brains. Alternatively, this could be indicative of an inhibitory signal during pregnancy, as suggested by studies on progesterone (see later) or a combination of factors.

Both prolactin and placental lactogen are thought to mediate their activity through the maternal prolactin receptor (Prlr) (Soares et al. 2007). Activation of the maternal Prlr during pregnancy is a key step in the induction of maternal caregiving, as shown in many studies where the receptor has been genetically ablated (Bridges et al. 1985; Brown et al. 2017; Lucas et al. 1998; Moltz et al. 1970; Shingo et al. 2003). Global heterozygous loss of *Prlr* in virgin female mice was reported to delay the acquisition of pup-induced maternal behaviour and in primiparous females resulted in maternal neglect and a failure to retrieve all pups in a pup retrieval task (Lucas et al. 1998). Importantly, acute deletion of *Prlr* from medial preoptic area of the hypothalamus results in a complete absence of maternal caregiving (Brown et al. 2017). An intact *Prlr* appears to also be required for pregnancy-related changes in neurogenesis that take place in the subventricular zone (Larsen and Grattan 2010; Shingo et al. 2003), and signalling via the same receptor may also be involved in pregnancy-related changes in neurogenesis in the subgranular zone, located within the hippocampus (Rolls et al. 2008; Walker et al. 2012) although this still needs to be demonstrated explicitly. These experiments identify signalling via *Prlr* as a key step in the priming of the maternal brain in preparation for caregiving.

Infusion experiments demonstrated that prolactin and Prl3d have a similar potency to stimulate maternal behaviour in rats (Bridges et al. 1997). In vivo, placental lactogens were proposed to enter the maternal brain through the choroid plexus, which accumulates the short and long isoforms of the prolactin receptor, potentially facilitating entry into the brain (Augustine et al. 2003; Bridges and Hays 2005; Grattan 2002; Pi et al. 2003; Torner et al. 2002) although recent studies suggest an alternative route via cerebral microvasculature (Brown et al. 2016). Placental lactogens have been experimentally shown to function by activating the Prlr in the medial preoptic area and paraventricular nucleus of the hypothalamus and the hippocampus (Bridges 1990; Kokay et al. 2006; Nogami et al. 2007; Torner et al. 2009; Vergara-Castaneda et al. 2016), regions important for maternal behaviour (Kimble et al. 1967; Stack et al. 2002). In addition to regions where prolactin activity is known to induce aspects of maternal behaviour, increased prolactin receptor signalling during lactation (i.e., the period of active maternal behaviour) is also observed in other key regions of the neural circuit underlying maternal behaviour, notably in the bed nucleus of the stria terminalis, medial amygdala, and ventromedial nucleus of the hypothalamus (Brown et al. 2011). Studies have yet to specifically manipulate Prlr signalling in these areas and examine behavioural effects.

In vivo, in rodents the maternal brain is exposed sequentially to prolactin, Prl3d, and then Prl3b1. Towards the end of pregnancy, in addition to Prl3b1 secreted from the placenta, there is co-exposure to pituitary-derived prolactin as levels rise again. The precise contribution of these sequential exposures to the priming of maternal behaviour has not been determined in vivo due to the lack of mouse models with disrupted placental expression of either *Prl3d1* or *Prl3b1*. A transgenic mouse carrying the entire *growth hormone* locus from human chromosome 17 has been generated, but while antibodies suggested expression of human placental lactogen in the labyrinth, potentially in the sinusoidal TGCs, no staining was evident in the spongiotrophoblast or *parietal* TGCs which are the main sources of Prl3b1 in rodents (Jin et al. 2009). Moreover, maternal behaviour has not been studied in this model. Even when mice modelling loss or gain in expression of *Prl3d1–3* or *Prl3b1* become available, one of the challenges of this work will be teasing apart the multi-fold roles of these hormones in pregnancy. Changes in the expression of these critical placental hormones will impact many different aspects of pregnancy, such as glucose management, which may have an indirect impact on maternal behaviour. However, it might be possible to combine models in which expression of either *Prl3d1–3* or *Prl3b1* is elevated in the placenta where the dam has a conditional heterozygous loss-of-function of *Prlr* restricted to the brain or, more specifically, the medial preoptic area. In such a model, the higher than normal expression of the placental hormones might compensate for the underactivity of Prlr restoring optimal maternal behaviour.

2.7 Steroid Hormones

Lactogenic hormones are not solely responsible for inducing changes in the maternal brain required for maternal caregiving. Steroid hormones, which are essential for the establishment and continuation of pregnancy, play a role in sensitising the maternal brain (Costa 2016; Napso et al. 2018; Newbern and Freemark 2011). Steroid hormones are synthesised from the substrate cholesterol though a series of enzymic reactions involving cytochromes P450 (CYP) and hydroxysteroid dehydrogenase (HSD) enzymes. The corpus luteum of the ovary, which in a non-pregnant female is a transient ovarian endocrine gland, is a major source of steroid hormones in non-pregnant females. In a rodent pregnancy, the prolactin surges initiated by mating maintain the corpus luteum in the first half of gestation (see Chap. 1). From mid-gestation, the placental lactogen activity of the placenta maintains the corpus luteum until the end of pregnancy, and therefore indirectly maintains steroid hormones synthesis (Galosy and Talamantes 1995; Soares and Talamantes 1982). This is one major difference between rodents and primates, as the primate placenta expresses many of the essential enzymes involved in steroidogenesis whereas many of the genes encoding these enzymes are either expressed at very low levels or not detectably expressed in the mature rodent placenta (Costa 2016; Napso et al. 2018; Newbern and Freemark 2011) (Fig. 2.4). In rodents and many other mammals, progesterone levels increase in the earlier stages of pregnancy and then rapidly decline just before parturition, associated with the initiation of labour. In contrast to progesterone, oestrogen levels rise rapidly just prior to term and then drop rapidly after birth.

The role of oestrogens in priming maternal care was demonstrated through combinations of surgeries involving removal of the uterus (removing all hormonal signals from the placenta), with or without removal of the ovary, from pregnant rats alongside oestrogen infusions (Bridges 2015; Rosenblatt and Siegel 1975; Siegel et al. 1978; Siegel and Rosenblatt 1975a, b, c, 1978). In these experiments, oestrogens were found to stimulate aspects of maternal behaviour in virgin rats when the ovary was removed and in pregnant female rats when the uterus was removed, including when oestrogen was targeted to the medial preoptic area (Fahrbach et al. 1986; Numan et al. 1977, Rosenblatt et al. 1994). Further evidence for the stimulatory role of oestrogens came from studying the impact of deficiency of *oestrogen receptor-alpha* localised to the medial preoptic area in mice, which resulted in increased latency to pup retrieval and reduced nursing and licking (Ribeiro et al. 2012). Progesterone, sometimes referred to as the "priming and timing" hormone, is thought to have a dual role in maternal caregiving, first acting to enhance the actions of oestrogen as levels increase over gestation (Bridges 1984), and then permitting the expression of maternal behaviour when progesterone levels decline at delivery (Bridges 2015). This inhibitory function of progesterone is suggested by experiments maintaining elevated progesterone levels at the end of pregnancy, which was found to delay maternal behaviour (Bridges et al. 1978; Numan 1978; Siegel and Rosenblatt 1975c, 1978), and the finding that the progesterone inhibitory effects were blocked by a progesterone receptor antagonist (Numan et al. 1999).

2.8 Other Placental Hormones and Hormone-Related Genes Expressed by the Placenta

2.8.1 Pregnancy-Specific Glycoproteins

Another distinct and complex family of hormones expressed in the placenta are the pregnancy-specific glycoproteins (Psgs) (Moore and Dveksler 2014). Psgs are related to carcinoembryonic antigen cell adhesion molecules, but are secreted rather than membrane-associated proteins. *Psgs* are highly expressed in the placenta (Fig. 2.4) but only in species where maternal blood comes into direct contact with foetal cells (haemochorial **placentation**), which include rodents and humans (McLellan et al. 2005). In mice all members of this multigene gene family are expressed by the spongiotrophoblast lineage (Wynne et al. 2006). The main function of these hormones are modifying the maternal immune system to prevent rejection of the foetus, and remodelling placental and maternal vasculature (Moore and Dveksler 2014). Currently there is no evidence that Psgs are involved either in priming maternal behaviour or in influencing foetal brain development. However, no study has sought to investigate these potential functions.

2.8.2 Insulin-Like Growth Factor 2

Insulin-like growth factor 2 is a paternally expressed imprinted gene (DeChiara et al. 1991) that encodes a growth factor hormone with a general role in promoting growth and development before birth (Baker et al. 1993; DeChiara et al. 1990). In the adult, Igf2 is important for brain plasticity, learning, and memory (Chen et al. 2011; Schmeisser et al. 2012) and the localised action of Igf2 regulates adult neurogenesis (Ferron et al. 2015). In the mouse placenta, *Igf2* is widely expressed with transcription initiated from four promoters. Three of these promoters drive expression in the foetal endothelium, TGCs, spongiotrophoblast, and, most highly, glycogen cells, while the fourth—called P0—drives *Igf2* expression specifically in labyrinthine cell layers (Constancia et al. 2002; Moore et al. 1997). Igf2 plays a key role in regulating placental growth particularly the proliferation of the glycogen cells and consequently placental glycogen biosynthesis (Esquiliano et al. 2009). Our global hormone profiling indicates that *Igf2* mRNA is expressed at tenfold lower levels than *Prl3b1* in the E14.5 placenta (Fig. 2.4). It is not known to what extent placentally derived Igf2 reaches either the maternal or foetal circulations but there is potential for placental Igf2 to have an indirect effect on foetal brain development. Mice with a targeted deletion of the placental-specific *Igf2* promoter, which are growth-restricted in utero, grow up to be anxious adults (Mikaelsson et al. 2013). Whether this is due to the growth restriction per se, or to the direct function of placental Igf2 hormone on foetal brain development, is not currently known.

2.8.3 Delta-Like 1

Delta-like 1 (*Dlk1*) is another paternally expressed imprinted gene with a role in both brain and placental development (da Rocha et al. 2008; Moon et al. 2002). *Dlk1* encodes a transmembrane domain protein that can be cleaved to generate a soluble form and can therefore function as a hormone (Smas et al. 1997). Loss of *Dlk1* in mice results in increased neural stem cell proliferation at early postnatal stages and a reduced numbers of quiescent neural stem cells and new neurons in the adult (Ferron et al. 2011). In the mouse placenta *Dlk1* is expressed in the foetal endothelium in the labyrinth. Foetally derived Dlk1 has been shown to reach the maternal circulation and influence maternal metabolism (Cleaton et al. 2016). At E14.5, *Dlk1* is expressed at approximately 140-fold lower levels than *Prl3b1* in whole placenta (Fig. 2.4). The relevance of placental Dlk1 for maternal behaviour and foetal brain development has not been explored but there is certainly potential to undertake studies that could be enlightening.

2.9 Low Level Placental Expression of Neuroendocrine Hormones Implicated in Maternal Behaviour

There is a high degree of experimental evidence that prolactin-related peptides and steroid hormones function in priming maternal behaviour in rodents through their direct action on the maternal brain. The placenta is a site of expression for a number of other genes known to contribute to the manifestation of maternal behaviour when expressed centrally in the brain, such as oxytocin, and also enzymes involved in the synthesis of serotonin, dopamine, and melatonin. However, the placenta is thought to express >70% of all coding genes at some level and these factors have not yet been demonstrated to function as hormones on the maternal brain when expressed from this peripheral site. Moreover, they represent highly unlikely candidates for the priming of maternal behaviour due to their exceedingly low levels of expression relative to the placental lactogen hormones, being expressed at 1000–10,000-fold lower levels (Fig. 2.4).

2.10 Placental Hormones and Foetal Brain Development

While it is unlikely that placentally derived hormones and enzymes involved in hormone synthesis expressed at relatively low levels in the placenta have the potential to influence maternal behaviour, it is possible that these factors may impact foetal brain development, as exemplified by two studies in mice, described below.

2.10.1 Serotonin

Serotonin, also known as 5-hydroxytryptamine (5-HT), is a monoamine neurotransmitter, synthesised mainly by the brain and the gut in the non-pregnant animal. Serotonin synthesis begins with the hydrolysis of L-tryptophan to 5-hydroxytryptophan (5-HTP) by tryptophan hydroxylase (Tph1/2). Aromatic L-amino acid decarboxylase (Aadc or Aaad), also known as DOPA decarboxylase (Ddc), then converts 5-HTP to 5-HT. Inactivation of 5-HT occurs through the action of mono-amine oxidases (MaoA/B), with conversion to 5-hydroxyindoleacetic acid (5-HIAA). Therefore, cells expressing both Tph1/2 and Aadc have the capacity to synthesise serotonin and those expressing MaoA/B have the capacity to clear serotonin. During pregnancy the genes required to synthesise serotonin are expressed in the rodent placenta from E10.5 until E18.5, and localised to the labyrinth (Bonnin et al. 2011; Bonnin and Levitt 2011; Wu et al. 2016). More precisely, both *Tph1* and *Aadc* are expressed in SynT-I layer but not SynT-II, and in the sinusoidal TGCs (Wu et al. 2016). In the same study, serotonin was detectable in the sinusoidal TGCs, SynT-I, SynT-II, and foetal blood space close to the endothelial cells from E12.5 onwards. Another study localised serotonin immunologically to the *parietal* TGCs underlying the decidua and to the glycogen cells in the junctional zone in addition to that detectable in the labyrinth (Mao et al. 2020). In the rat placenta, IHC was used to demonstrate that Tph is expressed in the labyrinth sinusoidal TGCs while Mao was localised to the syncytiotrophoblast cell layer (Abad et al. 2020). *Tph1* is the more highly expressed of the two tryptophan hydroxylase enzymes in the mouse placenta at E14.5 but it is still expressed at very low levels (>50,000 times lower than *Prl3b1* in our dataset) (Fig. 2.4). Moreover, *MaoA* is expressed 300-fold times higher than *Tph1* at this same timepoint suggesting that degradation of serotonin is an important feature of the placenta. While it is unlikely that the amount of serotonin synthesised by the placenta is sufficient to directly impact maternal behaviour, a role for placental serotonin in foetal brain development has been suggested. During foetal development, serotonin stimulates cell division, neuronal migration, cell differentiation, and synaptogenesis (Bonnin et al. 2011; Rosenfeld 2021). Foetal neurons provide a direct source of serotonin later in gestation but a study in mice found that serotonin could be detected in the foetal forebrain between E10.5 and E15.5, before the serotonin-producing foetal neurons developed (Bonnin and Levitt 2011). The source of early foetal serotonin was identified as the placenta.

2.10.2 Allopregnanolone

Allopregnanolone (ALLO) is a GABAergic neuro-steroid synthesised from progesterone. ALLO is thought to regulate a number of processes important for behaviour including neurogenesis, cell survival, synapse stabilisation, and myelination (Schumacher et al. 2014). The gene *Akr1c14*, which encodes the enzyme 3α-hydroxysteroid dehydrogenase (3α-HSD), involved in the synthesis of ALLO, is expressed in the mouse placenta, peaking in the second half of gestation (Vacher et al. 2021). This suggests that the placenta is a site for the synthesis of ALLO. In one study, in which ablation of the enzyme was localised to the trophoblast cells of the placenta, changes in the offspring's cerebellar myelination were reported alongside alterations in their behaviour with potential relevance to autism spectrum disorders. Most importantly, these changes could be rescued by administering ALLO or the selective GABAA agonist muscimol (Vacher et al. 2021). These data indicate that the endocrine function of the placenta directly influences foetal brain development, rather than simply supporting foetal brain growth through ensuring the provision of nutrients.

2.11 Beyond Individual Placental Hormones: Imprinted Genes and the Regulation of Placental Hormones in Mice

Definitive experimental proof that hormones produced by or dependent on the placenta are functionally important for maternal caregiving behaviour has been demonstrated in an in vivo system. Placental hormone levels can be manipulated *en mass* using mice carrying genetic modification of imprinted genes shown to regulate development of placental endocrine lineages (John 2017). **Genomic imprinting** describes parent-of-origin specific monoallelic gene expression and is dependent on epigenetic gene regulation (Surani 1998). Genomic imprinting may have evolved in response to the imbalance in parental investment to offspring imposed by viviparity (live birth), with the paternal silencing of genes in the offspring securing increased maternal investment in these offspring and maternal silencing protecting maternal fitness (Moore and Haig 1991; Tucci et al. 2019). Studies reporting the consequence of targeting individual genes for placental development identified both maternally and paternally expressed imprinted genes that function to regulate the number of placental endocrine cells that develop (John and Hemberger 2012; John 2013). Three classically maternally expressed imprinted genes function as negative regulators of key placental endocrine lineages. *Pleckstrin Homology Like Domain Family A Member 2* (*Phlda2*) (Frank et al. 2002; Salas et al. 2004; Tunster et al. 2010, 2016a), *Achaete-scute complex homolog 2* (*Ascl2* aka *Mash2*) (Guillemot et al. 1995; Oh-McGinnis et al. 2011; Tanaka et al. 1997; Tunster et al. 2016b), and *Cyclin dependent kinase inhibitor 1c* (*Cdkn1c*) (Takahashi et al. 2000; Tunster et al. 2011). These three genes function to limit the number of spongiotrophoblast cells that develop. *Ascl2* additionally limits the number of *parietal* TGCs and *Cdkn1c* is important for the proper differentiation of sinusoidal TGCs in the labyrinth. In apparent opposition to the maternally expressed imprinted genes, paternally expressed genes appear to function to increase the number of cells synthesising placental hormones. For example, loss-of-imprinting (two-fold expression) of the paternally expressed *Igf2* gene results in double the number of glycogen cells and more than double the number of *parietal* TGCs, although with no effect on the spongiotrophoblast (Esquiliano et al. 2009) while loss of expression of *Paternally expressed gene 3* (*Peg3*) results in a significant loss of both spongiotrophoblast and glycogen cells (Tunster et al. 2018). By genetically manipulating the expression of individual genes that function as regulators of the placental endocrine lineages in mice, it is possible to increase or decrease the number of cells synthesising placental hormones. These genetic manipulations provide a novel tool to formally demonstrate that signals from the placenta to the maternal brain prime maternal behaviour, because they allow a large number of hormones to be manipulated in a physiologically relevant manner. Moreover, by using different models there is potential to allow delineation of the endocrine lineages and, therefore, the placental hormone combinations that are most important, and which may play a redundant role.

As an example of this approach, the *Phlda2* mouse model was used to demonstrate that this gene influences the quality of maternal care that offspring receive before they are born, through changing the number of cells synthesising placental hormones (Creeth et al. 2018). *Phlda2* encodes a negative regulator of the spongiotrophoblast lineage. Two-fold expression of *Phlda2* (2x*Phlda2*), equivalent to loss-of-imprinting, results in a 40% loss of spongiotrophoblasts alongside decreased expression of all the hormones expressed by spongiotrophoblast cells detectable from as early as E10.5 of development (Tunster et al. 2010, 2016a). Conversely, loss-of-function (*Phlda2* knock-out; KO) has the opposite effect on the spongiotrophoblast with almost double the representation of this lineage in the mature mouse placenta, and increased expression of these same hormones (Tunster et al. 2010, 2016a). Consistent with the function of placental hormones in securing nutrients, 2x*Phlda2* in the conceptus (low placental hormones) results in foetal growth restriction followed by rapid catch-up growth once these mice have been born (Tunster et al. 2010, 2014). Consistent with the function of prolactin-related hormones in inducing maternal behaviour (Grattan 2012; Larsen and Grattan 2012), wildtype (WT) dams carrying 2x*Phlda2* embryos (low placental hormones) exhibit reduced nursing and grooming behaviour towards their pups whereas those carrying *Phlda2*KO embryos (high placental hormones) exhibit increased maternal caregiving (Creeth et al.). The experimental dams exposed to offspring with abnormal doses of *Phlda2* showed alterations to their brains before their pups were born formally demonstrating that some priming of the maternal brain occurred during gestation. Gene alterations were present in both the maternal hypothalamus and hippocampus including alterations in G protein-coupled receptor (GPCR) pathways, olfactory transduction pathways, and gonadotropin-releasing hormone signalling pathways all of which are implicated in maternal care. This study also involved a fostering experiment whereby genetically WT pups were given to WT dams to nurse, either dams that had carried WT embryos in utero (control group) or dams that had been exposed to higher levels of placental hormones through carrying all *Phlda2*KO embryos (experimental group). In this scenario, WT dams exposed to placental endocrine insufficiency were nursing WT pups. Enhanced maternal caregiving was still evident in dams from the experimental group despite the considerable disruption that fostering involves. There were two critical aspects to this study. First, the WT genetic status of the dam was maintained, in order to isolate the function of *Phlda2* in the offspring achieved through the process of recipient transfer. In this specific experiment, all the dams were genetically WT but it is possible to transfer any combination of genetic modification to the dams and offspring (Box 2.3). The second key aspect to this study was the fostering (Box 2.4). Newborn pups are known to influence their dams through smell, sucking, and vocalisation. Consequently, it is possible that alterations in behaviour are not due to programming in pregnancy (Potter et al. 2019). By providing WT pups to the dams exposed during pregnancy to mutant placenta, it was possible to formally demonstrate that the effect of maternal behaviour occurred before birth (prenatal programming). However, while dams exposed to higher levels of hormones spent more time on pup-focused behaviours and their own selfcare, their performance in the nest-building task was significantly poorer than in the control scenario. Instead, dams exposed to low levels of placental hormones excelled at this task. In addition, dams exposed to higher levels of hormones were slower to retrieve their pups. It was therefore not simply the case that hormones synthesised by the placental spongiotrophoblast lineage function uniformly to enhance all aspects of maternal behaviour.

Box 2.3: Recipient Transfer

One of the challenges with studying maternal behaviour in genetically altered mice is distinguishing between the action of genes in the mother and those in the foetus since, in most cases, it necessary to breed from genetically altered females in order to generate genetically modified offspring. Recipient transfer of embryos is a technique that allows a distinction to be made. This procedure involves the superovulation or natural mating of virgin females, usually aged 5–7 weeks, which are then mated with suitable studs to generate fertilised embryos of a particular genetic status. Embryos are flushed from oviducts into a suitable media (ex. M2 media from Sigma) 2 days after a discernible plug on E1.5 and incubated (ex. KSOM from Specialty Media) at 37 °C/5% CO_2 under embryo tested mineral oil for 1–2 h before embryos are transferred by bilateral oviduct transfer into E0.5 pseudopregnant females aged 5–7 weeks to establish a surrogate pregnancy. The number of embryos transferred is dependent on the strain of mice and also, potentially, the genetic modification. It is usual to transfer E1.5 embryos into E0.5 recipient dams but embryos can be transferred at a later stage when they are blastocysts, which may enable non-surgical transfer. Any combination of genetic modification to the dams and offspring can be generated by this approach.

Box 2.4: Fostering

To distinguish between the prenatal and postnatal contribution of a specific genetic modification, newborn pups can be fostered between relevant dams. Pups must be born on the same day and fostering performed on the day of birth. For a natural mating, several pairings of the relevant genotypes are set up and plug checked in order to generate pregnant females that will deliver on the same day. Consider how many pairings will be needed for a particular strain to maximise the chances of dams delivering on the same day. For example, fully WT matings can be set up to generate WT pups and matings between genetically modified mice set up to generate experimental dams with genetically altered pups. Alternatively, control and experimental dams can be generated by recipient transfer on the same day, which will then deliver on the same day. Pregnant dams are visually monitored twice daily from E18.5 for the presence of newborn pups without disturbing the dam. Once pups are observed, they are removed from the nest by an experimenter with gloved hands. Pups are counted and can be weighed. Litter A is then rubbed with some bedding from Dam B and pups placed in cage with Dam B. Pups from litter B are rubbed with some bedding from Dam A and pups then placed in cage with Dam A. The inclusion of non-cross fostered litters in the experiment controls for the impact of fostering. Particularly important for studies of maternal behaviour, it is also important to ensure litters of a normal size for the particular strain are used. The average litter size is 6–7 for a 129 strain and 7–8 for a BL6 strain.

The *Phlda2* mouse model is not the only model in which the placental endocrine lineages have been shown to influence maternal behaviour. Loss-of-function of *Peg3* results in a substantial loss of both the spongiotrophoblast and the glycogen cell lineages (Tunster et al. 2018). WT dams carrying litters of all *Peg3*KO foetuses also display atypical maternal behaviour (McNamara et al. 2018). When pregnant dams were first transferred from their home cage into Phenotyper cages at E14.5, these dams moved significantly less than WT dams carrying WT foetuses particularly evident during the first hour. RNA sequence analysis did not identify global changes to the maternal brain transcriptomes of the hypothalamus or hippocampus before birth. However, this approach does not exclude changes within specific regions or cell populations. No differences in anxiety-like behaviour were detected before birth, but WT dams carrying and caring for *Peg3*KO offspring were highly anxious after the birth and considerably slower to retrieve their pups. *Peg3* loss of function was also found to cause a deficit in pup vocalisations. As the pre- and postnatal contribution to the mother's behaviour of loss-of-function of *Peg3* in the offspring was not distinguished through a fostering experiment, maternal anxiety and slower pup retrieval cannot be confidentially attributed to the placental changes reported in this model. However, it is of interest to note that low levels of prolactin in mice are linked to elevated anxiety (Larsen and Grattan 2010) and in human pregnancy, low placental lactogen at term is significantly associated with maternally reported anxiety symptoms (Sumption et al. 2020).

Despite these caveats, both studies formally demonstrate that manipulating the number of cells in the placenta that synthesise placental hormones can alter the way in which mothers behave during pregnancy (McNamara et al. 2018) and after she has given birth (Creeth et al. 2018), which is completely consistent with the observations that placental lactogens play a major role in the priming of maternal caregiving in mice. There are some improvements with the experimental design that can now be made. In both models, the expression of a large number of hormones was altered. It may be possible to undertake further manipulations in these models to try and identify specific placental hormones, for example, by reintroducing individual hormones into the 2x*Phlda2* model (low placental hormones) to restore their level to normal or by manipulating the expression of receptors, such as *Prlr*, in the maternal brain in the *Phlda2*KO model.

2.12 Relevance to Human Pregnancy and Translational Implications

One of the greatest challenges of reproductive studies is translating findings between experimental models and humans. Mice and rats provide highly effective systems in which individual hormones and genes can be precisely manipulated, and the consequences for either maternal behaviour or behaviour in the offspring studied. However, the placenta is the most diverse tissue that mammals possess, with differences both in the recognisable cell types and the nature and timing of expression of placental hormones. While both human and rodent placenta are haemochorial with maternal blood in direct contact with trophoblast cells, the human placenta is structurally distinct (Fig. 2.6). At maturity, the human placenta consists of multiple placental villi which are composed of an inner core of extraembryonic mesenchymal cells, Hoffbauer cells, and foetal blood vessels covered by a single layer of cytotrophoblast cells, which remain proliferative, and an outer layer of multinucleated syncytiotrophoblast (Hemberger et al. 2020; Soares 2004; Turco and Moffett 2019) (Fig. 2.6). At the tip of these villi, individual cytotrophoblast cells invade into decidua to form extravillous trophoblast within the maternal decidua, with some similarity to the invasive spiral artery TGCs of rodents (Hemberger et al. 2020; Woods et al. 2018). The major site of placental hormone synthesis in humans is the syncytiotrophoblast, which is in direct contact with maternal blood (Burton and Fowden 2015; Burton and Jauniaux 2015) (Fig. 2.6). While the human placenta has traditionally been described as possessing three main types of trophoblast (cytotrophoblast, syncytiotrophoblast cells, and extravillous trophoblast), recent single cell RNA-sequencing analysis has identified 14 subtypes of placental cells from first and second trimester placenta (Liu et al. 2018). Three subtypes of cytotrophoblast cells and five subtypes of extravillous trophoblast have been suggested. In addition to this further classification of cell types, this study identified the extravillous trophoblast as a second source of

Fig. 2.6 Diagrammatic representation of the human placenta. The foetally-derived components of the human placenta consist of foetal blood vessels and trophoblast cell types organised into finger-like structures called villi, which are bathed in maternal blood. The primary sites of hormone synthesis are the syncytiotrophoblast, which covers the surface of placental villi, and the extravillous trophoblast, which migrate into the maternal uterus to modify maternal blood vessels, with some similarity to the mouse spiral artery giant trophoblast cells. Both cell types synthesise placental lactogens

placental peptide hormones (Liu et al. 2018). To summarise, key differences between the human and rodent placenta are the nature of the placental endocrine lineages. In rodents the spongiotrophoblast and TGCs are the major endocrine lineages, with TGCs undergoing endoreduplication and gene amplification but remaining as single cells. In humans, the syncytiotrophoblast cells and, to some extent, the extravillous trophoblast are the major endocrine lineages. The **human syncytiotrophoblast** is generated by fusion into vast tracks of syncytium and these cells have both a transport and an endocrine function. In the rodent placenta, there are similarly named multinucleated structures located in the labyrinth called syncytiotrophoblast, but these principally function in nutrient transport. Despite these key differences in structure and development, the placental endocrine lineages in both humans and rodents are in direct contact with maternal blood, which means that placental hormones are secreted directly into the maternal circulation.

2.13 Human Placental Hormones

As in the rodent placenta, the peptide hormones most highly expressed in the human placenta belong to the prolactin/growth hormone gene family and the PSG family (Fig. 2.7). Human placental lactogen arose from duplications centred on the ancestral GH gene copy, generating four variants expressed in the placenta (Newbern and Freemark 2011). This is in contrast to rodents, where the ancestral prolactin gene underwent extensive gene duplication. In humans, two identical placental lactogen peptides are encoded by *CHORIONIC SOMATOMAMMOTROPIN HORMONE 1* and *2* (aka *HPL-A* and *HPL-B*). There is a related non-coding pseudogene *CSHL1* (aka *HPL-L*). The 4th gene family member expressed in the placenta is *GROWTH HORMONE 2* (aka GH-V), encoding placental growth hormone (Handwerger and Freemark 2000). Human *GH1*, *CSHL1*, *CSH1*, *GH2*, and *CSH2* are all located within a single locus on human chromosome 17 and human *PROLACTIN* is located on chromosome 6 (Fig. 2.5c). Placental lactogen is synthesised in extraordinary amounts by the syncytiotrophoblast and extravillous trophoblast lineages throughout pregnancy with the rate of secretion reaching 1–3 g/day towards term, exceeding that of any other peptide hormone (Kaplan et al. 1968). Human studies suggest that the majority of placental lactogen is released into the maternal circulation with approximately 0.05% transferred to the developing foetus (Linnemann et al. 2000), with 5–7 μg/ml and 20–50 ng/ml in maternal and foetal blood, respectively, at term (Newbern and Freemark 2011; Walker et al. 1991). In contrast to rodents, where prolactin is supressed during pregnancy, it is important to note that in humans there is a concurrent rise in prolactin and placental lactogen throughout pregnancy (Aghaeepour et al. 2018). Placental growth hormone, also expressed from the syncytiotrophoblast, is detectable in the maternal bloodstream from 12 weeks until term. Members of prolactin/growth hormone gene family have stronger affinities for their cognate receptors, but not exclusively, which means there may be some functional crossover (Handwerger and Freemark 2000; Jomain et al. 2007; Lowman et al. 1991) (Fig. 2.5c). Prolactin signals exclusively via PRLR whereas growth hormone can signal via both GHR and PRLR, placental lactogen has higher affinity for PRLR with very weak affinity for GHR, and placental growth hormone has high affinity for GHRs, with negligible affinity for PRLR (Haig 2008). Another notable difference between primates and rodents is the source of steroid hormones. In rodents, the ovary remains the primary source of steroid hormones but in human pregnancy, enzymes for the synthesis of steroid hormones are expressed by the syncytiotrophoblast of the placenta at high levels (Fig. 2.7). Through the expression of steroid enzymes, the human placenta becomes a direct source of progesterone and oestrogens for most of pregnancy, though still being reliant on androgen precursors for the production of oestrogens. Again, in contrast to rodents, in which levels of progesterone decline before delivery, progesterone blood levels increase throughout a human pregnancy peaking during the last 4 weeks of gestation and decreasing only after delivery of the placenta. Oestrogen levels also increase significantly during human pregnancy and decline rapidly when the placenta is delivered. Human chorionic gonadotropin is the first placental hormone detectable in the human mother's blood very soon after conception. Human chorionic gonadotropin primarily functions to stimulate the ovaries to synthesise steroid hormones in the early stages of pregnancy, but chorionic gonadotropin beta genes have not been found in mammals other than primates (Maston and Ruvolo 2002). Thus, a major hormone absolutely required for human pregnancy is not produced during gestation in rodents. Humans and other primates uniquely express placental corticotrophin-releasing hormone, which floods the maternal circulation during the second and third trimester reaching peak levels near term (Petraglia et al. 1987). Corticotrophin-releasing hormone functions to stimulate maternal production of cortisol, known to influence both maternal and offspring behaviour (Thomson 2013). Thus, there are a number of key differences between rodents and human placental hormones, which must be considered when interpreting studies on rodents.

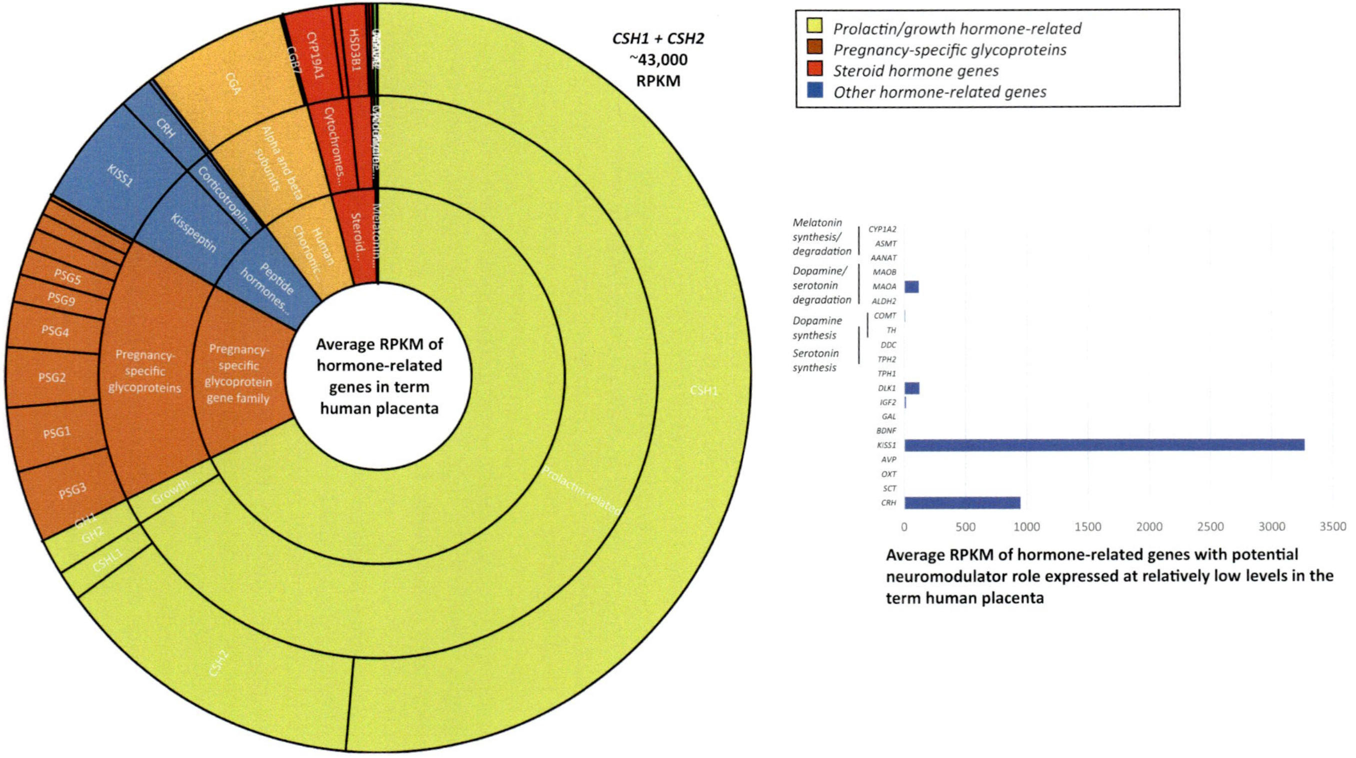

Fig. 2.7 Global hormone transcriptional profiling of the term human placenta. Pie chart shows relative expression levels of key peptide hormone and hormone-related genes in the human placenta determined from total RNA sequencing of term placenta performed to a depth of 30–32 Mb reads per sample; counts averaged from 40 placental samples, 16 male and 24 female. The outermost circle of the pie chart provides individual gene names while the inner circle indicates gene families. Genes expressed at considerably lower levels in comparison to *CSH1/2* are also shown in the bar graph. *CSH1* and *CSH2*, which combined encode human placental lactogen, are the most highly expressed peptide hormones in the placenta at approaching 43,000 RPKM in this dataset. RPKM: reads per kilobase of transcript, per million mapped reads

2.14 Perspectives

2.14.1 A Placental Origin for Maternal Mood Disorders and Neurodevelopmental Disorders?

The traditional focus of placental studies in humans has been in relation to pregnancy complications such as foetal growth restriction, preeclampsia, and gestational diabetes. Where longer term behavioural outcomes for children have been studied, the focus has been on the role of the placenta in transmitting damaging factors from the mother to her developing child. For example, there has been considerable focus on maternal cortisol which might be transmitted to offspring impacting their later life health (Glover 2015). Cortisol is released from the adrenal gland under conditions of acute or chronic stress. In pregnancy, the placenta normally prevents excess cortisol reaching the foetus through the action of the enzyme 11beta-hydroxysteroid dehydrogenase type 2 but there may be conditions where the placenta fails to provide an effective barrier such as when this enzyme is downregulated (O'Donnell et al. 2012). Despite this focus on the barrier function of the placenta and differences in placental hormones between humans and rodents, there is a growing body of evidence that the placenta plays a much more active role in influencing maternal mood in pregnancy and neurodevelopmental outcomes for offspring through the action of placental hormones.

2.14.2 Mood Disorders and Placental Endocrine Insufficiency

Human mothers are at high risk of developing mood symptoms in pregnancy with one in seven women reporting clinically concerning symptoms of depression and one in four women reporting concerning levels of anxiety (Heron et al. 2004; Janssen et al. 2018; Lockwood Estrin et al. 2019). Depression in pregnancy is commonly comorbid with anxiety (Glover 2014) and both these mood disorders have both been linked to a higher risk of low birth weight and atypical behavioural outcomes (Janssen et al. 2016b). Hormonal changes represent prime candidates linking pregnancy to the higher vulnerability for depression because their levels change so dramatically driven by the placenta (Burton and Fowden 2015). Studies in rodents highlighted the importance of prolonged exposure to oestrogen, progesterone, and the lactogenic hormones prolactin and placental lactogens, in the induction of maternal care behaviours (Creeth et al. 2019). In two independent animal models, placental endocrine insufficiency has been shown to result in maternal neglect of pups (Creeth et al. 2018) and maternal anxiety (McNamara et al. 2018), again implicating a placental origin for maternal mood disorders. In human pregnancy, postnatal depression (Abou-Saleh et al. 1998; Groer and Morgan 2007) and anxiety (Asher et al. 1995) have been linked to low maternal serum prolactin. Low expression of placental lactogen has been reported in term placenta from women with clinically diagnosed depression in pregnancy and those self-reporting in depressive symptoms at term (Janssen et al. 2016a). Low serum placental lactogen at term has been linked to symptoms of postnatal depression and anxiety up to 10 weeks after birth (Sumption et al. 2020). While these associations do not demonstrate a causal relationship, it is important to note that this study found that the association between serum placental lactogen and mood was stronger when samples from women prescribed selective serotonin reuptake inhibitors (SSRIs) were removed from the dataset. SSRIs are commonly prescribed in pregnancy to manage mood symptoms (Dubovicky et al. 2017). SSRIs function to increase serotonin levels in the brain by blocking the serotonin negative feedback loop, and this has been associated with hyperprolactinemia in rodents (Gil-ad et al. 1976; Kamberi et al. 1971) and in humans (Goodnick et al. 2000). One explanation for the stronger relationship between mood symptoms and placental lactogens is that SSRIs are functioning to elevate prolactin levels locally, compensating for the reduced levels of placental lactogen from the placenta. It will be important to measure both prolactin and placental lactogen over gestation in this vulnerable population of women. Furthermore, the focus in animal models has been to study maternal behaviour, with little attention paid to quantifying depressive-like behaviours, whereas studies in human mothers have focused principally on depression and anxiety rather than maternal caregiving behaviour. Considerably more work is required to establish whether there is a functional link between placental endocrine insufficiency and changes in the mothers' caregiving. However, both depression and anxiety in human pregnancy are linked to poor mother-infant bonding and it is tempting to speculate that placental endocrine insufficiency in a human pregnancy impairs maternal caregiving, as it does in mice (Creeth and John 2020) (Fig. 2.8).

2.14.3 The Human Placenta and Neurodevelopmental Disorders

While there is a dearth of data on maternal mood and behaviour in the context of placental hormones, there is much evidence that the function of the placenta is critically important for healthy neurodevelopment in offspring (Janssen et al. 2016b; Kratimenos and Penn 2019), for example, in cases of under/overnutrition in pregnancy, where the foetal brain is exposed to suboptimal levels of nutrition at critical stages of brain development or when the placental barrier breaks down, exposing the foetus to damaging maternal factors. With the development of high throughput and sensitive techniques, it is now possible to generate proteomic profiles of both maternal and cord blood serum. In one such study, a comparison made between protein concentrations in foetal blood as it leaves the placenta through the umbilical vein with the levels in foetal blood as it returns from the foetus to the placenta via the umbilical artery revealed 341 placental hormones reaching the foetal circulation (Michelsen et al. 2019). Again, determining the functional significance of individual hormones presents significant challenges.

In summary, the placenta is an endocrine organ that contributes to the priming of maternal caregiving in rodents and potentially also humans. While in vivo experiments are required to fully tease apart the functional contribution of individual placental hormones to maternal caregiving, both overdevelopment and

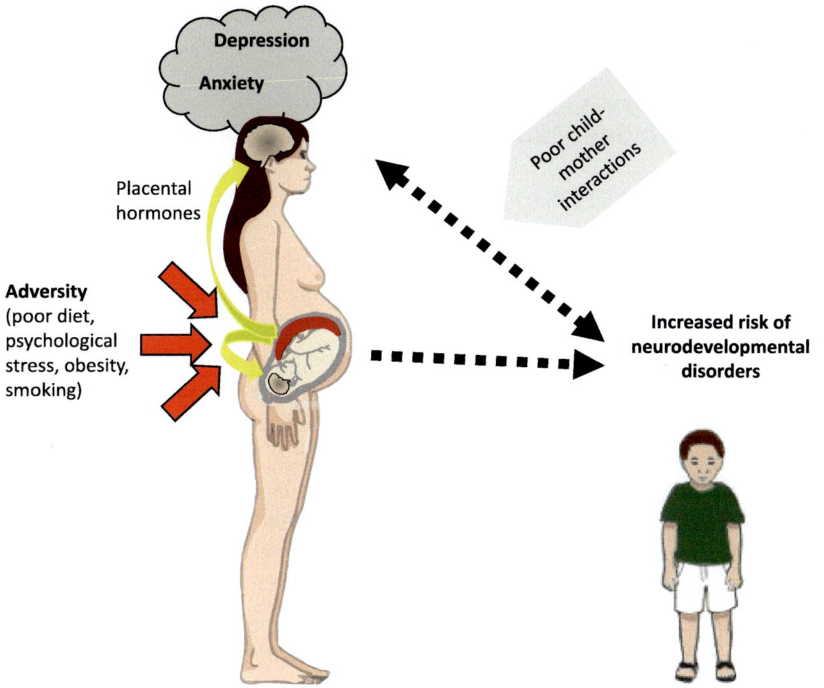

Fig. 2.8 Placental endocrine insufficiency. Animal models of prenatal adversity report alterations in the placenta consistent with defects in the synthesis of placental hormones. Placental endocrine insufficiency prevents adaptation of the maternal brain required for motherhood in rodents, which may manifest as symptoms of depression and anxiety in a human pregnancy, and poor mother–child interactions, increasing the risk of neurodevelopmental and metabolic disorders in children. Placental endocrine insufficiency may also directly impact neurodevelopment in utero through reduced nutrient supply resulting in foetal growth restriction or through the direct action of placental hormones on the developing foetal brain. Figure adapted from Creeth and John (2020)

underdevelopment of the placental endocrine lineages can contribute to changes in the behaviour of the mother towards her offspring, with likely consequences for their lifelong health. The placenta can influence foetal brain development indirectly either through allowing the transmission of damaging factors from the mother or by driving foetal growth restriction to alter later life health of offspring. Finally, expression of neuromodulators by the placenta, albeit in some cases at exceedingly low levels, may directly influence developmental processes in the foetal brain. For these reasons, we should study the neuroendocrine function of the placenta in considerably more detail. In addition, it is imperative that we identify factors that disrupt the expression of placental hormones, as has been reported for human obesity and placental lactogen expression (Cattini et al. 2020; Jin et al. 2018), as placental endocrine insufficiency could provide a mechanism linking prenatal adversity to negative behavioural outcomes for both mothers and their offspring (Creeth and John 2020).

Acknowledgements With thanks to members of Preg_lab past and present, and Hannah Tyson for her comments on the chapter. Research relevant to this chapter was funded by Medical Research

Council (*MR/M013960/1*) and *Biotechnology and Biological Sciences Research Council* (*BB/J015156; BB/P002307/1*; BB/P008623/1).

Key References: See Main List for Reference Details

Bridges and Freemark (1995). First report that (human) placental lactogen induces maternal behaviour in non-pregnant rats.

Simmons et al. (2008). Characterisation of the spatial and temporal expression of prolactin family members in the mouse placenta. This work identified the spongiotrophoblast as a major endocrine lineage.

Tunster et al. (2016a). First demonstration that the maternally expressed imprinted gene *Phlda2* is a key negative regulator of the **mouse spongiotrophoblast** lineage, which is the major endocrine lineage of the mouse placenta. Loss-of-function of *Phlda2* increased the size of this lineage and the global expression of key placental hormones, including mouse placental lactogen.

Creeth et al. (2018). First demonstration that maternal behaviour is modulated by the dosage of *Phlda2* in the mouse foetus. Genetically wild type dams exposed to higher levels of *Phlda2* in their offspring (lower placental hormones) were more focused on nest building while those exposed to lower levels of *Phlda2* in their offspring (higher placental hormones) were more focused on nurturing and grooming their pups.

Bonnin et al. (2011). First experimental demonstration that the mouse placenta is a source of serotonin for the foetal brain, with the potential to influence foetal brain development and later-life behaviour.

Sumption et al. (2020). Reported low placental lactogen associated with symptoms of postnatal depression and anxiety.

References

Abad C, Karahoda R, Kastner P, Portillo R, Horackova H, Kucera R, Nachtigal P, Staud F (2020) Profiling of tryptophan metabolic pathways in the rat fetoplacental unit during gestation. Int J Mol Sci 21(20):7578

Abou-Saleh MT, Ghubash R, Karim L, Krymski M, Bhai I (1998) Hormonal aspects of postpartum depression. Psychoneuroendocrinology 23:465–475

Aghaeepour N, Lehallier B, Baca Q, Ganio EA, Wong RJ, Ghaemi MS, Culos A, El-Sayed YY, Blumenfeld YJ, Druzin ML, Winn VD, Gibbs RS, Tibshirani R, Shaw GM, Stevenson DK, Gaudilliere B, Angst MS (2018) A proteomic clock of human pregnancy. Am J Obstet Gynecol 218:347 e1–347 e14

Ain R, Dai G, Dunmore JH, Godwin AR, Soares MJ (2004) A prolactin family paralog regulates reproductive adaptations to a physiological stressor. Proc Natl Acad Sci U S A 101:16543–16548

Alam SM, Ain R, Konno T, Ho-Chen JK, Soares MJ (2006) The rat prolactin gene family locus: species-specific gene family expansion. Mamm Genome 17:858–877

Asher I, Kaplan B, Modai I, Neri A, Valevski A, Weizman A (1995) Mood and hormonal changes during late pregnancy and puerperium. Clin Exp Obstet Gynecol 22:321–325

Augustine RA, Kokay IC, Andrews ZB, Ladyman SR, Grattan DR (2003) Quantitation of prolactin receptor mRNA in the maternal rat brain during pregnancy and lactation. J Mol Endocrinol 31:221–232

Baker J, Liu J-P, Robertson EJ, Efstratiadis A (1993) Role of insulin-like growth factors in embryonic and post natal growth. Cell 75:73–82

Bao L, Tessier C, Prigent-Tessier A, Li F, Buzzio OL, Callegari EA, Horseman ND, Gibori G (2007) Decidual prolactin silences the expression of genes detrimental to pregnancy. Endocrinology 148:2326–2334

Barkley MS, Bradford GE, Geschwind II (1978) The pattern of plasma prolactin concentration during the first half of mouse gestation. Biol Reprod 19:291–296

Bonnin A, Levitt P (2011) Fetal, maternal, and placental sources of serotonin and new implications for developmental programming of the brain. Neuroscience 197:1–7

Bonnin A, Goeden N, Chen K, Wilson ML, King J, Shih JC, Blakely RD, Deneris ES, Levitt P (2011) A transient placental source of serotonin for the fetal forebrain. Nature 472:347–350

Bouillot S, Rampon C, Tillet E, Huber P (2006) Tracing the glycogen cells with protocadherin 12 during mouse placenta development. Placenta. 27(8):882–888. https://doi.org/10.1016/j.placenta.2005.09.009. Epub 2005 Nov 2. PMID: 16269175

Bredy TW, Grant RJ, Champagne DL, Meaney MJ (2003) Maternal care influences neuronal survival in the hippocampus of the rat. Eur J Neurosci 18:2903–2909

Bridges RS (1984) A quantitative analysis of the roles of dosage, sequence, and duration of estradiol and progesterone exposure in the regulation of maternal behavior in the rat. Endocrinology 114:930–940

Bridges RS (1990) Endocrine regulation of parental behavior in rodents. Oxford University Press, New York

Bridges RS (2015) Neuroendocrine regulation of maternal behavior. Front Neuroendocrinol 36:178–196

Bridges RS, Freemark MS (1995) Human placental lactogen infusions into the medial preoptic area stimulate maternal behavior in steroid-primed, nulliparous female rats. Horm Behav 29:216–226

Bridges RS, Hays LE (2005) Steroid-induced alterations in mRNA expression of the long form of the prolactin receptor in the medial preoptic area of female rats: Effects of exposure to a pregnancy-like regimen of progesterone and estradiol. Brain Res Mol Brain Res 140:10–16

Bridges RS, Ronsheim PM (1990) Prolactin (PRL) regulation of maternal behavior in rats: bromocriptine treatment delays and PRL promotes the rapid onset of behavior. Endocrinology 126:837–848

Bridges RS, Scanlan VF (2005) Maternal memory in adult, nulliparous rats: effects of testing interval on the retention of maternal behavior. Dev Psychobiol 46:13–18

Bridges RS, Rosenblatt JS, Feder HH (1978) Serum progesterone concentrations and maternal behavior in rats after pregnancy termination: behavioral stimulation after progesterone withdrawal and inhibition by progesterone maintenance. Endocrinology 102:258–267

Bridges RS, DiBiase R, Loundes DD, Doherty PC (1985) Prolactin stimulation of maternal behavior in female rats. Science 227:782–784

Bridges RS, Numan M, Ronsheim PM, Mann PE, Lupini CE (1990) Central prolactin infusions stimulate maternal behavior in steroid-treated, nulliparous female rats. Proc Natl Acad Sci U S A 87:8003–8007

Bridges RS, Robertson MC, Shiu RP, Friesen HG, Stuer AM, Mann PE (1996) Endocrine communication between conceptus and mother: placental lactogen stimulation of maternal behavior. Neuroendocrinology 64:57–64

Bridges RS, Robertson MC, Shiu RP, Sturgis JD, Henriquez BM, Mann PE (1997) Central lactogenic regulation of maternal behavior in rats: steroid dependence, hormone specificity, and behavioral potencies of rat prolactin and rat placental lactogen I. Endocrinology 138:756–763

Brown RS, Herbison AE, Grattan DR (2011) Differential changes in responses of hypothalamic and brainstem neuronal populations to prolactin during lactation in the mouse. Biol Reprod 84:826–836

Brown RS, Wyatt AK, Herbison RE, Knowles PJ, Ladyman SR, Binart N, Banks WA, Grattan DR (2016) Prolactin transport into mouse brain is independent of prolactin receptor. FASEB J 30:1002–1010

Brown RSE, Aoki M, Ladyman SR, Phillipps HR, Wyatt A, Boehm U, Grattan DR (2017) Prolactin action in the medial preoptic area is necessary for postpartum maternal nursing behavior. Proc Natl Acad Sci U S A 114:10779–10784

Bu P, Alam SM, Dhakal P, Vivian JL, Soares M (2016) A prolactin family paralog regulates placental adaptations to a physiological stressor in the mouse. Biol Reprod 94(5):107

Burton GJ, Fowden AL (2015) The placenta: a multifaceted, transient organ. Philos Trans R Soc Lond B Biol Sci 370:20140066

Burton GJ, Jauniaux E (2015) What is the placenta? Am J Obstet Gynecol 213:S6 e1, S6–8

Cameron NM, Fish EW, Meaney MJ (2008) Maternal influences on the sexual behavior and reproductive success of the female rat. Horm Behav 54:178–184

Carter AM (2012) Evolution of placental function in mammals: the molecular basis of gas and nutrient transfer, hormone secretion, and immune responses. Physiol Rev 92:1543–1576

Carter AM (2018) Comparative placentation. In: Skinner MK (ed) Encyclopedia of reproduction. Academic, Cambridge, pp 129–136

Carter AM (2020) Animal models of human pregnancy and placentation: alternatives to the mouse. Reproduction 160:R129–R143

Carter AM, Enders AC (2004) Comparative aspects of trophoblast development and placentation. Reprod Biol Endocrinol 2:46

Cattini PA, Jin Y, Jarmasz JS, Noorjahan N, Bock ME (2020) Obesity and regulation of human placental lactogen production in pregnancy. J Neuroendocrinol 32(11):e12859

Champagne FA, Francis DD, Mar A, Meaney MJ (2003) Variations in maternal care in the rat as a mediating influence for the effects of environment on development. Physiol Behav 79:359–371

Champagne FA, Curley JP, Keverne EB, Bateson PP (2007) Natural variations in postpartum maternal care in inbred and outbred mice. Physiol Behav 91:325–334

Chen DY, Stern SA, Garcia-Osta A, Saunier-Rebori B, Pollonini G, Bambah-Mukku D, Blitzer RD, Alberini CM (2011) A critical role for IGF-II in memory consolidation and enhancement. Nature 469:491–497

Cindrova-Davies T, Jauniaux E, Elliot MG, Gong S, Burton GJ, Charnock-Jones DS (2017) RNA-seq reveals conservation of function among the yolk sacs of human, mouse, and chicken. Proc Natl Acad Sci U S A 114:E4753–E4761

Cleaton MA, Dent CL, Howard M, Corish JA, Gutteridge I, Sovio U, Gaccioli F, Takahashi N, Bauer SR, Charnock-Jones DS, Powell TL, Smith GC, Ferguson-Smith AC, Charalambous M (2016) Fetus-derived DLK1 is required for maternal metabolic adaptations to pregnancy and is associated with fetal growth restriction. Nat Genet 48(12):1473–1480

Coan PM, Conroy N, Burton GJ, Ferguson-Smith AC (2006) Origin and characteristics of glycogen cells in the developing murine placenta. Dev Dyn 235:3280–3294

Constancia M, Hemberger M, Hughes J, Dean W, Ferguson-Smith A, Fundele R, Stewart F, Kelsey G, Fowden A, Sibley C, Reik W (2002) Placental-specific IGF-II is a major modulator of placental and fetal growth. Nature 417:945–948

Costa MA (2016) The endocrine function of human placenta: an overview. Reprod Biomed Online 32:14–43

Creeth HDJ, John RM (2020) The placental programming hypothesis: Placental endocrine insufficiency and the co-occurrence of low birth weight and maternal mood disorders. Placenta 98:52–59

Creeth HDJ, McNamara GI, Tunster SJ, Boque-Sastre R, Allen B, Sumption L, Eddy JB, Isles AR, John RM (2018) Maternal care boosted by paternal imprinting in mammals. PLoS Biol 16:e2006599

Creeth HDJ, McNamara GI, Isles AR, John RM (2019) Imprinted genes influencing the quality of maternal care. Front Neuroendocrinol 53:100732

Cross JC (2006) Placental function in development and disease. Reprod Fertil Dev 18:71–76

Curley JP, Champagne FA (2016) Influence of maternal care on the developing brain: Mechanisms, temporal dynamics and sensitive periods. Front Neuroendocrinol 40:52–66
da Rocha ST, Edwards CA, Ito M, Ogata T, Ferguson-Smith AC (2008) Genomic imprinting at the mammalian Dlk1-Dio3 domain. Trends Genet 24:306–316
DeChiara TM, Efstratiadis A, Robertson EJ (1990) A growth-deficiency phenotype in heterozygous mice carrying an insulin-like growth factor II gene disrupted by targeting. Nature 345:78–80
DeChiara TM, Robertson EJ, Efstratiadis A (1991) Parental imprinting of the mouse insulin-like growth factor II gene. Cell 64:849–859
Dobolyi A, Grattan DR, Stolzenberg DS (2014) Preoptic inputs and mechanisms that regulate maternal responsiveness. J Neuroendocrinol 26:627–640
Dubovicky M, Belovicova K, Csatlosova K, Bogi E (2017) Risks of using SSRI / SNRI antidepressants during pregnancy and lactation. Interdiscip Toxicol 10:30–34
Dulac C, O'Connell LA, Wu Z (2014) Neural control of maternal and paternal behaviors. Science 345:765–770
Esquiliano DR, Guo W, Liang L, Dikkes P, Lopez MF (2009) Placental glycogen stores are increased in mice with H19 null mutations but not in those with insulin or IGF type 1 receptor mutations. Placenta 30:693–699
Fahrbach SE, Morrell JI, Pfaff DW (1986) Identification of medial preoptic neurons that concentrate estradiol and project to the midbrain in the rat. J Comp Neurol 247:364–382
Ferron SR, Charalambous M, Radford E, McEwen K, Wildner H, Hind E, Morante-Redolat JM, Laborda J, Guillemot F, Bauer SR, Farinas I, Ferguson-Smith AC (2011) Postnatal loss of Dlk1 imprinting in stem cells and niche astrocytes regulates neurogenesis. Nature 475:381–385
Ferron SR, Radford EJ, Domingo-Muelas A, Kleine I, Ramme A, Gray D, Sandovici I, Constancia M, Ward A, Menheniott TR, Ferguson-Smith AC (2015) Differential genomic imprinting regulates paracrine and autocrine roles of IGF2 in mouse adult neurogenesis. Nat Commun 6:8265
Frank D, Fortino W, Clark L, Musalo R, Wang W, Saxena A, Li CM, Reik W, Ludwig T, Tycko B (2002) Placental overgrowth in mice lacking the imprinted gene Ipl. Proc Natl Acad Sci U S A 99:7490–7495
Franks B, Curley JP, Champagne FA (2011) Measuring variations in maternal behavior: relevance for studies of mood and anxiety. In: Gould TD (ed) Mood and anxiety related phenotypes in mice. Humana, Totowa
Freeman ME, Neill JD (1972) The pattern of prolactin secretion during pseudopregnancy in the rat: a daily nocturnal surge. Endocrinology 90:1292–1294
Freyer C, Renfree MB (2008) The mammalian yolk sac placenta. J Exp Zool Part B 312(6):545–554
Furuta M, Bridges RS (2005) Gestation-induced cell proliferation in the rat brain. Brain Res Dev Brain Res 156:61–66
Gallagher JM, Nephew BC, Poirier G, King JA, Bridges RS (2019) Estrogen receptor-alpha knockouts and maternal memory in nulliparous rats. Horm Behav 110:40–45
Galosy SS, Talamantes F (1995) Luteotropic actions of placental lactogens at midpregnancy in the mouse. Endocrinology 136:3993–4003
Gasperowicz M, Surmann-Schmitt C, Hamada Y, Otto F, Cross JC (2013) The transcriptional co-repressor TLE3 regulates development of trophoblast giant cells lining maternal blood spaces in the mouse placenta. Dev Biol 382:1–14
Georgiades P, Ferguson-Smith AC, Burton GJ (2002) Comparative developmental anatomy of the murine and human definitive placentae. Placenta 23:3–19
Gil-ad I, Zambotti F, Carruba MO, Vicentini L, Muller EE (1976) Stimulatory role for brain serotoninergic system on prolactin secretion in the male rat. Proc Soc Exp Biol Med 151:512–518
Glover V (2014) Maternal depression, anxiety and stress during pregnancy and child outcome; what needs to be done. Best practice & research. Clin Obstet Gynaecol 28:25–35
Glover V (2015) Prenatal stress and its effects on the fetus and the child: possible underlying biological mechanisms. Adv Neurobiol 10:269–283
Goodnick PJ, Chaudry T, Artadi J, Arcey S (2000) Women's issues in mood disorders. Expert Opin Pharmacother 1:903–916

Grattan DR (2002) Behavioural significance of prolactin signalling in the central nervous system during pregnancy and lactation. Reproduction 123:497–506
Grattan D (2012) A mother's brain knows. J Neuroendocrinol 23:1188–1189
Groer MW, Morgan K (2007) Immune, health and endocrine characteristics of depressed postpartum mothers. Psychoneuroendocrinology 32:133–139
Guernsey MW, Chuong EB, Cornelis G, Renfree MB, Baker JC (2017) Molecular conservation of marsupial and eutherian placentation and lactation. Elife 6:e27450
Guillemot F, Caspary T, Tilghman SM, Copeland NG, Gilbert DJ, Jenkins NA, Anderson DJ, Joyner AL, Rossant J, Nagy A (1995) Genomic imprinting of *Mash2*, a mouse gene required for trophoblast development. Nat Genet 9:235–242
Gunnet JW, Freeman ME (1983) The mating-induced release of prolactin: a unique neuroendocrine response. Endocr Rev 4:44–61
Haig D (2008) Placental growth hormone-related proteins and prolactin-related proteins. Placenta 29 Suppl A:S36–S41
Handwerger S, Freemark M (2000) The roles of placental growth hormone and placental lactogen in the regulation of human fetal growth and development. J Pediatr Endocrinol Metab 13:343–356
Hannibal RL, Baker JC (2016) Selective amplification of the genome surrounding key placental genes in trophoblast giant cells. Curr Biol 26:230–236
Hemberger M, Hanna CW, Dean W (2020) Mechanisms of early placental development in mouse and humans. Nat Rev Genet 21:27–43
Heron J, O'Connor TG, Evans J, Golding J, Glover V, ALSPAC Study Team (2004) The course of anxiety and depression through pregnancy and the postpartum in a community sample. J Affect Disord 80:65–73
Horseman ND, Zhao W, Montecino-Rodriguez E, Tanaka M, Nakashima K, Engle SJ, Smith F, Markoff E, Dorshkind K (1997) Defective mammopoiesis, but normal hematopoiesis, in mice with a targeted disruption of the prolactin gene. EMBO J 16:6926–6935
Hu D, Cross JC (2011) Ablation of Tpbpa-positive trophoblast precursors leads to defects in maternal spiral artery remodeling in the mouse placenta. Dev Biol 358:231–239
Janssen AB, Capron LE, O'Donnell K, Tunster SJ, Ramchandani PG, Heazell AE, Glover V, John RM (2016a) Maternal prenatal depression is associated with decreased placental expression of the imprinted gene PEG3. Psychol Med 46:2999–3011
Janssen AB, Kertes DA, McNamara GI, Braithwaite EC, Creeth HD, Glover VI, John RM (2016b) A role for the placenta in programming maternal mood and childhood behavioural disorders. J Neuroendocrinol 28(8)
Janssen AB, Savory KA, Garay SM, Sumption L, Watkins W, Garcia-Martin I, Savory NA, Ridgway A, Isles AR, Penketh R, Jones IR, John RM (2018) Persistence of anxiety symptoms after elective caesarean delivery. BJPsych Open 4:354–360
Jin Y, Lu SY, Fresnoza A, Detillieux KA, Duckworth ML, Cattini PA (2009) Differential placental hormone gene expression during pregnancy in a transgenic mouse containing the human growth hormone/chorionic somatomammotropin locus. Placenta 30:226–235
Jin Y, Vakili H, Liu SY, Menticoglou S, Bock ME, Cattini PA (2018) Chromosomal architecture and placental expression of the human growth hormone gene family are targeted by prepregnancy maternal obesity. Am J Physiol Endocrinol Metab 315:E435–E445
John RM (2013) Epigenetic regulation of placental endocrine lineages and complications of pregnancy. Biochem Soc Trans 41:701–709
John RM (2017) Imprinted genes and the regulation of placental endocrine function: Pregnancy and beyond. Placenta 56:86–90
John R, Hemberger M (2012) A placenta for life. Reprod Biomed Online 25:5–11
Jomain JB, Tallet E, Broutin I, Hoos S, van Agthoven J, Ducruix A, Kelly PA, Kragelund BB, England P, Goffin V (2007) Structural and thermodynamic bases for the design of pure prolactin receptor antagonists: X-ray structure of Del1-9-G129R-hPRL. J Biol Chem 282:33118–33131
Kamberi IA, Mical RS, Porter JC (1971) Effects of melatonin and serotonin on the release of FSH and prolactin. Endocrinology 88:1288–1293

Kaplan SL, Gurpide E, Sciarra JJ, Grumbach MM (1968) Metabolic clearance rate and production rate of chorionic growth hormone-prolactin in late pregnancy. J Clin Endocrinol Metab 28:1450–1460
Kim SK, Hwang IK, Yoo KY, Yoo DY, Bae E, Lee CH, Choi JH, Choi JW, Seong JK, Yoon YS, Won MH (2010) Pregnancy inhibits cell proliferation and neuroblast differentiation without neuronal damage in the hippocampal dentate gyrus in C57BL/6N mice. Brain Res 1315:25–32
Kimble DP, Rogers L, Hendrickson CW (1967) Hippocampal lesions disrupt maternal, not sexual, behavior in the albino rat. J Comp Physiol Psychol 63:401–407
Knox K, Leuenberger D, Penn AA, Baker JC (2011) Global hormone profiling of murine placenta reveals Secretin expression. Placenta 32:811–816
Kohl J, Dulac C (2018) Neural control of parental behaviors. Curr Opin Neurobiol 49:116–122
Kohl J, Babayan BM, Rubinstein ND, Autry AE, Marin-Rodriguez B, Kapoor V, Miyamishi K, Zweifel LS, Luo L, Uchida N, Dulac C (2018) Functional circuit architecture underlying parental behaviour. Nature 556:326–331
Kokay IC, Bull PM, Davis RL, Ludwig M, Grattan DR (2006) Expression of the long form of the prolactin receptor in magnocellular oxytocin neurons is associated with specific prolactin regulation of oxytocin neurons. Am J Physiol Regul Integr Comp Physiol 290:R1216–R1225
Kratimenos P, Penn AA (2019) Placental programming of neuropsychiatric disease. Pediatr Res 86:157–164
Ladyman SR, Hackwell ECR, Brown RSE (2020) The role of prolactin in co-ordinating fertility and metabolic adaptations during reproduction. Neuropharmacology 167:107911
Larsen CM, Grattan DR (2010) Prolactin-induced mitogenesis in the subventricular zone of the maternal brain during early pregnancy is essential for normal postpartum behavioral responses in the mother. Endocrinology 151:3805–3814
Larsen CM, Grattan DR (2012) Prolactin, neurogenesis, and maternal behaviors. Brain Behav Immun 26:201–209
Latham N, Mason G (2004) From house mouse to mouse house: the behavioural biology of free-living Mus musculus and its implications in the laboratory. Appl Anim Behav Sci 86:261–289
Linnemann K, Malek A, Sager R, Blum WF, Schneider H, Fusch C (2000) Leptin production and release in the dually in vitro perfused human placenta. J Clin Endocrinol Metab 85:4298–4301
Liu Y, Fan X, Wang R, Lu X, Dang YL, Wang H, Lin HY, Zhu C, Ge H, Cross JC, Wang H (2018) Single-cell RNA-seq reveals the diversity of trophoblast subtypes and patterns of differentiation in the human placenta. Cell Res 28:819–832
Lockwood Estrin G, Ryan EG, Trevillion K, Demilew J, Bick D, Pickles A, Howard LM (2019) Young pregnant women and risk for mental disorders: findings from an early pregnancy cohort. BJPsych Open 5:e21
Lonstein JS (2007) Regulation of anxiety during the postpartum period. Front Neuroendocrinol 28:115–141
Lowman HB, Cunningham BC, Wells JA (1991) Mutational analysis and protein engineering of receptor binding determinants in human placental lactogen. J Biol Chem 266:10982–10988
Lucas BK, Ormandy CJ, Binart N, Bridges RS, Kelly PA (1998) Null mutation of the prolactin receptor gene produces a defect in maternal behavior. Endocrinology 139:4102–4107
Mao J, Jain A, Denslow ND, Nouri MZ, Chen S, Wang T, Zhu N, Koh J, Sarma SJ, Sumner BW, Lei Z, Sumner LW, Bivens NJ, Roberts RM, Tuteja G, Rosenfeld CS (2020) Bisphenol A and bisphenol S disruptions of the mouse placenta and potential effects on the placenta-brain axis. Proc Natl Acad Sci U S A 117:4642–4652
Maston GA, Ruvolo M (2002) Chorionic gonadotropin has a recent origin within primates and an evolutionary history of selection. Mol Biol Evol 19:320–335
McLellan AS, Zimmermann W, Moore T (2005) Conservation of pregnancy-specific glycoprotein (PSG) N domains following independent expansions of the gene families in rodents and primates. BMC Evol Biol 5:39
McNamara GI, Creeth HDJ, Harrison DJ, Tansey KE, Andrews RM, Isles AR, John RM (2018) Loss of offspring Peg3 reduces neonatal ultrasonic vocalizations and increases maternal anxiety in wild-type mothers. Hum Mol Genet 27:440–450

Meaney MJ (2001) Maternal care, gene expression, and the transmission of individual differences in stress reactivity across generations. Annu Rev Neurosci 24:1161–1192
Menzies BR, Pask AJ, Renfree MB (2011) Placental expression of pituitary hormones is an ancestral feature of therian mammals. Evodevo 2:16
Michelsen TM, Henriksen T, Reinhold D, Powell TL, Jansson T (2019) The human placental proteome secreted into the maternal and fetal circulations in normal pregnancy based on 4-vessel sampling. FASEB J 33:2944–2956
Mikaelsson MA, Constancia M, Dent CL, Wilkinson LS, Humby T (2013) Placental programming of anxiety in adulthood revealed by Igf2-null models. Nat Commun 4:2311
Miller WL, Eberhardt NL (1983) Structure and evolution of the growth hormone gene family. Endocr Rev 4:97–130
Moltz H, Robbins D (1965) Maternal behavior of primiparous and multiparous rats. J Comp Physiol Psychol 60:417–421
Moltz H, Lubin M, Leon M, Numan M (1970) Hormonal induction of maternal behavior in the ovariectomized nulliparous rat. Physiol Behav 5:1373–1377
Moon YS, Smas CM, Lee K, Villena JA, Kim KH, Yun EJ, Sul HS (2002) Mice lacking paternally expressed Pref-1/Dlk1 display growth retardation and accelerated adiposity. Mol Cell Biol 22:5585–5592
Moore T, Dveksler GS (2014) Pregnancy-specific glycoproteins: complex gene families regulating maternal-fetal interactions. Int J Dev Biol 58:273–280
Moore T, Haig D (1991) Genomic imprinting in mammalian development: a parental tug-of-war. TIG 7:45–49
Moore T, Constancia M, Zubair M, Bailleul B, Feil R, Sasaki H, Reik W (1997) Multiple imprinted sense and antisense transcripts, differential methylation and tandem repeats in a putative imprinting control region upstream of mouse Igf2. Proc Natl Acad Sci U S A 94:12509–12514
Napso T, Yong HEJ, Lopez-Tello J, Sferruzzi-Perri AN (2018) The role of placental hormones in mediating maternal adaptations to support pregnancy and lactation. Front Physiol 9:1091
Nephew BC, Bridges RS (2008) Arginine vasopressin V1a receptor antagonist impairs maternal memory in rats. Physiol Behav 95:182–186
Newbern D, Freemark M (2011) Placental hormones and the control of maternal metabolism and fetal growth. Curr Opin Endocrinol Diab Obes 18:409–416
Ng RK, Dean W, Dawson C, Lucifero D, Madeja Z, Reik W, Hemberger M (2008) Epigenetic restriction of embryonic cell lineage fate by methylation of Elf5. Nat Cell Biol 10:1280–1290
Nogami H, Hoshino R, Ogasawara K, Miyamoto S, Hisano S (2007) Region-specific expression and hormonal regulation of the first exon variants of rat prolactin receptor mRNA in rat brain and anterior pituitary gland. J Neuroendocrinol 19:583–593
Noirot E (1969) Serial order of maternal responses in mice. Anim Behav 17:547–550
Numan M (1978) Progesterone inhibition of maternal behavior in the rat. Horm Behav 11:209–231
Numan M, Young LJ (2016) Neural mechanisms of mother-infant bonding and pair bonding: Similarities, differences, and broader implications. Horm Behav 77:98–112
Numan M, Rosenblatt JS, Komisaruk BR (1977) Medial preoptic area and onset of maternal behavior in the rat. J Comp Physiol Psychol 91:146–164
Numan M, Roach JK, del Cerro MC, Guillamon A, Segovia S, Sheehan TP, Numan MJ (1999) Expression of intracellular progesterone receptors in rat brain during different reproductive states, and involvement in maternal behavior. Brain Res 830:358–371
Ocampo Daza D, Larhammar D (2018) Evolution of the receptors for growth hormone, prolactin, erythropoietin and thrombopoietin in relation to the vertebrate tetraploidizations. Gen Comp Endocrinol 257:143–160
O'Donnell KJ, Bugge Jensen A, Freeman L, Khalife N, O'Connor TG, Glover V (2012) Maternal prenatal anxiety and downregulation of placental 11beta-HSD2. Psychoneuroendocrinology 37:818–826
Oh-McGinnis R, Bogutz AB, Lefebvre L (2011) Partial loss of Ascl2 function affects all three layers of the mature placenta and causes intrauterine growth restriction. Dev Biol 351:277–286

Perez-Garcia V, Fineberg E, Wilson R, Murray A, Mazzeo CI, Tudor C, Sienerth A, White JK, Tuck E, Ryder EJ, Gleeson D, Siragher E, Wardle-Jones H, Staudt N, Wali N, Collins J, Geyer S, Busch-Nentwich EM, Galli A, Smith JC, Robertson E, Adams DJ, Weninger WJ, Mohun T, Hemberger M (2018) Placentation defects are highly prevalent in embryonic lethal mouse mutants. Nature 555:463–468

Petraglia F, Sawchenko PE, Rivier J, Vale W (1987) Evidence for local stimulation of ACTH secretion by corticotropin-releasing factor in human placenta. Nature 328:717–719

Pi X, Zhang B, Li J, Voogt JL (2003) Promoter usage and estrogen regulation of prolactin receptor gene in the brain of the female rat. Neuroendocrinology 77:187–197

Potter HG, Ashbrook DG, Hager R (2019) Offspring genetic effects on maternal care. Front Neuroendocrinol 52:195–205

Prigent-Tessier A, Tessier C, Hirosawa-Takamori M, Boyer C, Ferguson-Gottschall S, Gibori G (1999) Rat decidual prolactin. Identification, molecular cloning, and characterization. J Biol Chem 274:37982–37989

Rai A, Cross JC (2014) Development of the hemochorial maternal vascular spaces in the placenta through endothelial and vasculogenic mimicry. Dev Biol 387:131–141

Rawn SM, Huang C, Hughes M, Shaykhutdinov R, Vogel HJ, Cross JC (2015) Pregnancy hyperglycemia in prolactin receptor mutant, but not prolactin mutant, mice and feeding-responsive regulation of placental lactogen genes implies placental control of maternal glucose homeostasis. Biol Reprod 93:75

Ribeiro AC, Musatov S, Shteyler A, Simanduyev S, Arrieta-Cruz I, Ogawa S, Pfaff DW (2012) siRNA silencing of estrogen receptor-alpha expression specifically in medial preoptic area neurons abolishes maternal care in female mice. Proc Natl Acad Sci U S A 109:16324–16329

Riddle O, Lahr EL, Bates RW (1935) Maternal behavior induced in virgin rats by prolactin. Proc Soc Exp Biol Med 32:730–734

Rolls A, Schori H, London A, Schwartz M (2008) Decrease in hippocampal neurogenesis during pregnancy: a link to immunity. Mol Psychiatry 13:468–469

Rosenblatt JS (1967) Nonhormonal basis of maternal behavior in the rat. Science 156:1512–1514

Rosenblatt JS, Siegel HI (1975) Hysterectomy-induced maternal behavior during pregnancy in the rat. J Comp Physiol Psychol 89:685–700

Rosenblatt JS, Wagner CK, Morrell JI (1994) Hormonal priming and triggering of maternal behavior in the rat with special reference to the relations between estrogen receptor binding and ER mRNA in specific brain regions. Psychoneuroendocrinology 19:543–552

Rosenfeld CS (2021) The placenta-brain-axis. J Neurosci Res 99:271–283

Ross C, Boroviak TE (2020) Origin and function of the yolk sac in primate embryogenesis. Nat Commun 11:3760

Salas M, John R, Saxena A, Barton S, Frank D, Fitzpatrick G, Higgins MJ, Tycko B (2004) Placental growth retardation due to loss of imprinting of Phlda2. Mech Dev 121:1199–1210

Scanlan VF, Byrnes EM, Bridges RS (2006) Reproductive experience and activation of maternal memory. Behav Neurosci 120:676–686

Schmeisser MJ, Baumann B, Johannsen S, Vindedal GF, Jensen V, Hvalby OC, Sprengel R, Seither J, Maqbool A, Magnutzki A, Lattke M, Oswald F, Boeckers TM, Wirth T (2012) IkappaB kinase/nuclear factor kappaB-dependent insulin-like growth factor 2 (Igf2) expression regulates synapse formation and spine maturation via Igf2 receptor signaling. J Neurosci 32:5688–5703

Schumacher M, Mattern C, Ghoumari A, Oudinet JP, Liere P, Labombarda F, Sitruk-Ware R, De Nicola AF, Guennoun R (2014) Revisiting the roles of progesterone and allopregnanolone in the nervous system: resurgence of the progesterone receptors. Prog Neurobiol 113:6–39

Seegal RF, Denenberg VH (1974) Maternal experience prevents pup-killing in mice induced by peripheral anosmia. Physiol Behav 13:339–341

Shingo T, Gregg C, Enwere E, Fujikawa H, Hassam R, Geary C, Cross JC, Weiss S (2003) Pregnancy-stimulated neurogenesis in the adult female forebrain mediated by prolactin. Science 299:117–120

Siegel HI, Rosenblatt JS (1975a) Estrogen-induced maternal behavior in hysterectomized-overiectomized virgin rats. Physiol Behav 14:465–471

Siegel HI, Rosenblatt JS (1975b) Latency and duration of estrogen induction of maternal behavior in hysterectomized-ovariectomized virgin rats: effects of pup stimulation. Physiol Behav 14:473–476
Siegel HI, Rosenblatt JS (1975c) Progesterone inhibition of estrogen-induced maternal behavior in hysterectomized-ovariectomized virgin rats. Horm Behav 6:223–230
Siegel HI, Rosenblatt JS (1978) Duration of estrogen stimulation and progesterone inhibition of maternal behavior in pregnancy-terminated rats. Horm Behav 11:12–19
Siegel HI, Doerr HK, Rosenblatt JS (1978) Further studies on estrogen-induced maternal behavior in hysterectomized-ovariectomized virgin rats. Physiol Behav 21:99–103
Simmons DG, Fortier AL, Cross JC (2007) Diverse subtypes and developmental origins of trophoblast giant cells in the mouse placenta. Dev Biol 304:567–578
Simmons DG, Rawn S, Davies A, Hughes M, Cross JC (2008) Spatial and temporal expression of the 23 murine prolactin/placental lactogen-related genes is not associated with their position in the locus. BMC Genomics 9:352
Smas CM, Chen L, Sul HS (1997) Cleavage of membrane-associated pref-1 generates a soluble inhibitor of adipocyte differentiation. Mol Cell Biol 17:977–988
Smiley KO, Ladyman SR, Gustafson P, Grattan DR, Brown RSE (2019) Neuroendocrinology and adaptive physiology of maternal care. Curr Top Behav Neurosci 43:161–210
Soares MJ (2004) The prolactin and growth hormone families: pregnancy-specific hormones/cytokines at the maternal-fetal interface. Reprod Biol Endocrinol 2:51
Soares MJ, Talamantes F (1982) Gestational effects on placental and serum androgen, progesterone and prolactin-like activity in the mouse. J Endocrinol 95:29–36
Soares MJ, Konno T, Alam SM (2007) The prolactin family: effectors of pregnancy-dependent adaptations. Trends Endocrinol Metab 18:114–121
Stack EC, Balakrishnan R, Numan MJ, Numan M (2002) A functional neuroanatomical investigation of the role of the medial preoptic area in neural circuits regulating maternal behavior. Behav Brain Res 131:17–36
Stolzenberg DS, Champagne FA (2016) Hormonal and non-hormonal bases of maternal behavior: The role of experience and epigenetic mechanisms. Horm Behav 77:204–210
Stolzenberg DS, Rissman EF (2011) Oestrogen-independent, experience-induced maternal behaviour in female mice. J Neuroendocrinol 23:345–354
Sumption LA, Garay SM, John RM (2020) Low serum placental lactogen at term is associated with postnatal symptoms of depression and anxiety in women delivering female infants. Psychoneuroendocrinology 116:104655
Surani MA (1998) Imprinting and the initiation of gene silencing in the germ line. Cell 93:309–312
Swart JM, Grattan DR, Ladyman SR, Brown RSE (2021) Changes in maternal motivation across reproductive states in mice: A role for prolactin receptor activation on GABA neurons. Horm Behav 135:105041
Szyf M, Weaver IC, Champagne FA, Diorio J, Meaney MJ (2005) Maternal programming of steroid receptor expression and phenotype through DNA methylation in the rat. Front Neuroendocrinol 26:139–162
Takahashi K, Kobayashi T, Kanayama N (2000) p57(Kip2) regulates the proper development of labyrinthine and spongiotrophoblasts. Mol Hum Reprod 6:1019–1025
Tanaka M, Gertsenstein M, Rossant J, Nagy A (1997) Mash2 acts cell autonomously in mouse spongiotrophoblast development. Dev Biol 190:55–65
Terkel J, Rosenblatt JS (1972) Humoral factors underlying maternal behavior at parturition: cross transfusion between freely moving rats. J Comp Physiol Psychol 80:365–371
Theiler K (1989) The house mouse: atlas of embryonic development. Springer, Berlin
Thomson M (2013) The physiological roles of placental corticotropin releasing hormone in pregnancy and childbirth. J Physiol Biochem 69:559–573
Torner L, Toschi N, Nava G, Clapp C, Neumann ID (2002) Increased hypothalamic expression of prolactin in lactation: involvement in behavioural and neuroendocrine stress responses. Eur J Neurosci 15:1381–1389

Torner L, Karg S, Blume A, Kandasamy M, Kuhn HG, Winkler J, Aigner L, Neumann ID (2009) Prolactin prevents chronic stress-induced decrease of adult hippocampal neurogenesis and promotes neuronal fate. J Neurosci 29:1826–1833

Tucci V, Isles AR, Kelsey G, Ferguson-Smith AC, Erice Imprinting G (2019) Genomic imprinting and physiological processes in mammals. Cell 176:952–965

Tunster SJ, Tycko B, John RM (2010) The imprinted Phlda2 gene regulates extraembryonic energy stores. Mol Cell Biol 30:295–306

Tunster SJ, Van de Pette M, John RM (2011) Fetal overgrowth in the Cdkn1c mouse model of Beckwith-Wiedemann syndrome. Dis Model Mech 4:814–821

Tunster SJ, Van De Pette M, John RM (2014) Isolating the role of elevated Phlda2 in asymmetric late fetal growth restriction in mice. Dis Model Mech 7:1185–1191

Tunster SJ, Creeth HD, John RM (2016a) The imprinted Phlda2 gene modulates a major endocrine compartment of the placenta to regulate placental demands for maternal resources. Dev Biol 409:251–260

Tunster SJ, McNamara GI, Creeth HD, John RM (2016b) Increased dosage of the imprinted Ascl2 gene restrains two key endocrine lineages of the mouse Placenta. Dev Biol 418:55–65

Tunster SJ, Boqué-Sastre R, McNamara GI, Hunter SM, Creeth HDJ, John RM (2018) Peg3 Deficiency Results in Sexually Dimorphic Losses and Gains in the Normal Repertoire of Placental Hormones. Front Cell Dev Biol. 6:123. https://doi.org/10.3389/fcell.2018.00123. PMID: 30320110; PMCID: PMC6170603

Turco MY, Moffett A (2019) Development of the human placenta. Development 146(22)

Vacher CM, Lacaille H, O'Reilly JJ, Salzbank J, Bakalar D, Sebaoui S, Liere P, Clarkson-Paredes C, Sasaki T, Sathyanesan A, Kratimenos P, Ellegood J, Lerch JP, Imamura Y, Popratiloff A, Hashimoto-Torii K, Gallo V, Schumacher M, Penn AA (2021) Placental endocrine function shapes cerebellar development and social behavior. Nat Neurosci 24:1392–1401

Vergara-Castaneda E, Grattan DR, Pasantes-Morales H, Perez-Dominguez M, Cabrera-Reyes EA, Morales T, Cerbon M (2016) Prolactin mediates neuroprotection against excitotoxicity in primary cell cultures of hippocampal neurons via its receptor. Brain Res 1636:193–199

Walker WH, Fitzpatrick SL, Barrera-Saldana HA, Resendez-Perez D, Saunders GF (1991) The human placental lactogen genes: structure, function, evolution and transcriptional regulation. Endocr Rev 12:316–328

Walker TL, Vukovic J, Koudijs MM, Blackmore DG, Mackay EW, Sykes AM, Overall RW, Hamlin AS, Bartlett PF (2012) Prolactin stimulates precursor cells in the adult mouse hippocampus. PLoS One 7:e44371

Weaver IC, Cervoni N, Champagne FA, D'Alessio AC, Sharma S, Seckl JR, Dymov S, Szyf M, Meaney MJ (2004) Epigenetic programming by maternal behavior. Nat Neurosci 7:847–854

Weber EM, Olsson AS (2008) Maternal behaviour in Mus musculus sp.: An ethological review. Appl Anim Behav Sci 114:1–22

Woods L, Perez-Garcia V, Hemberger M (2018) Regulation of placental development and its impact on fetal growth-new insights from mouse models. Front Endocrinol (Lausanne) 9:570

Wu HH, Choi S, Levitt P (2016) Differential patterning of genes involved in serotonin metabolism and transport in extra-embryonic tissues of the mouse. Placenta 42:74–83

Wynne F, Ball M, McLellan AS, Dockery P, Zimmermann W, Moore T (2006) Mouse pregnancy-specific glycoproteins: tissue-specific expression and evidence of association with maternal vasculature. Reproduction 131:721–732

Zhang Q, Hao J, Li G (2019) Deletion of Prl7d1 causes placental defects at mid-pregnancy in mice. Mol Reprod Dev 86:696–713

Adaptations of the Maternal Brain to Meet the Increased Energy Demands of Pregnancy and Lactation

3

Sharon R. Ladyman and Barbara C. Woodside

Abstract

In adult mammals, peripheral signals of energy balance act on central circuits to match energy intake with energy expenditure. The unique metabolic challenges of pregnancy and lactation, however, require alterations in this energy balance equation. In this chapter we describe the multiple behavioral and physiological changes in energy intake and expenditure that accompany pregnancy and lactation and enable the mother to meet the nutritional requirements of her young. We then discuss how secretion of metabolic hormones and neuropeptides modulating energy balance are changed in pregnancy and lactation as well as how alterations in sensitivity of the brain to metabolic hormones facilitates the increased food intake required of these reproductive states. Finally, the contribution of the distinct hormonal profiles of pregnancy and lactation to the maternal adaptations of the brain allowing for these changes in energy homeostasis systems is described.

Keywords

Pregnancy · Lactation · Food intake · Hypothalamus · Leptin · Insulin

S. R. Ladyman (✉)
Centre for Neuroendocrinology, Department of Anatomy, University of Otago, Dunedin, New Zealand

Maurice Wilkins Centre for Molecular Biodiscovery, University of Auckland, Auckland, New Zealand
e-mail: sharon.ladyman@otago.ac.nz

B. C. Woodside
Center for Studies in Behavioral Neurobiology, Concordia University, Montreal, QC, Canada

P. J. Brunton, D. R. Grattan (eds.), *Neuroendocrine Regulation of Mammalian Pregnancy and Lactation*, Masterclass in Neuroendocrinology 15,
https://doi.org/10.1007/978-3-031-51138-7_3

3.1 Introduction

Mammalian females are the sole source of nutrients for their offspring from conception until weaning and adaptations that allow female mammals to provide these nutrients begin early after conception, before any direct investment in embryonic growth, and persist at least until the young begin to eat independently. Although supplying food to the young is the primary function of both pregnancy and lactation, the context in which this occurs is quite different. In pregnancy, the increasing mass of the developing fetuses, growth and development of maternal tissues, and accumulation of energy reserves in preparation for lactation all combine to increase maternal body mass and place both behavioral and physiological constraints on multiple systems. Metabolic systems, in particular, have to adapt to the need for increased food intake in the presence of **positive energy balance**. The birth of young relieves some of these constraints but presents an array of new challenges. As with pregnancy, the most energetically costly of these is providing young with the nutrients necessary for their growth as well as their ever-increasing **energy expenditure**. In addition, because **altricial young** require not only food but also warmth, grooming, and protection, **postpartum** maternal care entails a wide range of young-directed behaviors. In contrast to pregnancy, in lactation the nutrients directed to the young are transported away from the mother, creating a **negative energy balance** and presenting a different set of inputs to the neural circuits controlling energy balance. These changing demands of pregnancy and lactation result in adaptations within the central neural circuits controlling energy balance and the metabolic signals impinging upon them that are, at least in part, orchestrated by reproductive hormones.

3.2 Meeting the Energy Requirements of Pregnancy and Lactation

The increase in body weight that accompanies pregnancy (reviewed in Khant Aung et al. 2020) primarily reflects fetal growth but some organs such as the liver, pancreas, mammary gland, and uterus also increase in size. In many species, there is also an accumulation of fat mass. Women with a pre-pregnancy **body mass index (BMI)** between 20 and 25 would be expected to gain approximately 3 kg of adipose tissue across pregnancy and, in mice, small yet significant increases in fat mass can be detected as early as day 7 of pregnancy, with fat accumulation peaking in late pregnancy. Skeletal calcium also increases during pregnancy (Brommage and Baxter 1988) and in early pregnancy there are increases in protein content that are used in the last trimester to support rapid fetal growth (Naismith and Morgan 1976). In total, approximately 15% of pregnancy weight gain is attributable to increases in maternal body weight per se. Overall increased caloric expenditure in pregnancy and lactation in women has been estimated at an additional 350 kcal/day in the second trimester and 500 kcal/day for the last trimester (Catalano et al. 1998; Butte et al. 2004) and for breastfeeding a single infant 626 kcal/day (Butte and King 2005). The energetic cost of lactation increases dramatically with the number of young nursed and in small rodents, such as mice nursing multiple young, the daily energy requirement at peak lactation is about 56% greater than that of non-reproducing females (Krol and Speakman 2003).

Numerous behavioral and physiological strategies are used to meet these energetic demands, perhaps the most obvious of which are changes in food intake (see Fig. 3.1). In both laboratory rodents and humans, food intake peaks in the last third of pregnancy, when there is the most rapid growth of the offspring. Late pregnant rats and mice eat 20–30% more than in pre-pregnancy and this can rise to about 350% of their pre-pregnancy intake by peak lactation (day 14 postpartum) (Woodside et al. 2008). The amount of food eaten is not the only change in feeding behavior across a reproductive episode, as timing of meals as well as type of food eaten all change to match day-by-day alterations in the energetic demand of fetal growth and milk production. Interestingly, not all species increase food intake to meet pregnancy energy demands (reviewed in Khant Aung et al. 2020). For example, the golden hamster does not significantly increase food intake during pregnancy and instead uses up fat resources to meet the demands of the growing conceptus. Even in species that show increases in food intake in lactation, nutrients stored in pregnancy can be important. Fat accumulated in pregnancy is an easily utilizable source of calories for the initial stages of lactation, reducing foraging demands and allowing the dam to spend most of her time with her young.

Usage of ingested nutrients is facilitated by morphological changes in the gastrointestinal (GI) tract, including increases in overall stomach mass, increased height of the villi and thickness of gastric mucosa, and lengthening of the small intestine and colon, all of which are evident in late pregnancy and peak in late lactation (Johnson et al. 2019). One key nutrient required in both pregnancy and lactation is calcium. Pregnancy is associated with increased calcium absorption in the small intestine, and in both female rodents and women there is an increase in calcium reabsorption in the kidney during lactation (Kovacs 2001). Calcium from bone is also utilized for later milk production, particularly in situations where there is a shortage of food or when the dietary calcium is deficient. While stored nutrients can protect against negative consequences of variations in energy intake during lactation, it is not a failsafe device in lactating women. Short-term reductions in caloric intake can impact milk output in some situations of negative energy balance although this is an area that needs further investigation (Lovelady 2004).

The obligatory increases in energy expenditure seen across pregnancy and lactation are offset in part by decreases in energy investment in other processes. Ovarian cyclicity is suppressed immediately post-conception and in many mammals, lactation is also associated with a period of **anovulation** (McNeilly 2001) or, in species that exhibit a postpartum ovulation, a delay in implantation (Lamming 1978), both of which serve to decrease the overlap in investment between current offspring and the next. In rodents, voluntary physical activity not directed toward foraging or caring for the offspring is reduced in pregnancy prior to any constraint due to weight gain and this decrease persists in lactation (Gamo et al. 2013; Ladyman et al. 2018a). Brown adipose tissue (BAT) thermogenesis is also decreased in lactation resulting both from a decrease in its sympathetic activation and reduced levels of proteins involved in the thermogenic process, UCP1 and UCP3 (reviewed in Krol and Speakman 2019). The extent to which decreases in physical activity and BAT

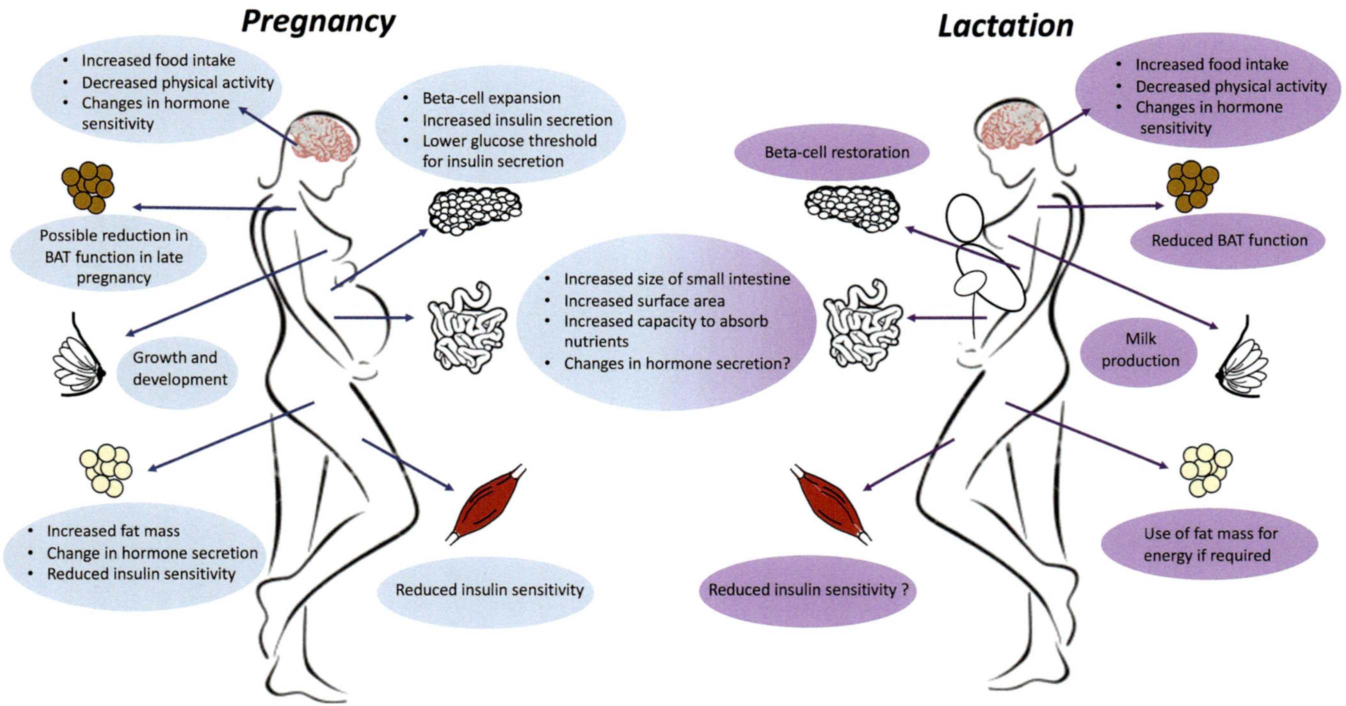

Fig. 3.1 Adaptations in metabolic tissues in pregnancy and lactation. In both pregnancy and lactation, metabolically relevant tissues undergo adaptations to meet the increased energy demands of these states. Some of the changes in pregnancy, such as increased energy storage in the form of adipose tissue and development of the mammary gland, are to prepare for the demands of lactation. Others, such as those in the pancreas, are specific to pregnancy and are quickly restored to normal in lactation. There are some tissues, such as the brain and intestine, and possibly brown adipose tissue (BAT) and muscle, where modifications are seen in both pregnancy and lactation, although whether they result from the same mechanisms in both these states is not always clear

thermogenesis generalize across species is not clear, but in humans at least, diet-induced thermogenesis seems to be unaffected by lactation (Spaaij et al. 1994).

In sum, providing nutrition for offspring, both before and after birth, entails a wide array of physiological and behavioral adaptations, many of which are common to pregnancy and lactation. How these adaptations may be affected by food availability and the degree to which they influence female metabolism in the long term are discussed in Boxes 3.1 and 3.2. When food is plentiful in pregnancy these adaptations result in energy intake overtaking energy expenditure allowing for a state of positive energy balance to develop. In contrast, in lactation, channeling of available resources to milk production results in the loss of body fat and development of negative energy balance in spite of dramatic increases in energy intake. Underlying these adaptations is a specific pattern of hormonal and neuropeptide activity.

Box 3.1: Effect of Under- and Over-Nutrition on Pregnancy and Lactation

Maternal under- and over-nutrition are both considered among the early life adverse events that can program offspring for later health risks. Studies of both overall food restriction and dietary protein restriction during pregnancy in rodents suggest that fetal growth is maintained as far as possible by utilization of maternal nutrient stores (Naismith et al. 1982; Leon and Woodside 1983) and funneling available nutrients to the offspring. Preferential use of available nutrition by fetuses is facilitated by adaptations in the placenta such that even though placental weight may itself be decreased in undernourished mothers, its capacity for amino acid and glucose transport is maintained (Burton 2012). This cannot fully compensate for low energy resources, however, and fetal growth in the later stages of pregnancy falls further behind that of controls. Food restriction during lactation results in an increased channeling of maternal resources to offspring resulting in a greater fat loss in the mother. In spite of this, milk volume is typically reduced, as is milk lactose content, resulting in slower growth of the young (Wattez et al. 2015).

Given the energetic costs of pregnancy and lactation one might expect that increased body weight would confer an advantage. However, women with obesity are more likely than those within a healthy weight range to experience pregnancy complications, such as spontaneous preterm birth, difficult deliveries, and to give birth to offspring that are large for gestational age. Birth weight is highly correlated with placental weight and both the higher circulating levels of insulin and proinflammatory cytokines as well as the lower adiponectin concentrations typical of obesity have been associated with increased placental growth. There is also a tendency for overall insulin sensitivity to decrease to a greater extent in women with obesity (Catalano and Shankar 2017). Obesity has also been associated with problems with lactation (Bever Babendure et al. 2015). In women, this is often seen as a delay in onset and/or early termination of breast feeding and in experimental studies in rodents as a reduction in litter growth. It has been suggested that this reflects, at least in part, a reduction in the ability of suckling to stimulate prolactin secretion (Rasmussen 2007).

Box 3.2: Long-Term Metabolic Consequences of Adaptations of Pregnancy and Lactation

Although it is generally acknowledged that pregnancy and lactation entail major adaptations in multiple physiological systems, the consequences of these changes for maternal biology following weaning of the offspring have not been widely investigated (Duarte-Guterman et al. 2019). In terms of metabolic health, pregnancy commonly leads to long-term weight retention, and women regularly report that they are unable to return to their pre-pregnancy bodyweight (Harris et al. 1997; Hollis et al. 2017). Women who have undergone at least one pregnancy have an increased risk of obesity and metabolic syndrome compared to women who have never been pregnant (Gunderson et al. 2009). Some studies have suggested that lactation reduces the risk for developing metabolic syndrome later in life, although a recent meta-analysis found no clear positive effect of lactation on this outcome (Torris and Bjornnes 2020). One of the factors that most likely contributes to differences in outcome is the mother's choice of diet itself. For example, a Swedish study found that women who snacked more than three times a day did not lose weight over lactation whereas those that did not snack did lose weight (Ohlin and Rossner 1996).

Having a child can lead to major changes in lifestyle (reviewed in McKinley et al. 2018), making it challenging to tease apart the impact of pregnancy and lactation per se on the biological processes underlying maternal metabolism. Work from laboratory rodents can investigate the biological effects independent of lifestyle changes, and in mice that have experienced one pregnancy and successfully nursed pups till weaning, body weight remains higher than in age-matched controls for at least 6 months (Ladyman et al. 2018b; Teixeira et al. 2019b). While there are no changes in food intake, increased levels of fat mass, increased size of the small intestine, reductions in daily physical activity, and increased susceptibility of glucose regulation to the negative consequence of a high fat diet have all been demonstrated (Casirola and Ferraris 2003; Ladyman et al. 2018b; Teixeira et al. 2019b). Interestingly, mice lacking the signaling molecule STAT5 in the brain do not show many of the effects of pregnancy and lactation on body weight retention (Teixeira et al. 2019b). Prolactin and growth hormone both act through the JAK/STAT5 signaling pathway and are responsible for the increase in STAT5 activity in the brain during pregnancy, suggesting that either or both of these hormones contributes to the long-term effects of pregnancy on body weight. Further work is required to investigate this possibility directly.

3.3 Metabolic Signals in Pregnancy and Lactation

Given the extraordinary energetic costs of pregnancy and lactation, it is not surprising that it is associated with changes in patterns of metabolic hormones as well as central expression of the neuropeptides that modulate energy balance systems. A brief overview of the key neural circuits and hormones (Fig. 3.2) modulating energy balance is given in Box 3.3. Changes in expression of the neuropeptides modulating food intake are observed in pregnancy and lactation. In both states the changes are such as to favor **orexigenic** over **anorectic** pathways and as might be expected, this is seen to a greater extent in lactation than in pregnancy. In rodent models, anorectic pro-opiomelanocortin (POMC) mRNA has been shown to be relatively constant or reduced during pregnancy while orexigenic neuropeptide Y (NPY) and agouti-related peptide (AGRP) mRNA are either similarly constant or increased (reviewed in Khant Aung et al. 2020). In lactation hypothalamic expression of the anorectic peptides POMC and cocaine- and amphetamine-regulated transcript (CART) are decreased, suggesting limited suppression of food intake by these peptides. Opioid peptides, prodynorphin and proenkephalin, are increased in lactation and their orexigenic properties may contribute to the hyperphagia of lactation. Some orexigenic neuropeptides do not show increased expression during lactation, for example, galanin mRNA is lower in lactation and hypocretin/orexin has expression levels during lactation similar to those in virgin rats, suggesting these neuropeptides do not play a major contributing role in driving lactational-hyperphagia (reviewed in Woodside 2016). There is an increase in AGRP mRNA expression by Day 5 postpartum and of NPY mRNA expression by Day 10 (Suzuki et al. 2014), and eliminating AGRP neurons by treatment with diphtheria toxin in adult mice abolishes both basal food intake and lactation-induced hyperphagia (Suzuki et al. 2014).

Box 3.3: Overview of Peripheral Signals and Neural Circuits Controlling Energy Balance

The neural circuits controlling energy balance involve neurons in multiple sites, including various hypothalamic nuclei, ventral tegmental area (VTA), parabrachial nucleus (PBN), and hindbrain, but most attention has focused on the arcuate nucleus of the hypothalamus (ARC) and in particular on two neural populations, the NPY/AGRP- and POMC/CART-expressing neurons. There are many detailed reviews covering this area, and for more information with emphasis on the maternal brain we recommend Woodside et al. (2012). Stimulation of the POMC/CART neurons reduces food intake and increases energy expenditure and the anorectic effects of POMC are mediated in large part by its cleavage product, alpha-MSH, acting at MC3/MC4 receptors. Stimulation of the NPY/AGRP neurons increases both food intake and conservation of energy. Both NPY and AGRP have orexigenic effects but AGRP may have the greater effect on food intake and NPY on suppression of sympathetic activation. The orexigenic effects of AGRP are mediated through its

(continued)

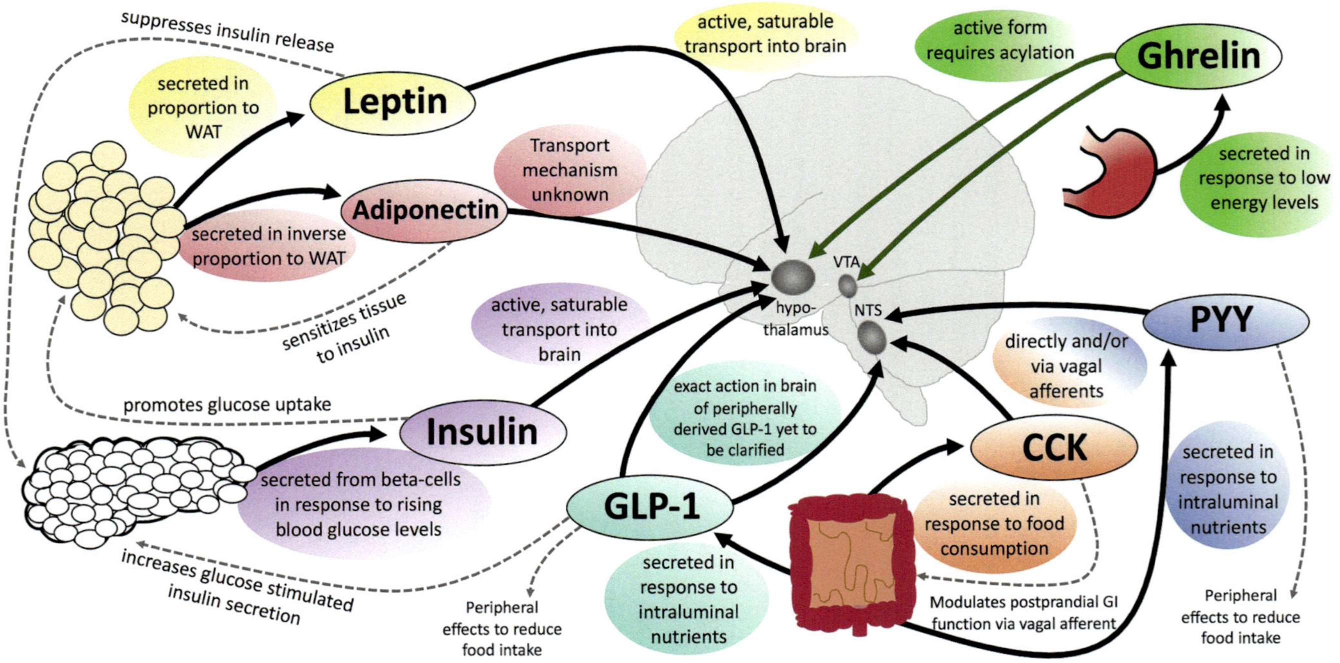

Fig. 3.2 Hormones regulating food intake. Key hormones involved in regulating food intake are shown, along with where in the periphery they are secreted from and their main targets in the brain to influence food intake. Most of these hormones suppress food intake, except ghrelin, which stimulates food intake (this difference is indicated by green arrows). All of these hormones have other functions within the body, some of which are shown (grey dashed arrows). Key notes about secretion, transport into the brain or other features are indicated for each hormone by similar colored bubbles, however there is still a vast number of unanswered questions regarding function, sites of action, and entry into the brain for all hormones. CCK, cholecystokinin; PYY, peptide YY; GLP-1, glucagon-like peptide; NTS, nucleus of the solitary tract; VTA, ventral tegmental area; WAT, white adipose tissue

Box 3.3 (continued)

action as an endogenous antagonist of alpha-MSH at MC3/MC4 receptors. These two populations interact, such that activation of the NPY/AGRP neurons increases inhibitory GABA inputs onto the POMC/CART neurons. Both populations are targeted, either directly or indirectly, by a wide array of metabolic signals arising from various tissues in the periphery, including white adipose tissue, the pancreas, and the digestive tract (see Fig. 3.2).

Key anorectic hormones involved in **energy homeostasis** are leptin, insulin, GLP-1, CCK, PYY, and adiponectin. While these hormones have a range of metabolic functions, including regulating glucose homeostasis, they are all involved in suppressing food intake by actions in **both** the brain and the periphery (Fig. 3.2). Ghrelin stimulates appetite through its actions in the hypothalamus and also through modulation of the reward pathways in the VTA. There is still a lot we do not know about the actions of these hormones on food intake, including how they access the brain. Some of these hormones, notably CCK and GLP-1, are also synthesized centrally, and it is not yet clear whether peripheral or central contributions of these hormones make the greater contribution to food intake. It is proposed that many of them enter by active transport systems and/or access at locations where the blood brain barrier is particularly leaky, such as the median eminence or area postrema. Vagal afferents arising from the GI tract express receptors for these hormones and are also a major route through which they influence central circuits controlling energy balance.

One might expect that the alterations in central neuropeptide expression and increased food intake during pregnancy and lactation would be accompanied by an increase in orexigenic signals and a decrease in anorectic signals from the periphery, but as shown in Fig. 3.3, this does not necessarily take place. Circulating **leptin** levels rise in pregnancy, with a decrease just before parturition (Fig. 3.4). Increases in circulating leptin are augmented by higher secretion rates from adipose tissue, production from the placenta, and/or increased leptin binding proteins that may reduce leptin clearance (reviewed in Khant Aung et al. 2020) and the relative contribution of each of these sources varies across species and throughout pregnancy. At the beginning of lactation, maternal circulating leptin levels in rats may be at least as high as that of non-lactating females but as fat stores are utilized, leptin levels decrease, so that in rats, at peak lactation (Day 15 postpartum) both fat stores and leptin levels are approximately half those seen in age-matched virgin females (Woodside et al. 2000) (Fig. 3.4). In women, leptin levels in lactation are lower than during pregnancy (Andersson-Hall et al. 2020). Overall, these results indicate a dissociation between leptin concentrations and appetite in these reproductive states, such that pregnancy has elevated leptin concentrations and small yet significant increases in food intake while lactation has similar leptin concentrations to the non-pregnant, non-lactating state but much greater food intake.

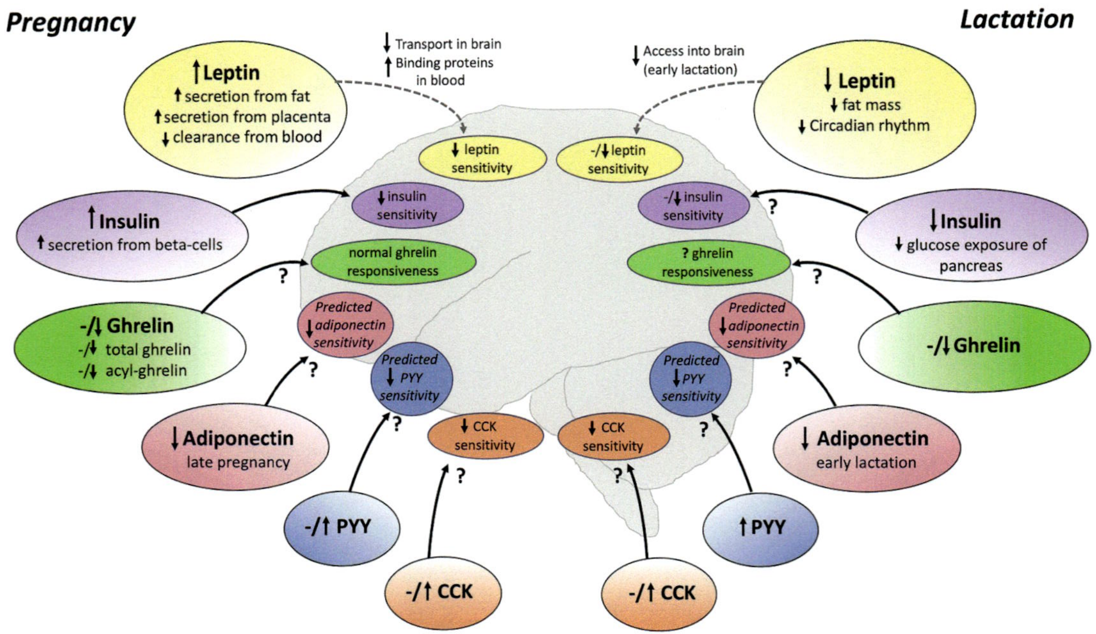

Fig. 3.3 Change in hormone level and sensitivity of the brain to peripheral signals of energy balance during pregnancy and lactation. As pregnancy and lactation progress, the increase in food intake does not always correspond to the concentration and secretion pattern of anorectic and orexigenic peripherally released hormones. In the cases of leptin and insulin, where levels increase in pregnancy, the brain becomes insensitive to the satiety actions of these hormones. There is some evidence that this insensitivity remains during lactation, despite the low levels of both leptin and insulin at this time. The underlying mechanisms of these changes in sensitivity are yet to be fully elucidated. For many of the other peripheral signals of energy balance, there is very little data even describing the physiological changes in secretion during pregnancy and lactation, let alone addressing whether there are changes in the brain's sensitivity to these factors. Here, we have condensed the information regarding hormone levels during pregnancy and lactation to its simplest form and made predictions of likely changes in the brain. CCK, cholecystokinin; PYY, peptide YY

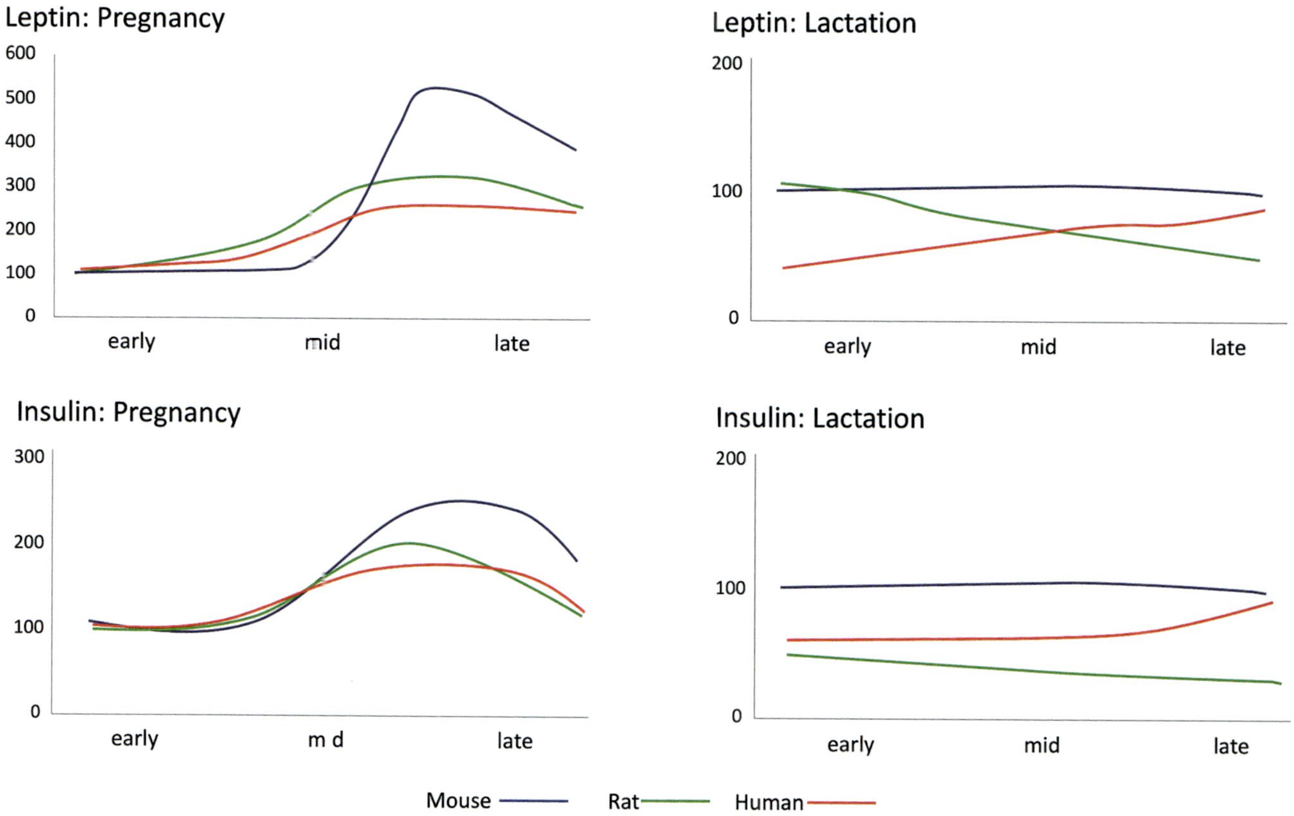

Fig. 3.4 Schematic representation of changing levels of leptin and insulin through pregnancy and lactation for mice, rats, and humans. Leptin and insulin are perhaps the most well-characterized metabolic hormones during pregnancy and lactation, yet there is still a lack of knowledge of exact secretion patterns over these reproductive states. Levels are expressed as an approximate percentage of average pre-pregnancy levels

Insulin levels, like those of leptin, gradually increase over pregnancy, in humans and rodents (reviewed in Ladyman and Brooks 2021) (Fig. 3.4). In particular, during pregnancy glucose-stimulated insulin secretion is elevated, the threshold for β-cells to release insulin is lowered, and this is accompanied by an increase in β-cell mass (Baeyens et al. 2016). This increased β-cell mass is rapidly reduced in the first few days postpartum largely as a result of a reduction in β-cell size and a decrease in proliferation (Takahashi et al. 2020). In contrast to pregnancy, lactating rodents are hypoinsulinemic, but this is associated with changes in insulin sensitivity and increased usage of glucose for milk production (reviewed in Ladyman and Brooks 2021) (Fig. 3.3).

In spite of the increase in adiposity during pregnancy, in humans, **adiponectin** levels are relatively stable until they decrease in the last trimester (Fig. 3.3), perhaps because of supplementation from placental adiponectin production (Pheiffer et al. 2021). Soon after birth in humans, maternal adiponectin levels are low compared to pregnancy, although by 3–6 months after birth the levels are comparable to pregnancy (Andersson-Hall et al. 2020).

There is very little data examining the hormones secreted from the gastrointestinal tract during pregnancy and lactation (Fig. 3.3). For **cholecystokinin** (CCK), what evidence there is suggests either an increase or no change in levels compared to non-pregnant controls, in both humans and dogs (reviewed in Clarke et al. 2021). In lactation, circulating concentrations of CCK increase in rats, although cerebrospinal fluid (CSF) levels, which are a likely indicator of levels accessing the brain, remain similar to those of nonlactating rats (Linden 1989). A number of studies indicate that **peptide YY** (PYY) levels remain stable during pregnancy (Johnson et al. 2019), although others have demonstrated, at least in rats, that PYY increases as gestation progresses, with peak levels in the last week of pregnancy (Tovar et al. 2004). In lactating rodents, colon levels of PYY and **glucagon-like peptide** (GLP-1) are increased, and this has been attributed to the increased size of this structure rather than food intake itself (Johnson et al. 2019). Circulating PYY levels were found to be higher in breastfeeding women than in matched bottle-feeding controls between 3 and 6 months postpartum (Vila et al. 2015).

There is some inconsistency between studies about how or whether plasma concentrations of the orexigenic hormone **ghrelin** change in pregnancy and lactation (Fig. 3.3). A recent study in mice demonstrated no change in total ghrelin but lower levels of its active form, acyl-ghrelin, in mid-pregnancy (Kaur et al. 2020). The number of stomach cells that show immunoreactivity for ghrelin, together with ghrelin concentrations within the stomach, increase at parturition but most studies have found either no change or lower circulating levels during lactation (Shibata et al. 2004; Johnson et al. 2019). Decreased circulating levels of acyl-ghrelin were observed in lactating rats on Day 14 postpartum, coincident with peak food intake and milk production, and levels were restored 24 h after removal of the litter (Abizaid et al. 2008). Since release of ghrelin is controlled, at least in part, by the presence of food in the stomach, it is possible that the increase in food intake seen during lactation itself constrains ghrelin release, as might the decrease in sympathetic activity that accompanies lactation.

Although the high levels of anorectic signals and unaltered levels of orexigenic signals at some stages of pregnancy and lactation described above and in Fig. 3.3 are inconsistent with the need for increased food intake, they reflect other functions of these hormones. For example, leptin plays a role in both implantation and placentation as well as in **angiogenesis** and hence the growth and development of the fetus (Pérez-Pérez et al. 2018). Stable levels of adiponectin may be key in ensuring appropriate blood glucose levels and glucose sensitivity in pregnancy. Indeed, mice deficient in adiponectin develop a gestational diabetic phenotype during pregnancy with impaired glucose tolerance, higher basal glucose levels, lower insulin levels, and impaired β-cell expansion (Qiao et al. 2021). Similarly, increased insulin secretion in pregnancy is vital for glucose regulation and the need of the fetus for energy, mostly in the form of glucose, to be transferred across the placenta in a near-continuous manner. Utilization of fat stores during lactation might be facilitated by low levels of circulating ghrelin, which favors carbohydrate metabolism and the protection of fat stores.

The apparent conflict between a requirement for the peripheral functions of these signals and their central effects on energy balance pathways regulating food intake and energy expenditure may be resolved by changes in the responsiveness of these pathways to circulating signals (Fig. 3.3). In rodents, central leptin insensitivity develops during pregnancy to prevent leptin-induced satiety despite the increasing levels of leptin, so that by the last trimester of pregnancy central leptin administration no longer produces a decrease in food intake. Multiple mechanisms underlie central leptin insensitivity, including impaired entry of leptin into the brain and reduced sensitivity of first-order leptin responsive neurons, perhaps because of a decrease in leptin receptor levels (reviewed in Khant Aung et al. 2020). Leptin signals through the JAK/STAT3 pathway and expression of SOCS-3, a negative regulator of the JAK/STAT3 signaling pathway, is elevated in the hypothalamus during pregnancy (Nagaishi et al. 2014), suggesting downregulation of this signaling cascade. Phosphorylation of the signaling molecule STAT3 is a vital step in this signaling cascade and although leptin-induced pSTAT3 in the arcuate nucleus is unchanged in pregnancy, it is attenuated within the ventromedial nucleus of the hypothalamus (VMH), a region of the medial basal hypothalamus that is involved in the regulation of appetite and blood glucose levels. Reduced leptin receptor mRNA expression has also been detected in the VMH during pregnancy (reviewed in Khant Aung et al. 2020).

There is also evidence for central leptin insensitivity in lactation, despite the lower levels of plasma leptin. For example, Suzuki et al. demonstrated that lactating rats were less sensitive to intravenous leptin administration than nonlactating females (Suzuki et al. 2010) and chronic subcutaneous administration of leptin produced only a short-term decrease in food intake (Woodside et al. 2000). Furthermore, downregulation of leptin receptors within the VMH in rats on Day 15 of lactation has been observed (Brogan et al. 2000; Xu et al. 2009b). On day 9–10 of lactation neither orexigenic peptide expression in the arcuate nucleus nor food intake were suppressed by chronic peripheral infusions of leptin designed to mimic the plasma levels seen in nonlactating rats, although this treatment was able to increase POMC mRNA within the arcuate nucleus (Xu et al. 2009a).

Similar to leptin, insulin-induced intracellular signaling is attenuated during pregnancy (reviewed in Ladyman and Brooks 2021). Using the phosphorylation of AKT as a marker of insulin-activated intracellular signaling pathways, it has been shown that when insulin levels are higher in pregnant than nonpregnant rats, pAKT levels in the arcuate nucleus are lower in pregnant rats. Moreover, the ability of acute exogenous insulin administration to induce pAKT in the arcuate nucleus and VMH is reduced in pregnancy, but not in other areas of the hypothalamus. The mechanisms underlying these changes are as yet unclear but do not seem to reflect changes in insulin transport and/or reductions in insulin receptor density. Recently, increased insulin degradation in the brain has been proposed to underlie the decreased ability of insulin to activate signaling pathways in the brain during pregnancy, although further work is required to confirm this hypothesis (reviewed in Ladyman and Brooks 2021).

In contrast to pregnancy, insulin sensitivity seems to be intact at the level of POMC and NPY/AGRP neurons in the ARC during lactation, since reversing the decrease in circulating insulin levels typical of lactation was able to increase POMC and decrease both NPY and AGRP mRNA expression in this area (Xu et al. 2009a). Interestingly, however, this did not result in a decrease in food intake, suggesting that other brain areas and/or signals downstream of the POMC and NPY/AGRP neurons in the ARC may be changed to meet the energetic challenges of lactation. For example, lactation in rats is associated with an increase in NPY expression within the non-compact zone of the dorsal medial hypothalamus (DMH), an area where such expression is typically only seen in genetically obese animals. Neither leptin nor insulin treatment downregulated lactation-induced NPY expression in this area (Xu et al. 2009a), suggesting that it may be important for lactational hyperphagia. It is also possible that the dissociation between effects on central neuropeptide expression and food intake are a result of changes in the effectiveness of downstream signals. As noted earlier, the anorectic effects of stimulation of POMC neurons are largely dependent on the actions of alpha-MSH. The ability of alpha-MSH to induce a satiety response and to stimulate neurons are both impaired during pregnancy in the rat (reviewed in Khant Aung et al. 2020), suggesting changes in responsivity at multiple levels of the neural circuits mediating energy balance.

Little is known about changes in sensitivity to the anorectic effects of PYY and GLP-1 across a reproductive episode but in both pregnant and lactating rats the feeding response to exogenous CCK is reduced (Linden 1989; Ladyman et al. 2011) and meal size and duration are increased, supporting the concept of reduction in sensitivity to meal-terminating factors (Ladyman et al. 2018a). Moreover, the response to CCK in the brain and brainstem, as reflected in c-fos activation, is diminished by mid-pregnancy (Ladyman et al. 2011).

As expected, in contrast to sensitivity to the anorectic hormones, sensitivity to ghrelin is maintained in pregnancy and is actually increased during lactation. Chronic administration of ghrelin, either via mini-pump or multiple daily injections during the last week of pregnancy, increased both food intake and fetal growth and in lactation chronic ghrelin administration increased milk production (Nakahara et al. 2003). In mid-lactation, rats are more sensitive to the ability of acute ghrelin administration to increase food intake than nonlactating females (Woodside et al. 2012) and in late lactation there is an increase in ghrelin receptor expression within the hypothalamus, hippocampus, and VTA (Woodside et al. 2012).

As reviewed in this section, the metabolic requirements of pregnancy and lactation both entail changes in peripheral signals and neural peptides modulating energy balance, as well as the ability of those peripheral signals to affect central pathways (Fig. 3.3). Consistent with the fact that in pregnancy an increase in food intake has to occur under conditions in which the mother's body mass is increasing whereas the hyperphagia of lactation is associated with negative energy balance, the profile of both peripheral signals and central modulators, as well as their relationship, differs in these two states.

3.4 Contributions of Reproductive Hormones to Metabolic Adaptations in Pregnancy and Lactation

Pregnancy and lactation are each associated with a specific pattern of pituitary (prolactin, oxytocin), ovarian (estrogen, progesterone), and adrenal (progesterone, glucocorticoids) hormone release (see Fig. 3.5), which contribute not only to the growth and development of the fetus and/or milk production and delivery but also, themselves, act as metabolic signals. The dual roles of these hormones make it difficult to differentiate the direct effects of manipulation of these hormones on metabolic parameters and effects that are mediated by changes in fetal growth and/or milk delivery. Methodological approaches used to differentiate between these two pathways are described in Box 3.4.

Box 3.4: Dissecting the Direct Effects of Hormones on Food Intake, Independent of Their Effects on Milk Production During Lactation

As would be expected with the high energy requirement for milk production, the level of milk produced and food intake during lactation are highly related. Two key rat models have been used to dissect the contribution of hormones secreted in response to suckling, in particular that of prolactin, in the hyperphagia of lactation, independent of their indirect effects on hyperphagia due to the energy drain of milk production (Fig. 3.6). The first model completely eliminates the energy drain of milk production by preventing milk from being delivered to the offspring yet maintains the suckling stimulus from offspring and thereby the hormonal profile of lactation. Before pregnancy, rats undergo surgery to have the **galactophores** (tubes that carry the milk from the site of production to the nipple) cut, then after recovery they are mated, undergo pregnancy, and give birth. This model is based on the knowledge that for milk production to continue, milk needs to be removed from the site of production, the alveoli in the mammary glands, to the nipple for release to offspring. Without milk removal, the milk-producing alveoli will distend then atrophy.

(continued)

Box 3.4 (continued)
In the galactophore-cut (GC) dams milk production is not stimulated, since milk is not being delivered to offspring, yet offspring will continue to suckle, as rat pups will continue this behavior even if no milk is delivered. Pups do need to be switched between their GC mother and a foster mother who provides the milk they need. The continued suckling drives the same hormonal changes in the GC dam as it does in a milk-producing mother, thus allowing assessment of the role of the hormonal changes in the presence of no energy drain. GC dams eat more than non lactating females, and the level of hyperphagia is related to the size of the suckling litter (reviewed in Woodside et al. 2012). A second rat model used to investigate this interaction is an acute resuckling model. In this model, pups are removed from their mothers for 48 h. Once the pups are reunited with their mother they will suckle and stimulate the hormone secretions associated with suckling, yet there is a 24-h period before the onset of full milk production, thus allowing an assessment of changes induced by suckling alone. This model has shown that suckling alone is sufficient to alter neuropeptide expression involved in hyperphagia in the hypothalamus (Li et al. 1998).

Pituitary Hormones

Prolactin is named for its important role in milk production, yet prolactin is thought to have many other functions in the body during pregnancy and lactation. The exact secretion patterns of prolactin and placental lactogens during pregnancy are species-dependent, for example, in humans both prolactin and placental lactogens increase across gestation, while in laboratory rodents twice-daily surges of prolactin are initiated by the mating stimulus, and then as the placenta develops placental lactogens increase while prolactin secretion is suppressed (Fig. 3.5) (reviewed in Phillipps et al. 2020). Because most placental lactogens, like prolactin itself, act through the prolactin receptor (Prlr), Prlr activity increases in pregnancy (reviewed in Phillipps et al. 2020). In lactation, pituitary prolactin secretion is stimulated by suckling stimulation from the young but is independent of milk production (see Chap. 7). Prolactin levels rise very early postpartum with the onset of suckling and in rodents peak around Day 10 postpartum, before declining in late lactation (Walker et al. 1995). Prolactin is a major lactogenic hormone and in rodents, suppression of prolactin release eliminates milk production.

Prlr activation has been implicated in the stimulation of many aspects of metabolism, and primarily promotes a state of positive energy balance (reviewed in Lopez-Vicchi et al. 2020). In the periphery, Prlr activation is important in driving β-cell adaptations during pregnancy and in transgenic mice lacking Prlr in the pancreas a state of gestational diabetes develops during pregnancy, characterized by impaired glucose tolerance and attenuated β-cell expansion (Banerjee et al. 2016). In mice deficient in adiponectin, a gestational diabetic-like state also develops during pregnancy, and this has been shown to be due to low placental lactogen levels,

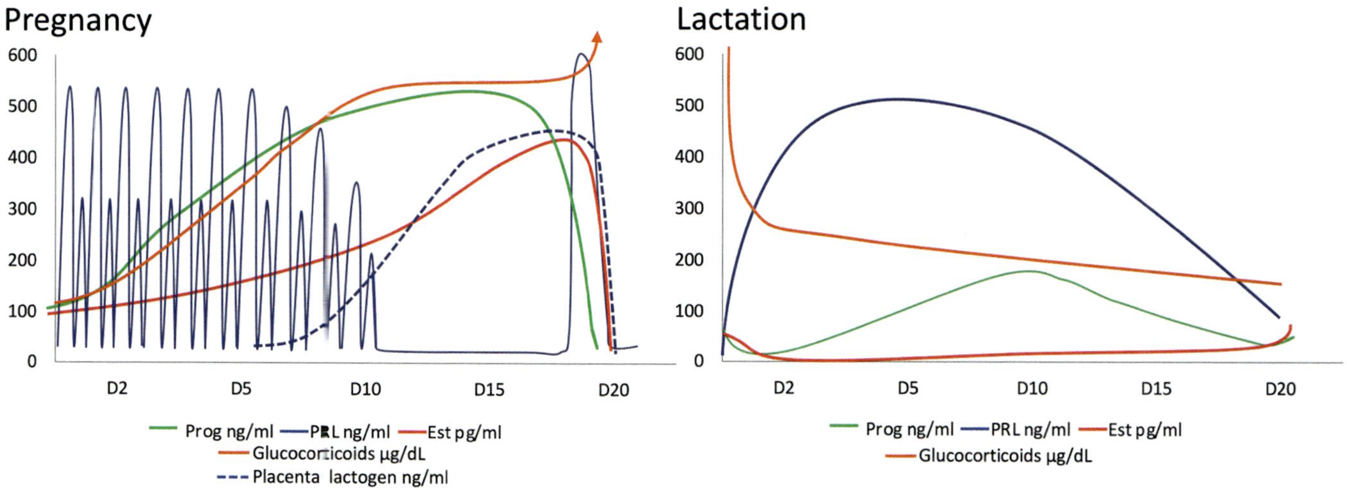

Fig. 3.5 Reproductive hormones in pregnancy and lactation. Schematic representation of changing reproductive hormones through pregnancy and lactation. Levels are expressed as an approximate percentage of average pre-pregnancy levels. These graphs are based on what is known in rats, as they are the most widely studied animal model in this field. The major difference in humans is that in early pregnancy instead of the twice daily prolactin surges seen in rats and mice, levels of both prolactin and placental lactogens rise across pregnancy

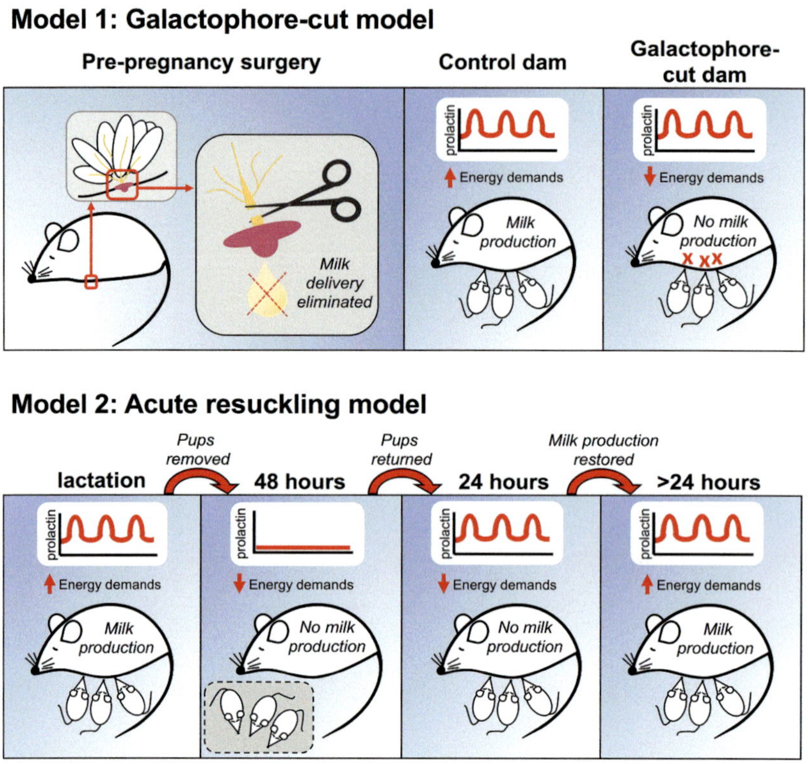

Fig. 3.6 Rat models used to dissect the direct effects of hormones on food intake independent of their effects on milk production during lactation. Two rat models have been used for this purpose, especially for the investigation of the role of prolactin. Model 1 (*top panel*) describes a galactophore-cut rat that can get pregnant and give birth, and then nurses offspring resulting in suckling-induced hormone changes, without milk being produced or transferred to the offspring. Model 2 (*bottom panel*) describes the acute resuckling model where the return of offspring after deprivation for 48 h leads to approximately 24 h where the suckling-induced hormone profile is stimulated but a delay in milk production means this is a time of low energy demands

suggesting interactions between these two hormones (Qiao et al. 2021). There is a large increase in Prlr in white adipose tissue of mice during lactation, and evidence from mice with a Prlr deficiency suggests that reduction of Prlr activation facilitates the differentiation of white adipose tissue into beige fat in mice on a high fat diet, resulting in an increase in energy expenditure (Auffret et al. 2012). Thus, it is possible that the increased Prlr in this tissue during lactation protects against a similar process occurring as a result of the increased food intake of lactation.

Central administration of prolactin increases food intake in estrous cycling rats and chronic central infusion of prolactin induces leptin resistance, which is likely to contribute to prolactin-induced elevated food intake (reviewed in Lopez-Vicchi et al. 2020). While the mechanisms underlying this are yet to be elucidated, there is evidence that the high levels of Prlr activation drive a suppression of leptin-induced activation of the JAK/STAT3 intracellular signaling pathway in regions of the hypothalamus involved in appetite regulation (reviewed in Khant Aung et al. 2020). The degree of colocalization of leptin receptors and prolactin receptors in these areas is not high (Nagaishi et al. 2014), suggesting that this effect of prolactin may be indirect. Chronic Prlr activation results in increased food intake and leptin resistance which is consistent with the increased food intake and leptin resistance typical of late pregnancy.

Because of its critical role in milk production, examining the effects of Prlr on the hyperphagia of lactation independent of changes in milk production has been complex. This problem has been addressed using the two models described in Box 3.4. Results of these studies suggest that maternal food intake increases with energetic demand but is not completely reduced to baseline levels when that energetic demand is eliminated, as long as stimulation from the young is maintained (reviewed in Woodside et al. 2012; Woodside 2016). Suppressing prolactin in these females reduces their food intake to that of estrous cycling females and simultaneous intracerebroventricular administration of prolactin restores it, suggesting that central Prlr activation contributes to the hyperphagia of lactation. Further, in contrast to intact lactating females, female rats that are suckled but not delivering milk (GC dams) do not utilize fat stores postpartum and their circulating leptin levels are higher than those of nonlactating rats. In spite of this they still show prolactin-dependent hyperphagia, suggesting that prolactin stimulation can overcome leptin's ability to decrease food intake in lactation, as it does in pregnancy (Woodside 2016).

Oxytocin is released into blood circulation via projections from oxytocinergic neurons in the PVN and supraoptic nucleus (SON) to the posterior pituitary. Pulsatile oxytocin release into the periphery plays an important role in uterine contractions during parturition and in milk release during lactation (see Chaps. 5 and 6). In addition to its peripheral action oxytocin is released centrally and acts distally as a neurotransmitter as well as locally as a **neurohormone**. During pregnancy, oxytocin neuron activity is suppressed, to prevent oxytocin release until it is required for birth. In lactation oxytocin is released both centrally and peripherally in response to suckling and stimulates **milk let-down**. Both peripheral and central oxytocin administration decrease food intake in male and virgin female rodents (McCormack et al. 2020). In pregnant rats, however, neither central administration of oxytocin nor an oxytocin receptor antagonist alters food intake (Douglas et al. 2007). As yet the effects of oxytocin on food intake in lactation have not been investigated. The mechanisms mediating the suppression of the effects of oxytocin on food intake in pregnancy are not clear. Recent evidence suggests that females are less responsive to oxytocin than males, especially when estrogen levels are high (Liu et al. 2020). Such an association is consistent with decreased anorectic effects of oxytocin in late pregnancy when estrogen is high, but is not consistent with the hyperphagia shown in lactation when oxytocin is high and estrogen is low (Fig. 3.5). Prolactin inhibits oxytocin neurons during pregnancy (Augustine et al. 2017), which might contribute to the effect, but central oxytocin release is increased, at least in some brain areas, during lactation (Menon et al. 2018), suggesting a specific decrease in the ability of oxytocin to modulate food intake (McCormack et al. 2020).

Growth hormone (GH) is secreted in a pulsatile manner, and elegant collection of tiny amounts of blood from the tails of mice every 10 min over 6 h across different stages of pregnancy, combined with ultra-sensitive ELISA, has provided detailed analysis of GH secretion patterns in mice pregnancies. While the pulsatile nature of secretion remains, basal levels of GH increase across pregnancy (Gatford et al. 2017). In humans, GH also increases in pregnancy, however it is a placentally produced variant of GH and does not show a pulsatile release pattern (Eriksson et al. 1989). GH acts throughout the body to promote growth, by increasing cell proliferation and promoting lipolysis of adipose tissue to mobilize energy stores. Recent work has shown that GH acts within the hypothalamus to drive increases in food intake and decrease energy expenditure, thus also promoting a state of positive energy balance (Furigo et al. 2019). In terms of pregnancy, studies using transgenic mice with specific deletion of GH receptors (GH-R) from neurons have shown that without the interaction of GH with GH-R in the brain, there is an attenuation of pregnancy-induced increase in food intake and fat mass accumulation, highlighting a key role for GH in promoting these maternal metabolic adaptations (Teixeira et al. 2019a). Consistent with the notion that fat mass accumulation in pregnancy may be important for subsequent lactation this manipulation was also associated with a decrease in litter growth. In lactation GH secretion from the anterior pituitary, like that of prolactin, is increased by suckling stimulation. GH does not make a major contribution to milk production in early lactation (Flint et al. 1992) but does in late lactation, when the ability of suckling to stimulate prolactin release is reduced (Mattheij et al. 1984)

Steroid Hormones

Estrogen levels normally peak preceding ovulation in the menstrual/estrous cycle and this peak is suppressed post-conception. However, basal estrogen levels increase during pregnancy in humans and in rodents are significantly higher in late pregnancy (Fig. 3.5). The peak estrogen levels seen just prior to parturition drop precipitously immediately postpartum and remain low for the first half of lactation, increasing as suckling stimulation declines until ovulation resumes (Fig. 3.5) (Smith and Neill 1977). Estrogen is both a potent anorectic factor and an enhancer of energy expenditure. The effects of estrogen on food intake are mediated primarily through POMC neurons in the ARC while its effects on energy expenditure are mediated primarily through the VMH (reviewed in Xu and Lopez 2018). Thus, the relatively low levels of estrogen in early pregnancy and lactation may contribute to the increase in food intake and conservation of energy at these times. Estrogen also modulates the effects of other metabolic signals on food intake, facilitating the effects of anorectic signals such as CCK, leptin, and insulin while suppressing the orexigenic effects of ghrelin (reviewed in Xu and Lopez 2018). There is also some evidence that the anorectic effects of estrogen are suppressed during lactation (Fleming 1976). Given the high estrogen levels and increased food intake in late pregnancy, it is highly tempting to speculate that this suppression of the anorectic effect of estrogen has already developed during pregnancy and is maintained during lactation. Future work is needed to confirm this hypothesis.

Progesterone is vital for maintaining pregnancy and increases across gestation (reviewed in Solano and Arck 2019) (Fig. 3.5). In rodents, progesterone drops rapidly prior to parturition and while this does not seem to be the case in humans, a reduction in progesterone function is hypothesized to take place, with reductions in tissue sensitivity probably playing a major role. In those species, such as rats and mice, that experience a postpartum ovulation, the resulting corpora lutea are stimulated by the elevated prolactin to secrete progesterone. Whether this high circulating progesterone contributes to lactational hyperphagia in rodents has not been determined. Some authors have reported an increase in food intake in response to exogenous progesterone administration to estrous cycling female rodents, while others failed to observe this effect. Establishing a role for progesterone in lactational hyperphagia is complicated, because removing the ovaries, the primary source of progesterone in lactating rodents, leads to a compensatory increase in prolactin-stimulated adrenal progesterone secretion and removing the adrenal glands decreases milk production (Plucinski and Baldwin 1976). Approaches similar to those described in Box 3.4 may be useful in studying this issue (Woodside et al. 2008).

Glucocorticoids released from the adrenal gland in response to stimulation of the hypothalamic-pituitary-adrenal (HPA) axis play an important role in the regulation of glucose, working in opposition to insulin to increase blood glucose levels, and making it available as a ready source of energy. Glucocorticoids also stimulate food intake, at least in part by stimulating NPY/AGRP neurons and reorganizing synaptic inputs both to these neurons and POMC neurons in the ARC (Gyengesi et al. 2010). During pregnancy, glucocorticoid levels increase, and while binding proteins (corticosteroid-binding globulin: CBG) for them also increase, the level of availability, or free hormone level, is elevated compared to pre-pregnancy (Fig. 3.5) (reviewed in Solano and Arck 2019). Basal circulating glucocorticoid levels are also high throughout lactation whereas CBG levels are decreased (Pawluski et al. 2009). Glucocorticoids facilitate successful implantation, are critical for fetal development, particularly adrenal and lung development, and play a critical role in milk synthesis. In contrast to the increase in basal glucocorticoids in pregnancy and lactation, stress-induced HPA activation is reduced, as is corticotrophin-releasing hormone (CRH) expression in the paraventricular nucleus (PVN) (Slattery and Neumann 2008) (see Chap. 4). Stress-induced CRH release acutely suppresses appetite, while in the longer term glucocorticoids stimulate appetite (Sominsky and Spencer 2014). Thus, the increase in basal glucocorticoids coupled with a decrease in stress-induced CRH favors the orexigenic aspects of HPA activation over its anorectic effects.

3.5 Perspectives

In general, as can be seen from Fig. 3.5, the reproductive hormone profiles of pregnancy and lactation favor orexigenic over anorexigenic factors and Fig. 3.7 summarizes how peripheral signals and neural circuits controlling energy balance

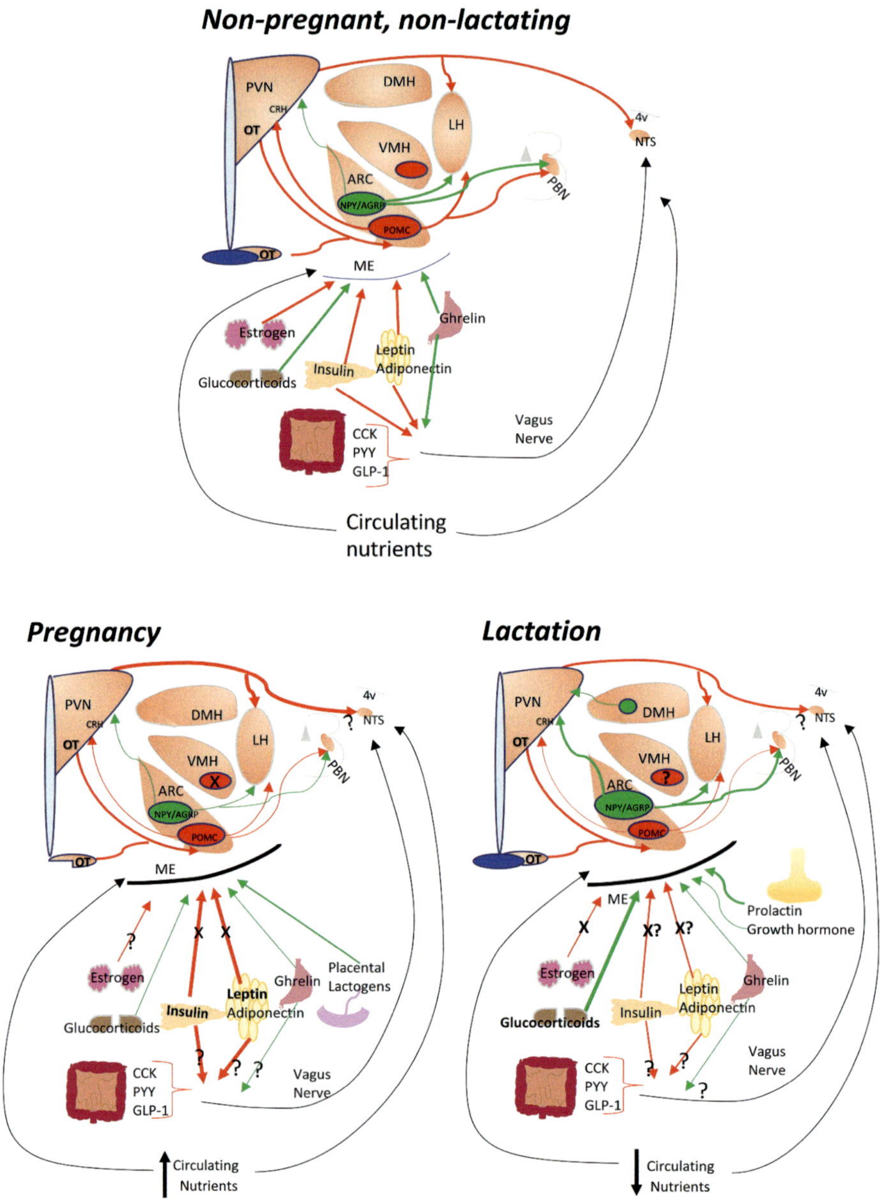

Fig. 3.7 Summary of maternal adaptations of peripheral signals and neural circuits controlling energy balance. *Top panel* shows key neural circuits that regulate food intake located in the forebrain, particularly the hypothalamus (ARC, VMH, DMH, LH, and PVN) and the hindbrain (PBN and NTS), with important projections between these regions indicated by arrows (green arrows indicate stimulatory effect, red arrows indicate inhibitory effect). Hormones involved in energy balance are released by peripheral tissues and can act either in the forebrain or the NTS via the vagal nerve to stimulate (green arrows) or inhibit (red arrows) food intake. Changes to this system are shown for pregnancy (*bottom left panel*) and lactation (*bottom right panel*). Thin arrows

change in pregnancy and lactation, as well as the contribution of key reproductive hormones. The high level of prolactin receptor activity that results from the release of lactogens from the placenta and suckling-induced release from the pituitary is key in facilitating the increased food intake of both pregnancy and lactation, not only through its own actions on food intake but also through its ability to alter responsivity to metabolic signals such as insulin and leptin. Prolactin is also implicated in the suppression of estrogen release during both pregnancy and lactation (Woodside et al. 2012) thereby reducing both the anorectic effects of estrogen itself and its ability to potentiate the effects of other anorectic hormones such as CCK.

Our knowledge of how metabolic hormone levels and the sensitivity of tissues to their actions during pregnancy and lactation is still limited. Going forward, understanding how these hormones and their actions are modulated during pregnancy and lactation may help to reveal the etiology of common pregnancy complications such as gestational diabetes and pre-eclampsia, as well as breast feeding problems, and to elucidate how parity can increase the long-term risk of developing metabolic syndrome and diabetes. With high rates of obesity in developed countries, understanding how the metabolic system adapts to pregnancy and lactation to promote healthy gestational weight gain and effective breastfeeding is of utmost importance, given the contribution of both of these factors to the long-term health of both mother and child.

Fig. 3.7 (continued) represent decreased concentration of hormones and reduced activation of neural projections, while bold text/thick arrows represent increased concentrations of hormones and increased activation of neural projections. "X" indicates a known development of insensitivity, such as the decrease in VMH neurons to anorectic hormones insulin and leptin during pregnancy. "?" indicates that these interactions are hypothesized to be affected but are yet to been specifically investigated. PVN, paraventricular nucleus; DMH, dorsomedial nucleus; LH, lateral hypothalamus; VMH, ventromedial nucleus; ARC, arcuate nucleus; NTS, nucleus of the solitary tract; ME, median eminence; PBN, parabrachial nucleus; 4v, 4th ventricle; CRH, corticotrophin-releasing hormone; OT, oxytocin; NPY, neuropeptide Y; AGRP, agouti-related peptide; CCK, cholecystokinin; PYY, peptide YY; GLP-1, glucagon-like peptide

Key References for Further Information

Woodside et al. (2012): In-depth review focused on maternal adaptations in neural circuits and metabolic hormones in lactation that allow the mother to meet the energy demands of providing milk for offspring.

Khant Aung et al. (2020): Recent review focused on mechanisms underlying the development of leptin and insulin insensitivity in the hypothalamus during pregnancy.

Duarte-Guterman et al. (2019): Summarizing what is known regarding the long-term physiological changes in females after undergoing pregnancy and lactation, including metabolic changes.

Catalano and Shankar (2017): Examines the impact of obesity on all aspects of pregnancy and lactation, from pre-conception to outcomes for the next generation.

Pheiffer et al. (2021): Review article exploring the interaction of adiponectin and common conditions in pregnancy, highlighting the potential role of metabolic hormones as biomarkers of pregnancy complications.

Teixeira et al. (2019a): Novel experimental results revealing a key role for central growth hormone actions in maternal metabolic adaptations during pregnancy.

References

Abizaid A, Schiavo L, Diano S (2008) Hypothalamic and pituitary expression of ghrelin receptor message is increased during lactation. Neurosci Lett 440(3):206–210

Andersson-Hall U, Svedin P, Svensson H, Lonn M, Mallard C, Holmang A (2020) Longitudinal changes in adipokines and free leptin index during and after pregnancy in women with obesity. Int J Obes (Lond) 44(3):675–683

Auffret J, Viengchareun S, Carré N, Denis RG, Magnan C, Marie PY, Muscat A, Fève B, Lombès M, N., B. (2012) Beige differentiation of adipose depots in mice lacking prolactin receptor protects against high-fat-diet-induced obesity. FASEB J 26(9):3728–3737

Augustine RA, Ladyman SR, Bouwer GT, Alyousif Y, Sapsford TJ, Scott V, Kokay IC, Grattan DR, Brown CH (2017) Prolactin regulation of oxytocin neurone activity in pregnancy and lactation. J Physiol 595(11):3591–3605

Baeyens L, Hindi S, Sorenson RL, German MS (2016) beta-Cell adaptation in pregnancy. Diabetes Obes Metab 18(Suppl 1):63–70

Banerjee RR, Cyphert HA, Walker EM, Chakravarthy H, Peiris H, Gu X, Liu Y, Conrad E, Goodrich L, Stein RW, Kim SK (2016) Gestational diabetes mellitus from inactivation of prolactin receptor and MafB in islet beta-cells. Diabetes 65(8):2331–2341

Bever Babendure J, Reifsnider E, Mendias E, Moramarco MW, Davila YR (2015) Reduced breastfeeding rates among obese mothers: a review of contributing factors, clinical considerations and future directions. Int Breastfeed J 10:21

Brogan RS, Grove KL, Smith MS (2000) Differential regulation of leptin receptor but not orexin in the hypothalamus of the lactating rat. J Neuroendocrinol 12(11):1077–1086

Brommage R, Baxter DC (1988) Elevated calcium, phosphorus, and magnesium retention in pregnant rats prior to the onset of fetal skeletal mineralization. J Bone Miner Res 3(6):667–672

Burton GJ (2012) The placenta and developmental programming: balancing fetal nutrient demands with maternal resource allocation. Placenta 33(3 Suppl):S23–S27

Butte NF, King JC (2005) Energy requirements during pregnancy and lactation. Public Health Nutr 8(7A):1010–1027
Butte NF, Wong WW, Treuth MS, Ellis KJ, O'Brian Smith E (2004) Energy requirements during pregnancy based on total energy expenditure and energy deposition. Am J Clin Nutr 79(6):1078–1087
Casirola DM, Ferraris RP (2003) Role of the small intestine in postpartum weight retention in mice. Am J Clin Nutr 78(6):1178–1187
Catalano PM, Shankar K (2017) Obesity and pregnancy: mechanisms of short term and long term adverse consequences for mother and child. BMJ 356:j1
Catalano PM, Roman-Drago NM, Amini SB, Sims EA (1998) Longitudinal changes in body composition and energy balance in lean women with normal and abnormal glucose tolerance during pregnancy. Am J Obstet Gynecol 179(1):156–165
Clarke GS, Gatford KL, Young RL, Grattan DR, Ladyman SR, Page AJ (2021) Maternal adaptions to food intake across pregnancy: central and peripheral mechanisms. Obesity 29(11):1813–1824
Douglas AJ, Johnstone LE, Leng G (2007) Neuroendocrine mechanisms of change in food intake during pregnancy: a potential role for brain oxytocin. Physiol Behav 91(4):352–365
Duarte-Guterman P, Leuner B, Galea LAM (2019) The long and short term effects of motherhood on the brain. Front Neuroendocrinol 53:100740
Eriksson L, Frankenne F, Eden S, Hennen G, Von Schoultz B (1989) Growth hormone 24-h serum profiles during pregnancy-lack of pulsatility for the secretion of the placental variant. Br J Obstet Gynaecol 96(8):949–953
Fleming AS (1976) Control of food intake in the lactating rat: role of suckling and hormones. Physiol Behav 17(5):841–848
Flint DJ, Tonner E, Beattie J, Panton D (1992) Investigation of the mechanism of action of growth hormone in stimulating lactation in the rat. J Endocrinol 134(3):377–383
Furigo IC, de Souza GO, Teixeira PDS, Guadagnini D, Frazao R, List EO, Kopchick JJ, Prada PO, Donato J Jr (2019) Growth hormone enhances the recovery of hypoglycemia via ventromedial hypothalamic neurons. FASEB J 33(11):11909–11924
Gamo Y, Troup C, Mitchell SE, Hambly C, Vaanholt LM, JR., S. (2013) Limits to sustained energy intake. XX. Body temperatures and physical activity of female mice during lactation. J Exp Biol 216(19):3751–3761
Gatford KL, Muhlhausler BS, Huang L, Sim PS, Roberts CT, Velhuis JD, Chen C (2017) Rising maternal circulating GH during murine pregnancy suggests placental regulation. Endocr Connect 6(4):260–266
Gunderson EP, Jacobs DR Jr, Chiang V, Lewis CE, Tsai A, Quesenberry CP Jr, Sidney S (2009) Childbearing is associated with higher incidence of the metabolic syndrome among women of reproductive age controlling for measurements before pregnancy: the CARDIA study. Am J Obstet Gynecol 201(2):177 e1–9
Gyengesi E, Liu Z-W, D'Agostino G, Gan G, Horvath TL, Gao X-B, Diano S (2010) Corticosterone regulates synaptic input organization of POMC and NPY-AgRP neurons in adult mice. Endocrinology 151(11):5395–5402
Harris HE, Ellison GT, Holliday M, Lucassen E (1997) The impact of pregnancy on the long-term weight gain of primiparous women in England. Int J Obes Relat Metab Disord 21(9):747–755
Hollis JL, Crozier SR, Inskip HM, Cooper C, Godfrey KM, Harvey NC, Collins CE, Robinson SM (2017) Modifiable risk factors of maternal postpartum weight retention: an analysis of their combined impact and potential opportunities for prevention. Int J Obes (Lond) 41(7):1091–1098
Johnson ML, Saffrey MJ, VJ., T. (2019) Gastrointestinal capacity, gut hormones and appetite change during rat pregnancy and lactation. Reproduction 157((5)):431–443
Kaur H, Muhlhausler BS, Sim PS, Page AJ, Li H, Nunez-Salces M, Clarke GS, Huang L, Wilson RL, Veldhuis JD, Chen C, Roberts CT, Gatford KL (2020) Pregnancy, but not dietary octanoic acid supplementation, stimulates the ghrelin-pituitary growth hormone axis in mice. J Endocrinol 245(2):327–342
Khant Aung Z, Grattan DR, Ladyman SR (2020) Pregnancy-induced adaptation of central sensitivity to leptin and insulin. Mol Cell Endocrinol 516:110933

Kovacs CS (2001) Calcium and bone metabolism in pregnancy and lactation. J Clin Endocrinol Metab 86(6):2344–2348
Krol E, Speakman JR (2003) Limits to sustained energy intake. VI. Energetics of lactation in laboratory mice at thermoneutrality. J Exp Biol 206(Pt 23):4255–4266
Krol E, Speakman JR (2019) Switching off the furnace: brown adipose tissue and lactation. Mol Aspects Med 68:18–41
Ladyman SR, Brooks VL (2021) Central actions of insulin during pregnancy and lactation. J Neuroendocrinol 33(4):e12946
Ladyman SR, Sapsford TJ, Grattan DR (2011) Loss of acute satiety response to cholecystokinin in pregnant rats. J Neuroendocrinol 23(11):1091–1098
Ladyman SR, Carter KM, Grattan DR (2018a) Energy homeostasis and running wheel activity during pregnancy in the mouse. Physiol Behav 194:83–94
Ladyman SR, Khant Aung Z, Grattan DR (2018b) Impact of pregnancy and lactation on the long-term regulation of energy balance in female mice. Endocrinology 159(6):2324–2336
Lamming G (1978) Control of ovulation. London, Butterworth, pp 335–353
Leon M, Woodside B (1983) Energetic limits on reproduction: maternal food intake. Physiol Behav 30(6):945–957
Li C, Chen P, Smith MS (1998) The acute suckling stimulus induces expression of neuropeptide Y (NPY) in cells in the dorsomedial hypothalamus and increases NPY expression in the arcuate nucleus. Endocrinology 139(4):1645–1652
Linden A (1989) Role of cholecystokinin in feeding and lactation. Acta Physiol Scand Suppl 585:i-vii, 1-49
Liu CM, Davis EA, Suarez AN, Wood RI, Noble EE, SE., K. (2020) Sex differences and estrous influences on oxytocin control of food intake. Neuroscience 447:63–73
Lopez-Vicchi F, De Winne C, Brie B, Sorianello E, Ladyman SR, Becu-Villalobos D (2020) Metabolic functions of prolactin: Physiological and pathological aspects. J Neuroendocrinol 32(11):e12888
Lovelady CA (2004) The impact of energy restriction and exercise in lactating women. Adv Exp Med Biol 554:115–120
Mattheij JA, Swarts HJ, Verstijnen CP (1984) The response of plasma prolactin to suckling during normal and prolonged lactation in the rat. Horm Res 20(4):261–268
McCormack SE, Blevins JE, Lawson EA (2020) Metabolic effects of oxytocin. Endocr Rev 41(2)
McKinley MC, Allen-Walker V, McGirr C, Rooney C, Woodside JV (2018) Weight loss after pregnancy: challenges and opportunities. Nutr Res Rev 31(2):225–238
McNeilly AS (2001) Lactational control of reproduction. Reprod Fertil Dev 13(7-8):583–590
Menon R, Grund T, Zoicas I, Althammer F, Fiedler D, Biermeier V, Bosch OJ, Hiraoka Y, Nishimori K, Eliava M, Grinevich V, Neumann ID (2018) Oxytocin signaling in the lateral septum prevents social fear during lactation. Curr Biol 28(7):1066–1078 e6
Nagaishi VS, Cardinali LI, Zampieri TT, Furigo IC, Metzger M, Donato J Jr (2014) Possible cross-talk between leptin and prolactin during pregnancy. Neuroscience 259:71–83
Naismith DJ, Morgan BL (1976) The biphasic nature of protein metabolism during pregnancy in the rat. Br J Nutr 36(3):563–566
Naismith DJ, Richardson DP, Pritchard AE (1982) The utilization of protein and energy during lactation in the rat, with particular regard to the use of fat accumulated in pregnancy. Br J Nutr 48(2):433–441
Nakahara K, Hayashida T, Nakazato M, Kojima M, Hosoda H, Kangawa K, N., M. (2003) Effect of chronic treatments with ghrelin on milk secretion in lactating rats. Biochem Biophys Res Commun 303(3):751–755
Ohlin A, Rossner S (1996) Factors related to body weight changes during and after pregnancy: the Stockholm Pregnancy and Weight Development Study. Obes Res 4(3):271–276
Pawluski JL, Charlier TD, Lieblich SE, Hammond GL, Galea LAM (2009) Reproductive experience alters corticosterone and CBG levels in the rat dam. Physiol Behav 96(1):108–114

Pérez-Pérez A, Toro A, Vilariño-García T, Maymó J, Guadix P, Dueñas JL, Fernández-Sánchez M, Varone C, Sánchez-Margalet V (2018) Leptin action in normal and pathological pregnancies. J Cell Mol Med 22(2):716–727
Pheiffer C, Dias S, Jack B, Malaza N, Adam S (2021) Adiponectin as a potential biomarker for pregnancy disorders. Int J Mol Sci 22(3)
Phillipps HR, Yip SH, Grattan DR (2020) Patterns of prolactin secretion. Mol Cell Endocrinol 502:110679
Plucinski T, Baldwin RL (1976) Effects of adrenalectomy and glucocorticoid therapy on enzyme activities in mammary and adipose tissues from lactating rats. J Dairy Sci 59(1):157–160
Qiao L, Saget S, Lu C, Hay WW Jr, Karsenty G, Shao J (2021) Adiponectin promotes maternal beta-cell expansion through placental lactogen expression. Diabetes 70(1):132–142
Rasmussen KM (2007) Association of maternal obesity before conception with poor lactational performance. Annu Rev Nutr 27:103–121
Shibata K, Hosoda H, Kojima M, Kangawa K, Makino Y, Makino I, Kawarabayashi T, Futagami K, Y., G. (2004) Regulation of ghrelin secretion during pregnancy and lactation in the rat: possible involvement of hypothalamus. Peptides 25(2):279–287
Slattery DA, Neumann ID (2008) No stress please! Mechanisms of stress hyporesponsiveness of the maternal brain. J Physiol 586(2):377–385
Smith MS, Neill JD (1977) Inhibition of gonadotropin secretion during lactation in the rat: relative contribution of suckling and ovarian steroids. Biol Reprod 17(2):255–261
Solano ME, Arck PC (2019) Steroids, pregnancy and fetal development. Front Immunol 10:3017
Sominsky L, Spencer SJ (2014) Eating behavior and stress: a pathway to obesity. Front Psychol 5:434
Spaaij CJ, van Raaij JM, de Groot LC, van der Heijden LJ, Boekholt HA, Hautvast JG (1994) Effect of lactation on resting metabolic rate and on diet- and work-induced thermogenesis. Am J Clin Nutr 59(1):42–47
Suzuki Y, Kurose Y, Takahashi H, Asakuma S, Azuma Y, Kobayashi S (2010) The differences in feeding-inhibitory responses to peripheral and central leptin between non-lactating and lactating rats. J Endocrinol 207(1):105–111
Suzuki Y, Nakahara K, Maruyama K, Okame R, Ensho T, Inoue Y, Murakami N (2014) Changes in mRNA expression of arcuate nucleus appetite-regulating peptides during lactation in rats. J Mol Endocrinol 52(2):97–109
Takahashi M, Miyatsuka T, Suzuki L, Osonoi S, Himuro M, Miura M, Katahira T, Wakabayashi Y, Fukunaka A, Nishida Y, Fujitani Y, Takeda S, Mizukami H, Itakura A, Watada H (2020) Biphasic changes in β-cell mass around parturition are accompanied by increased serotonin production. Sci Rep 10(1):4962–4974
Teixeira PDS, Couto GC, Furigo IC, List EO, Kopchick JJ, Donato J Jr (2019a) Central growth hormone action regulates metabolism during pregnancy. Am J Physiol Endocrinol Metab 317(5):E925–E940
Teixeira PDS, Ramos-Lobo AM, Furigo IC, Donato J (2019b) Brain STAT5 modulates long-term metabolic and epigenetic changes induced by pregnancy and lactation in female mice. Endocrinology 160(12):2903–2917
Torris C, Bjornnes AK (2020) Duration of lactation and maternal risk of metabolic syndrome: a systematic review and meta-analysis. Nutrients 12(9)
Tovar SA, Seoane LM, Caminos JE, Nogueiras R, Casanueva FF, Dieguez C (2004) Regulation of peptide YY levels by age, hormonal, and nutritional status. Obes Res 12(12):1944–1950
Vila G, Hopfgartner J, Grimm G, Baumgartner-Parzer SM, Kautzky-Willer A, Clodi M, Luger A (2015) Lactation and appetite-regulating hormones: increased maternal plasma peptide YY concentrations 3-6 months postpartum. Br J Nutr 114(8):1203–1208. https://doi.org/10.1017/S0007114515002536. Epub 2015 Aug 24. PMID: 26299586
Walker CD, Mitchell JB, Woodside BC (1995) Suppression of LH secretion in food-restricted lactating females: effects of ovariectomy and bromocryptine treatment. J Endocrinol 146(1):95–104

Wattez JS, Delmont A, Bouvet M, Beseme O, Goers S, Delahaye F, Laborie C, Lesage J, Foligné B, Breton C, Metges CC, Vieau D, Pinet F (2015) Maternal perinatal undernutrition modifies lactose and serotranferrin in milk: relevance to the programming of metabolic diseases? Am J Physiol Endocrinol Metab 308(5):E393–E401
Woodside B (2016) Mood, food and fertility: adaptations of the maternal brain. Compr Physiol 6(3):1493–1518
Woodside B, Abizaid A, Walker C (2000) Changes in leptin levels during lactation: implications for lactational hyperphagia and anovulation. Horm Behav 37(4):353–365
Woodside B, Augustine R, Naef L, Ladyman S, Grattan D (2008) Role of prolactin in the metabolic adaptations to pregnancy and lactation. In: Bridges RS (ed) The parental brain. Elsevier, Amsterdam
Woodside B, Budin R, Wellman MK, Abizaid A (2012) Many mouths to feed: The control of food intake during lactation. Front Neuroendocrinol 33(3):301–314
Xu Y, Lopez M (2018) Central regulation of energy metabolism by estrogens. Mol Metab 15:104–115
Xu J, Kirigiti MA, Grove KL, Smith MS (2009a) Regulation of food intake and gonadotropin-releasing hormone/luteinizing hormone during lactation: role of insulin and leptin. Endocrinology 150(9):4231–4240
Xu J, Kirigiti MA, Cowley MA, Grove KL, Smith MS (2009b) Suppression of basal spontaneous gonadotropin-releasing hormone neuronal activity during lactation: role of inhibitory effects of neuropeptide Y. Endocrinology 150(1):333–340

Regulation of the Hypothalamo-Pituitary-Adrenal Axis in Pregnancy and Lactation

4

Paula J. Brunton and Claire-Dominique Walker

Abstract

Physiological and behavioural responses to stress are critical to ensure individual survival in all species. This is even more important when the survival of the offspring depends on parental protection from predators and other environmental challenges. However, exaggerated maternal stress responses and enhanced glucocorticoid secretion can be deleterious to the development of the offspring, with negative impacts on the developing brain, as well as on the provision of maternal care and nourishment. Thus, in both late gestation and lactation, mothers exhibit a period of stress hyporesponsiveness that is regulated by both common and divergent mechanisms that are appropriate for the physiological, hormonal, and behavioural environment of pregnancy and lactation. This chapter will describe our current understanding of the mechanisms in the peripartum period that

Author websites:
https://www.ed.ac.uk/discovery-brain-sciences/our-staff/research-groups/dr-paula-brunton
https://douglas.research.mcgill.ca/claire-dominique-walker

P. J. Brunton (✉)
Centre for Discovery Brain Sciences, University of Edinburgh, Edinburgh, UK

Zhejiang University-University of Edinburgh Joint Institute, Zhejiang, P.R. China
e-mail: p.j.brunton@ed.ac.uk

C.-D. Walker
Department of Psychiatry, McGill University, Montreal, Canada

Douglas Mental Health University Institute, Montreal, Canada
e-mail: dominique.walker@douglas.mcgill.ca

P. J. Brunton, D. R. Grattan (eds.), *Neuroendocrine Regulation of Mammalian Pregnancy and Lactation*, Masterclass in Neuroendocrinology 15,
https://doi.org/10.1007/978-3-031-51138-7_4

underlie adaptations of the neuroendocrine stress system—the hypothalamo-pituitary-adrenal (HPA) axis. Changes in central brain structures regulating HPA axis activity and the role of steroid and peptide hormones in maintaining blunted responses to stress will be discussed. The importance of the offspring's presence and suckling stimulation on the maintenance of reduced stress responses in lactation will also be considered. Finally, implications of this research for postpartum depression and other mood disorders characterized by dysregulated stress responses in human mothers will be examined.

Keywords

Allopregnanolone · Leptin · Maternal behaviour · Opioids · Oxytocin · Peripartum · Prolactin · Steroid hormones · Stress

4.1 Introduction

The ability to mount an appropriate response when faced with a physiological challenge such as stress is critical for survival. Exposure to stress leads to a series of orchestrated responses involving complex **autonomic**, behavioural, and neuroendocrine adaptations that serve to promote restoration of **homeostasis**. However, equally important is an organism's ability to appropriately terminate a stress response, when the stress no longer poses a threat.

The hypothalamic-pituitary-adrenal (HPA) axis is the primary neuroendocrine system that regulates responses to stress. It consists of three components—the hypothalamus, a small region located at the base of the brain; the pituitary gland, situated just beneath the brain and attached to the hypothalamus by a narrow stalk (hypophyseal stalk); and the adrenal glands which are found immediately superior to the kidneys (Fig. 4.1). Stress activates neurosecretory neurons in the **parvocellular division of the hypothalamic paraventricular nucleus (pPVN)**, which synthesize and secrete corticotropin-releasing hormone (CRH). CRH is released from nerve terminals at the external zone of the median eminence into the hypophysial portal blood system that supplies the pituitary gland. CRH acts on CRH type 1 receptors (Crhr1), located on specialized cells of the anterior pituitary gland named corticotrophs, to stimulate the secretion of stored adrenocorticotropic hormone (ACTH) into the general circulation. Under normal conditions, CRH is the primary stimulator of ACTH secretion. However, in rats **pPVN** CRH neurons also synthesize arginine vasopressin (AVP), which is co-secreted with CRH at the median eminence (5, 6). AVP, signalling via AVP receptor-1b (Avpr1b) on pituitary corticotrophs, acts synergistically with CRH to enhance ACTH secretion. In turn, ACTH acts on the adrenal glands, where it rapidly stimulates the biosynthesis and secretion of glucocorticoids (primarily corticosterone in rats/mice and cortisol in humans) from the adrenal cortex. The principal action of glucocorticoids is to increase glucose availability by stimulating **gluconeogenesis** and promoting mobilization of energy

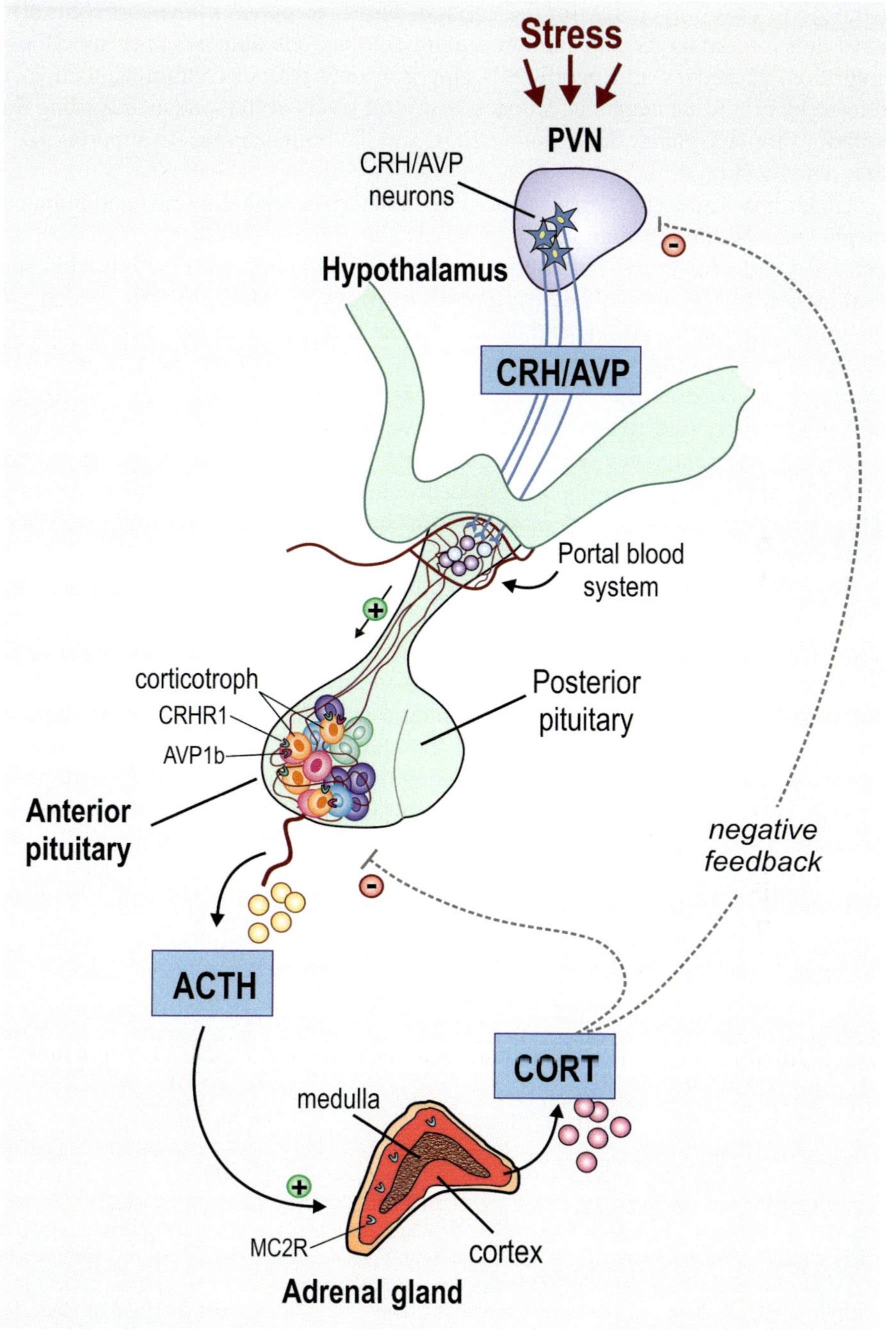

Fig. 4.1 The hypothalamo-pituitary-adrenal (HPA) axis. Stressful stimuli activate corticotropin-releasing hormone (CRH) neurons located in the medial parvocellular paraventricular nucleus (PVN) of the hypothalamus. These neurons co-express arginine vasopressin (AVP), and upon stimulation CRH and AVP are released into the hypophysial portal blood vessels at the median

from stores ensuring energy for the "fight or flight" response. Glucocorticoids also have anti-inflammatory and immunosuppressive actions and influence mood and cognition. Moreover, glucocorticoids play a critical role in restraining their own release by providing negative feedback at several levels of the system, including the anterior pituitary gland, the hypothalamus, and the hippocampus, to suppress HPA axis activity (Fig. 4.1).

Under non-stress (i.e., basal) conditions, mammals, including rats and humans, display a distinct **circadian** pattern of ACTH and glucocorticoid secretion, with the peak and nadir (or trough) of hormone secretion coinciding with the initiation and termination of the animal's active period, respectively. Hence in rats, which are nocturnal, circulating ACTH and corticosterone levels peak in the evening and are lowest in the morning, whereas in humans the opposite is observed. The pattern of increased glucocorticoid secretion at awakening likely reflects the anticipated demand for increased energy during the active period.

During late pregnancy and lactation, HPA axis responses to acute stress are markedly attenuated, meaning less glucocorticoid is secreted by the adrenal glands. This phenomenon has been demonstrated in several species, including rats, mice, and humans.

The physiological importance of a hyporesponsive HPA axis in pregnancy and lactation are multi-factorial. Firstly, exposure to stress during pregnancy has been consistently demonstrated to alter offspring physiology and behaviour, resulting in an increased risk of developing metabolic disorder, cardiovascular disease, anxiety, and mood disorders in later life. Hence, suppressed HPA axis responses to stress in pregnancy seemingly provide a mechanism to protect the foetuses from exposure to excessive maternal glucocorticoids. Secondly, glucocorticoids have **catabolic** actions, providing substrates for gluconeogenesis. Hence by limiting their release in response to stress, energy is conserved, ensuring adequate nutrients for the foetus(es), sufficient energy for the additional metabolic demands on the mother, and allowing surplus energy to be stored as fat in preparation for lactation. Finally, during the postpartum period, studies have shown that stress hormones can disrupt maternal behaviour, lactation, and milk transfer. Moreover, lactation is metabolically demanding, with increased energy required for milk production. Thus, minimizing activation of the maternal HPA axis to stressors is expected to limit the adverse impact of stress hormones on maternal care-giving and nourishment provision, maximizing the likelihood of successfully rearing the offspring.

Fig. 4.1 (continued) eminence from where they are transported to the anterior pituitary gland. CRH and AVP bind to their receptors (Crhr1and Avpr1b, respectively) located on corticotroph cells and stimulate ACTH secretion. ACTH enters the systemic circulation and acts on melanocortin-2 receptors (MC2R) in the adrenal cortex, where it triggers the synthesis and secretion of glucocorticoids (i.e., corticosterone/cortisol). Glucocorticoids regulate their own secretion through a negative feedback loop, acting on glucocorticoid receptors at various sites, including the anterior pituitary gland, hypothalamus, and hippocampus

4.2 Pregnancy

4.2.1 Basal Activity of HPA Axis in Pregnancy

4.2.1.1 Rodents

In early pregnancy in rats, basal secretion of ACTH and corticosterone at the circadian trough is comparable with the levels observed in virgin females (Waddell and Atkinson 1994; Atkinson and Waddell 1995). Likewise, gene expression for CRH, AVP, and glucocorticoid receptors (GR) in the pPVN, GR, and mineralocorticoid receptors (MR) in the hippocampus and pro-opiomelanocortin (POMC—the ACTH precursor) and ACTH content in the anterior pituitary gland are similar in rats in the early phase of pregnancy to those in virgin females (Johnstone et al. 2000; Ma et al. 2005). The circadian rhythm in ACTH secretion is maintained in early pregnancy in rats, but diminishes in the second half of pregnancy, due to a suppression in the **diurnal** peak (Atkinson and Waddell 1995) (Fig. 4.2a). In contrast, the circadian rhythm in corticosterone secretion persists throughout pregnancy in the rat (Fig. 4.2b). Absolute plasma corticosterone concentrations decline very early in pregnancy, remaining low until mid-pregnancy before rising progressively until term (around day 22 in the rat). This rise is associated with a significant increase in the trough corticosterone levels, with no increase in the amplitude of the rhythm (Atkinson and Waddell 1995). Increased corticosterone secretion at the end of pregnancy without a parallel increase in ACTH (Fig. 4.2) may result from increased oestrogen levels enhancing the sensitivity of the adrenal glands to stimulation by ACTH (Kitay et al. 1965). Furthermore, the sympathetic nervous system can increase glucocorticoid production through direct actions on the adrenal gland, independent of ACTH (Lowrance et al. 2016). However, it is not known whether increased sympathetic tone contributes to increased concentrations of corticosterone at the end of pregnancy.

During late pregnancy in the rat, basal activity of the HPA axis, at least at the levels of the hypothalamus and pituitary gland, is reduced. Gene expression for CRH in the pPVN and POMC in the anterior pituitary, CRH content in the median eminence and Crhr1 and Avpr1b binding are all significantly lower in late pregnancy compared to virgin females under basal conditions (Neumann et al. 1998; Johnstone et al. 2000; Ma et al. 2005), consistent with a reduction in the circadian peak of ACTH secretion observed in the later stages of rat pregnancy (Atkinson and Waddell 1995). More recently CRH gene expression in the PVN has also been demonstrated to be significantly lower under basal conditions in late pregnant mice (Gustafson et al. 2017a). However, it is unclear whether this leads to a reduction in the ACTH circadian peak, as it does in pregnant rats due to the difficulties in repeated blood sampling in mice, due to their size (Box 4.1).

Attenuated basal activity of the CRH neurons in late pregnancy may result from reduced excitatory input to the neurons in the pPVN, enhanced inhibitory input, or a combination of both. Basal CRH gene expression in the PVN is under **tonic** stimulatory influence from ascending **noradrenergic** and **adrenergic afferents** (Kiss

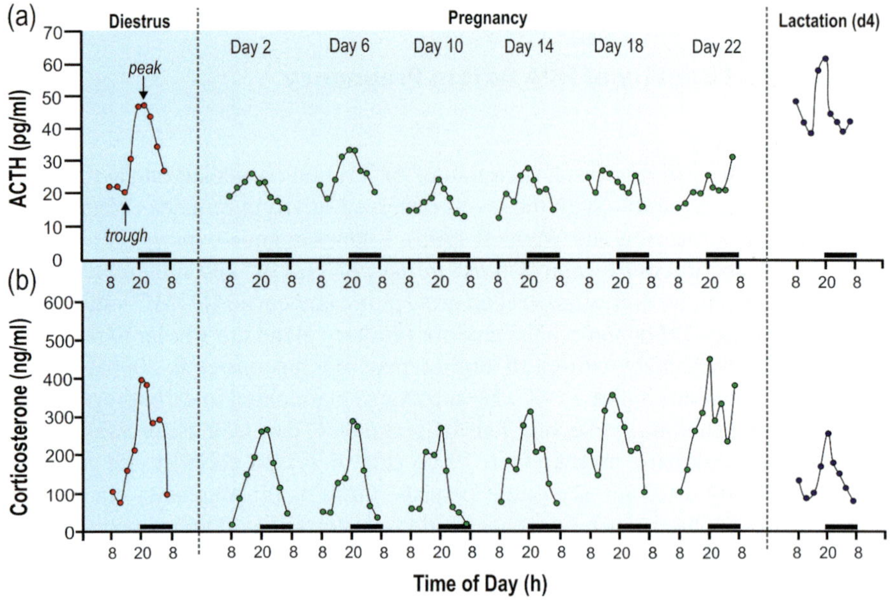

Fig. 4.2 Circadian variation in plasma ACTH and corticosterone concentrations in dioestrus females and during pregnancy and lactation in the rat. (**a**) Mesor levels (mid-point of the circadian range) of ACTH decline early in pregnancy and remain low across the remainder of pregnancy, due to a reduction in the amplitude of the circadian peak. ACTH secretion increases in lactation with marked increases in both peak and trough levels. (**b**) Compared with dioestrus, mesor corticosterone levels decline by day 2 of pregnancy, remaining low until day 10, before rising progressively reaching a peak on day 22 of pregnancy. This rise in corticosterone secretion is associated with a marked increase in trough levels, with no change in the amplitude of the rhythm. In contrast to ACTH secretion, corticosterone secretion declines between day 22 of pregnancy and day 4 of lactation with reductions in both peak and trough corticosterone levels. Data from Atkinson and Waddell (1995)

et al. 1996), meaning that reduced basal levels of CRH mRNA in late pregnancy could be a result of inhibition of these excitatory inputs to the PVN or these stimulatory inputs being counteracted by an inhibitory input.

Increased basal glucocorticoid secretion, together with up-regulated GR gene expression in the hippocampus at the end of pregnancy (Johnstone et al. 2000) might be expected to dampen activity of the CRH neurons and corticotrophs. However, given that the binding capacity of **corticosteroid binding globulin (CBG)**, which binds glucocorticoids in the bloodstream thereby modulating their availability to interact with target cells, is increased in late pregnancy (Douglas et al. 2003), an increase in corticosterone secretion into the blood may not necessarily result in increased glucocorticoid negative feedback. Indeed, **pharmacological adrenalectomy**, which blocks **endogenous** glucocorticoid synthesis, thereby eliminating their negative feedback actions on the HPA axis, results in a marked increase in ACTH

Box 4.1 Measuring and quantifying stress hormones

In order to determine circulating concentrations of ACTH and corticosterone, blood samples need to be collected. There are several options used in rodents, including sampling via an indwelling catheter, tail nicks, or blood collection at death—each with advantages and disadvantages. Venous catheters allow for repeated blood sampling, collection of larger volumes of blood, and fluid/red blood cell replacement to avoid hypovolemia/anaemia. However, indwelling catheters are invasive and require surgical skill for proper implantation and the avoidance of infection or thrombosis. These are frequently used in stress research in rats but are used much less frequently in mice, due to their smaller size. Nicking the tail vein is relatively straightforward and allows small volumes of blood to be collected from rats or mice, which is useful for determining plasma corticosterone concentrations, but typically larger volumes are required for measuring ACTH. The major drawback of this approach is that gentle restraint of the animal is typically required in order to perform the procedure which can in itself be stressful and lead to elevated levels of stress hormones. If samples are collected rapidly (i.e., within <3 min), ACTH and corticosterone concentrations are typically not elevated, however there is likely to be a lingering impact on stress hormones in any subsequent samples collected. Blood collection at the time of death is another option (e.g., trunk blood following decapitation or cardiac puncture). However, exposure to anaesthesia or carbon dioxide increases the secretion of stress hormones, confounding results. Hence, conscious decapitation is the most appropriate option for collecting blood for assessing stress hormone levels in terminal experiments. Radioimmunoassays (RIA), enzyme-linked immunosorbent assays (ELISA), or liquid/gas chromatography tandem mass spectrometry (LC/GC-MS/MS) are methods used to quantify plasma ACTH and/or corticosterone concentrations. RIAs and ELISAs are relatively straightforward to perform and kits are available commercially. Whereas, LC/GC-MS/MS methods require specialist instruments and expertise, however they offer greater sensitivity and specificity.

secretion. However, the magnitude of the increase is similar in virgin and late pregnant rats (Johnstone et al. 2000), suggesting that the HPA axis is similarly restrained by endogenous glucocorticoids. Nevertheless, activity of the enzyme 11β-hydroxysteroid dehydrogenase type 1, which reactivates glucocorticoids from their inactive metabolites (i.e., it catalyses the conversion of inert dehydrocorticosterone or cortisone into active corticosterone or cortisol, respectively), is increased in both the PVN and anterior pituitary gland in late pregnancy (Johnstone et al. 2000). This ostensibly augments local glucocorticoid generation and enhances glucocorticoid negative feedback actions on the HPA axis.

4.2.1.2 Humans

In women, basal concentrations of cortisol in saliva and serum are also higher in late pregnancy compared with early pregnancy (Sasaki et al. 1989; Lindsay and Nieman 2005; Obel et al. 2005). This is because glucocorticoids are essential for maturation of several foetal organs (e.g., the lungs and liver), which are critical for life outside the uterus after birth (Liggins 1994). In contrast to rats, however, the increase in basal cortisol secretion in late pregnant women is driven by placental CRH. From around 2 months of gestation, the placenta synthesizes and secretes CRH (which is indistinguishable from CRH of hypothalamic origin) into the maternal circulation (Frim et al. 1988; Goland et al. 1988; Sasaki et al. 1989). Placental CRH stimulates ACTH and cortisol secretion from the maternal pituitary and adrenal glands, respectively, leading to increased circulating cortisol levels during the third trimester (Goland et al. 1988; Sasaki et al. 1989). The diurnal variation in plasma ACTH and cortisol levels are maintained in pregnant women (Magiakou et al. 1996b). However, unlike hypothalamic CRH, placental CRH secretion does not exhibit a circadian rhythm and, importantly, it is stimulated, rather than inhibited, by cortisol (Magiakou et al. 1996b). Hence, placental CRH drives further cortisol secretion, resulting in a positive feedback loop that sustains CRH secretion at high levels in women during late pregnancy (Goland et al. 1988; Robinson et al. 1988; Sasaki et al. 1989). Reduced sensitivity of the HPA axis to glucocorticoid negative feedback is likely to contribute to elevated plasma concentrations of ACTH and cortisol in pregnant women during the last trimester, given the synthetic glucocorticoid, dexamethasone, is less effective in suppressing cortisol secretion at this time compared with non-pregnant women (Owens et al. 1987).

4.2.2 Stress-Induced Activity of HPA Axis in Pregnancy

In rodents, HPA axis responses to stress in early pregnancy are similar to those observed in non-pregnant females (Neumann et al. 1998). However, in late pregnancy, HPA axis responses to a range of both physical and psychological stressors (including restraint, forced swimming, immune challenge) are markedly attenuated, or even absent. Reduced ACTH and corticosterone/cortisol secretion following stress in late pregnancy have been demonstrated in rats (Neumann et al. 1998; Brunton et al. 2005), mice (Douglas et al. 2003), and women (Schulte et al. 1990; Hartikainen-Sorri et al. 1991; Kammerer et al. 2002). The mechanisms involved in women are unclear, though have been investigated extensively in rodents (see Sect. 4.3). It is unlikely that lower levels of circulating ACTH or corticosterone following stress are a result of enhanced metabolic clearance of these hormones during pregnancy, given that the clearance rate of corticosterone is similar between pregnant and non-pregnant animals (Waddell and Atkinson 1994). In rats, the reduction in HPA axis responsiveness to stress manifests around day 15 of pregnancy and persists until the end of pregnancy (Neumann et al. 1998), through **parturition** (Wigger et al. 1999) and lactation, until the pups are weaned (Walker et al. 1995; Windle et al. 1997b; Neumann et al. 2001).

4.3 Suppressed HPA Axis Responses in Late Pregnancy

4.3.1 Mechanisms Involved in Restraining HPA Axis Responses in Late Pregnancy

The mechanisms underlying hyporesponsiveness of the HPA axis to stressful stimuli in late pregnancy have mainly been investigated in rodents and involve adaptations at the level of the anterior pituitary gland corticotrophs, as well as adaptations in the brain (described below) (Neumann et al. 1998; Johnstone et al. 2000; Brunton et al. 2005, 2009).

As mentioned above, Crhr1 and Avpr1b binding in the anterior pituitary gland is lower in late pregnancy (Neumann et al. 1998; Ma et al. 2005). However, *in vitro* studies do not demonstrate reduced sensitivity of corticotrophs to CRH alone or in combination with AVP in late pregnancy (Ma et al. 2005). Instead, reduced secretion of the ACTH **secretagogues** evidently underlies attenuated ACTH responses to stress in late pregnancy. This is because although exogenous CRH or AVP administered alone is less effective in evoking ACTH secretion in late pregnant rats *in vivo*, the ACTH response to CRH and AVP administered together is indistinguishable from the response measured in virgin females (Ma et al. 2005). Moreover, while prior treatment with a Crhr1 antagonist or Avpr1a/b antagonist reduces stress-induced ACTH secretion in virgin rats, only the Crhr1 antagonist is effective in day 21 pregnant rats (Ma et al. 2005). Together these studies suggest that a lack of AVP (co-produced by parvocellular CRH neurons in the PVN) secretion into the portal blood underpins HPA axis hyporesponsiveness in late pregnancy.

There is also evidence for reduced activation in the neural circuits that regulate the HPA axis, including in the parvocellular neurons of the PVN themselves. In both rats and mice, stress-induced activation of the pPVN neurons is reduced compared with non-pregnant females. This is indicated by reduced gene expression for CRH, AVP, and the **immediate early genes** (used as a marker of recent neuronal activation), *c-fos*, and nerve growth factor inducible gene-B (NGFI-B or *Nur77*), following acute stress exposure (da Costa et al. 1996, 2001; Brunton and Russell 2003; Douglas et al. 2003; Brunton et al. 2005, 2006, 2009). This is not a result of augmented rapid negative feedback inhibition by glucocorticoids, as corticosterone is less efficient in restraining ACTH secretion induced by pharmacological adrenalectomy in late pregnant rats (Johnstone et al. 2000). Instead, the evidence indicates that reduced activation of pPVN neurons results from reduced excitatory drive from stress processing regions that provide afferent input to the pPVN. Stress-induced *c-fos* expression is lower in late pregnancy in extra-hypothalamic brain regions that are known to send afferent projections to the PVN and play a role in HPA axis activation, such as the medial amygdala (MeA) and lateral septum (da Costa et al.

1996). This suggests that modulation of afferent inputs to the PVN could, in part, explain attenuated HPA axis responses to stress in late pregnancy.

One such input that has been studied extensively is that from the brainstem, in particular the noradrenergic input arising from the A2 cells in the nucleus tractus solitarius (NTS) that project to the PVN (Cunningham and Sawchenko 1988; Sawchenko et al. 2000). Physical stressors such as forced swimming and immune challenge activate the cell bodies of these noradrenergic neurons in the NTS similarly in non-pregnant and late pregnant rats (Brunton et al. 2005). However, in contrast to virgin rats, noradrenaline release from the nerve terminals in the PVN is not stimulated by these stressors in late pregnancy. As a consequence, the CRH neurons are less excited and hence activation of the HPA axis is limited (Brunton et al. 2005; Douglas et al. 2005). Moreover, the HPA axis is evidently less sensitive to noradrenaline neurotransmission in late pregnant rats, as alpha1A-adrenoreceptor gene expression in the PVN is down-regulated and stimulation of a known brainstem noradrenergic input to the PVN using cholecystokinin results in markedly attenuated HPA axis responses (Douglas et al. 2005).

Studies have demonstrated that a central inhibitory endogenous opioid mechanism emerges in late pregnancy and inhibits stimulated noradrenaline release in the PVN, thereby minimizing HPA axis responses to stressors mediated via this pathway (Brunton et al. 2005; Douglas et al. 2005). ACTH and corticosterone secretory responses evoked by stressors, including restraint, forced swimming, social stress, cytokine challenge, or challenge with gastrointestinal or **orexigenic** peptides, which are all robust activators of the HPA axis in non-pregnant animals, are considerably diminished in late pregnancy (da Costa et al. 1996; Douglas et al. 1998; Neumann et al. 1998; Brunton and Russell 2003, 2010; Brunton et al. 2005, 2006; Windle et al. 2010). However, responses to some of these stressors (not all have been tested) can be restored by intravenous administration of the opioid receptor antagonist, naloxone, prior to the stimulus exposure (Douglas et al. 1998, 2005; Brunton et al. 2005; Bales et al. 2006). Notably, systemic administration of naloxone also reinstates a transcriptional response by the CRH neurons in the pPVN in late pregnant rats to the same levels as observed in non-pregnant females, indicating the inhibitory actions of endogenous opioids are exerted centrally (Brunton et al. 2005). Further support for this idea is the finding that when naloxone is infused directly into the PVN of late pregnant rats, immune challenge evokes noradrenaline release into the PVN, where it did not before (Brunton et al. 2005). Importantly, gene expression for the enkephalin precursor, proenkephalin-A (pENK-A), and the μ-opioid receptor are up-regulated in the NTS in late pregnancy (Brunton et al. 2005), and it is hypothesized that these NTS neurons are the same population that send excitatory noradrenergic projections to the PVN. Based on these studies, it is proposed that in late pregnancy, endogenous opioids act presynaptically on noradrenergic nerve terminals in the PVN to inhibit stimulated noradrenaline release, thereby diminishing excitatory drive to the CRH neurons. As mentioned above, the noradrenergic pathway from the NTS to the PVN plays a key role in relaying information relating to physical stressors to the CRH neurons (Sawchenko et al. 2000). If this is the case, it would provide a mechanism through which enkephalin, acting on up-regulated μ-opioid receptors located presynaptically on noradrenergic nerve terminals, could inhibit stressor-stimulated noradrenaline release in the PVN, hence limiting excitation of the CRH neurons in late pregnancy.

In contrast to rats, induction of an inhibitory endogenous opioid mechanism is not responsible for restraining neuroendocrine stress responses in mice during late pregnancy (Douglas et al. 2003), and the mechanisms underpinning HPA axis hyporesponsiveness in this species remain to be elucidated.

4.3.2 Hormonal Mechanisms Inducing HPA Axis Hyporesponsiveness in Late Pregnancy

Several studies have investigated the possible factor(s) that could be responsible for induction of the adaptations that occur in pregnancy to restrain HPA axis activity, with many focussing on endocrine factors, given that the hormonal milieu is dramatically altered in pregnancy.

4.3.2.1 Steroid Hormones: Oestradiol and Progesterone

The female sex steroids, oestradiol and progesterone, circulate in the blood at increased levels in pregnancy (Bridges 1984). In the rat, plasma oestradiol concentrations during the first week of pregnancy are similar to those measured during the **oestrus** phase of the **oestrous cycle**, but from around mid-gestation, oestradiol levels increase until term (Fig. 4.3a). Circulating progesterone concentrations rise in early pregnancy and remain at these increased levels from around day 4 to day 10 of pregnancy. There is another increase in progesterone levels from mid-pregnancy, reaching a peak around day 16–18. Progesterone levels then decline between days 20 and 21 and fall again further by day 22 prior to parturition (Fig. 4.3a).

Despite these increased levels of circulating sex steroids in late pregnancy, oestradiol and progesterone are not directly involved in suppressing HPA axis responses to stress at this time (Douglas et al. 2000). Studies investigating whether prolonged exposure to these steroids (by implanting subcutaneous capsules containing oestradiol or progesterone with or without progesterone withdrawal), to mimic the hormonal milieu of pregnancy in virgin female rats, report no impact on the HPA axis response to swim stress, nor is there any evidence to indicate that this regime induces increased endogenous inhibitory opioid tone over HPA axis responses to stress (Douglas et al. 2000). Similarly, administration of progesterone alone to non-pregnant female rats has no impact on the amplitude or profile of the ACTH secretory response to immune challenge (Brunton et al. 2009). Together, these studies indicate neither oestradiol nor progesterone are directly involved in the induction of HPA axis hyporesponsiveness in late pregnancy.

4.3.2.2 Steroid Hormones: Allopregnanolone

Increased secretion of progesterone in pregnancy leads to increased levels of allopregnanolone, a progesterone metabolite and neuroactive steroid, in the blood and brain (Fig. 4.3b). In the rat, plasma concentrations of allopregnanolone are

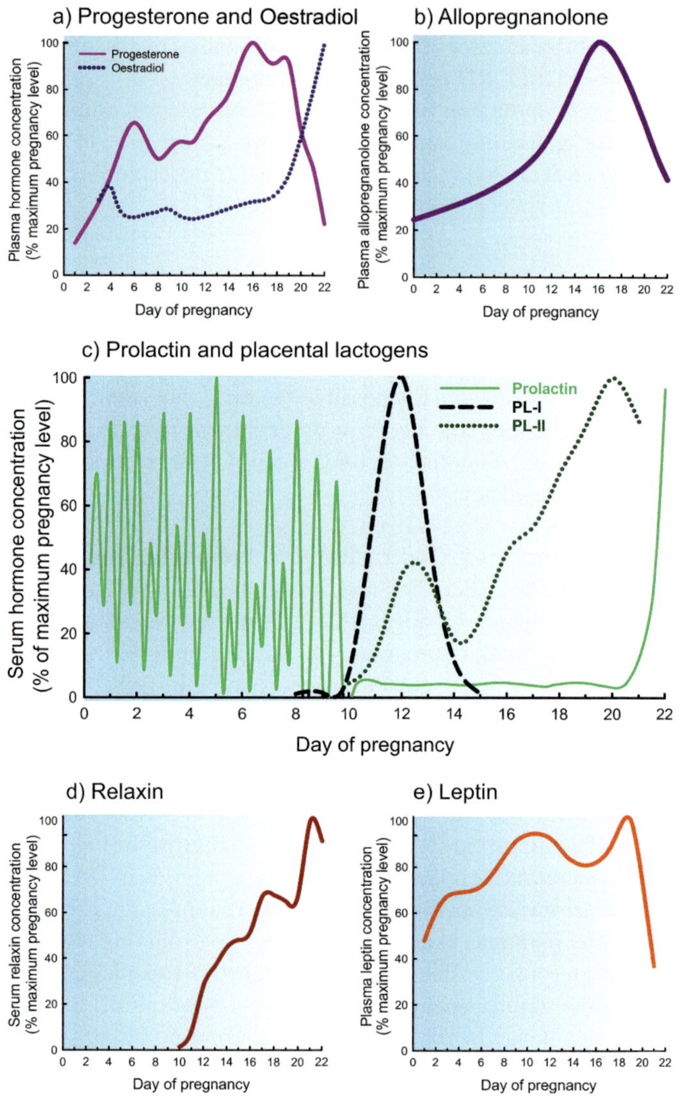

Fig. 4.3 Hormone profiles during pregnancy in the rat. Circulating concentrations of (**a**) progesterone and oestradiol; (**b**) allopregnanolone; (**c**) prolactin and placental lactogen I (PL-I) and II (PL-II); (**d**) relaxin; and (**e**) leptin in pregnant rats. In each case hormone concentrations are expressed as a percentage of the maximum levels found in pregnancy. Data are derived as follows: progesterone and oestradiol (Mann and Bridges 2001); allopregnanolone (Concas et al. 1998); prolactin (Smith and Neill 1976); placental lactogen (Robertson and Friesen 1981; Robertson et al. 1982); relaxin (Sherwood et al. 1980); and leptin (Kawai et al. 1997). (Please refer to Figs. 3.4 and 3.5, Chap. 3; Ladyman & Woodside for hormone profiles in lactation in the rat)

significantly increased by mid-pregnancy, and further increased to maximal levels on days 15–19 of pregnancy, before declining sharply by day 21 of pregnancy (Concas et al. 1998), while in the cerebral cortex, allopregnanolone concentrations are significantly higher than those in non-pregnant female rats by day 15 of pregnancy, peaking on day 19, before returning to non-pregnant levels on day 21 (Concas et al. 1998). Although the majority of allopregnanolone in the brain is likely to be of peripheral origin, the brain does express 5α-reductase and 3α-hydroxysteroid dehydrogenase (3αHSD), the steroidogenic enzymes necessary for the conversion of progesterone via dihydroprogesterone into allopregnanolone. Indeed, in late pregnancy, gene expression for 5α-reductase and 3αHSD is increased in the NTS and PVN, respectively, and hypothalamic 5α-reductase activity is also increased (Brunton et al. 2009), indicating the potential for enhanced allopregnanolone generation within the brain.

Treating male or virgin female rats with exogenous allopregnanolone suppresses HPA axis responses to stress (Patchev et al. 1996; Brunton et al. 2009). However, neither progesterone nor dihydroprogesterone impact the HPA axis response to stress in non-pregnant females (Brunton et al. 2009). This underlines the importance of up-regulation of the converting enzymes in the brain in late pregnancy. In contrast, administration of finasteride (an inhibitor of 5α-reductase) to late pregnant rats to prevent allopregnanolone synthesis reinstates ACTH and corticosterone secretory responses to immune challenge and up-regulates CRH gene transcription in the PVN (Brunton et al. 2009), indicating that allopregnanolone is necessary for maintaining suppressed HPA axis responses in late pregnancy.

The effect of allopregnanolone in restraining HPA axis stress responses in late pregnancy is evidently mediated through the actions of endogenous opioids, since treatment with allopregnanolone induces inhibitory opioid tone over HPA axis responses to stress in virgin rats (Brunton et al. 2009), similar to that observed in late pregnant rats (described above). Furthermore, allopregnanolone up-regulates pENK-A gene expression in the NTS of virgin rats, to a similar extent to that observed in late pregnancy, whereas finasteride has the opposite effect in pregnant rats, reducing pENK-A mRNA expression in the NTS to virgin levels (Brunton et al. 2009). Thus, in late pregnancy HPA axis hyporesponsiveness is driven, at least in part, by allopregnanolone acting in the NTS, where it induces and maintains the endogenous inhibitory opioid mechanism (described above) that limits stress-induced activation of the CRH neurons in the PVN (Fig. 4.4).

Another important point to note is that allopregnanolone acts as a positive allosteric modulator of the GABA$_A$ receptor, potentiating **GABA** inhibition by prolonging the opening time of chloride ion channels within GABA$_A$ receptors (Majewska 1990). The influx of chloride through activated GABA$_A$ receptors into neurons causes hyperpolarization and inhibits action potential (AP) generation. Given that CRH neurons in the PVN are under direct GABAergic control (Cullinan et al. 2008), increased allopregnanolone in the brain in pregnancy could conceivably

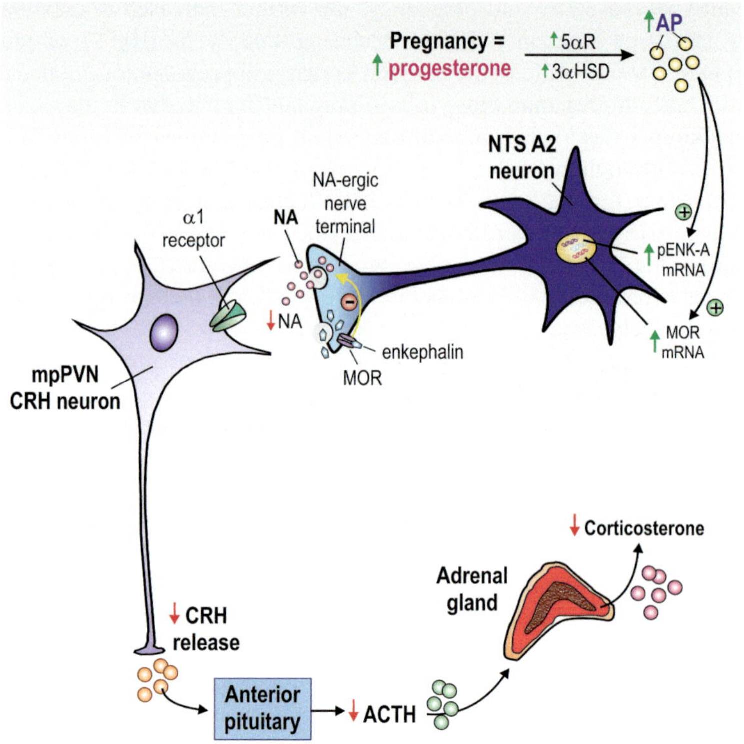

Fig. 4.4 Allopregnanolone-induced opioid mechanism involved in restraining HPA axis responses to stress in pregnancy. Brain and circulating levels of progesterone and its neuroactive metabolite, allopregnanolone (AP)—produced by the actions of 5α-reductase (5αR) and 3α-hydroxysteroid dehydrogenase (3αHSD)—are increased in pregnancy. Increased AP up-regulates proenkephalin-A (pENK-A) and μ-opioid receptor (MOR) mRNA expression in the nucleus tractus solitarii (NTS) of the brainstem. Noradrenergic A2 neurons in the nucleus NTS project to corticotropin-releasing hormone (CRH) neurons in the medial parvocellular paraventricular nucleus (mpPVN). Usually, noradrenaline (NA) released in the PVN acts via α1-adrenoceptors to excite the CRH neurons; however, in late pregnancy, immune challenge fails to evoke noradrenaline release from the terminals in the PVN. This is a result of increased opioid inhibition (by enkephalin) acting presynaptically on the up-regulated MOR, presumably on the noradrenergic nerve terminals in the PVN. A reduction in the NA-ergic excitatory drive to the CRH neurons leads to reduced stimulation of the HPA axis, reflected by reduced CRH mRNA expression in the mpPVN neurons, as well as attenuated ACTH and corticosterone secretion from the anterior pituitary gland and adrenal cortex, respectively

augment inhibitory GABA influence over the CRH neurons to suppress their response to stress, as has been demonstrated for the **magnocellular** oxytocin neurons in late pregnancy (Herbison 2001); however, this remains to be demonstrated.

4.3.2.3 Peptide Hormones: Oxytocin

Oxytocin is synthesized by **magnocellular** and parvocellular neurons of the hypothalamus, which project to either the posterior pituitary gland or to sites within the central nervous system. Oxytocin secreted into the bloodstream from the posterior pituitary gland acts on the uterus to stimulate uterine contractions at parturition, and acts on the mammary glands to promote milk ejection from the nipple during lactation (see Chap. 6). Oxytocin released from centrally projecting neurons facilitates social recognition and affiliative behaviours, including maternal behaviour and mother-infant bonding (Ross and Young 2009) (see Chap. 8).

In late pregnancy, oxytocin gene expression and immunoreactivity are significantly elevated in the hypothalamic supraoptic and paraventricular nuclei and oxytocin receptor expression is up-regulated in several brain regions involved in regulating the HPA axis (Young et al. 1997). In male and non-pregnant female rats, intracerebral administration of an oxytocin receptor antagonist significantly enhances basal and stress-induced ACTH and corticosterone secretion (Neumann et al. 2000a, b), indicating that endogenous oxytocin acting in the brain exerts an inhibitory action over the HPA axis. These findings led to the theory that the increased oxytocin synthesis in the brain in late pregnancy may contribute to HPA axis hyporesponsiveness. However, central administration of an oxytocin receptor antagonist fails to restore ACTH and corticosterone responses to stress in late pregnant rats, despite being effective in further elevating secretory responses in virgin females (Neumann et al. 2000a). These data indicate that the tonic inhibitory effect of endogenous oxytocin on HPA axis activity is absent in late pregnancy and therefore does not underpin suppressed neuroendocrine stress responses at this time.

4.3.2.4 Peptide Hormones: Prolactin and Placental Lactogens

The first half of pregnancy in the rat is characterized by twice-daily surges in prolactin secretion by the maternal anterior pituitary gland (Fig. 4.3c). These prolactin surges cease around day 10 of gestation, coinciding with an increase in the secretion of placental lactogen, of which there are two forms. Placental lactogen I (36–40 kDa) is luteotrophic and its secretion peaks around mid-gestation, whereas placental lactogen II (20–25 kDa) secretion increases progressively from around mid-gestation, reaching peak levels at term (Fig. 4.3c). During the final 24–36 h of pregnancy, prolactin secretion from the lactotrophs begins to increase once more, leading to a large nocturnal surge in plasma prolactin concentrations prior to parturition (Grattan et al. 2008). In addition, prolactin receptor expression in the choroid plexus is significantly increased during pregnancy (Augustine et al. 2003), providing a possible mechanism whereby increased levels of circulating prolactin may access the brain. However, despite studies demonstrating that centrally acting prolactin can attenuate HPA axis responses to stress in males, non-pregnant and lactating females (Torner et al. 2001, 2002; Donner et al. 2007), to date no studies have directly shown a role for prolactin or placental lactogen in restraining HPA axis responses to stress in pregnancy, though any action of prolactin is unlikely to be mediated through a direct action on the CRH neurons (Gustafson et al. 2017b, 2020).

4.3.2.5 Other Peptide Hormones: Leptin and Relaxin

Other peptide hormones, such as leptin, primarily secreted by **adipocytes**, and relaxin, secreted by the **corpus luteum**, are also synthesized and secreted by the placenta during pregnancy and circulate in the blood at increased concentrations (Fig. 4.3d, e). Leptin is generally considered to exert a negative influence on the HPA axis (Heiman et al. 1997) and studies have demonstrated that chronic leptin treatment prevents the prepartum activation of the foetal HPA axis in sheep (Yuen et al. 2004) and reduces stress-induced ACTH secretion and CRH gene expression in the PVN of neonatal rat pups (Oates et al. 2000). However, no studies have directly investigated whether increased circulating leptin contributes to the reduced responsiveness of the maternal HPA axis in late pregnancy. Similarly, although limited evidence suggests a role for the involvement of a relaxin-related peptide (relaxin-3) in the regulation of stress responses (Callander and Bathgate 2010), it is unclear whether this peptide potentiates or inhibits stress responses and to date no studies have investigated a role for peptides of the relaxin family in suppressed HPA axis responses to stress in pregnancy.

4.4 Parturition

Transitioning from gestation to postpartum is associated with large changes in hormone levels and the initiation of maternal care. The mother is now focussed on the well-being of her offspring and threats that can affect her offspring's survival, rather than on other stressors less relevant to the **dyad**. At parturition in rodents, plasma concentrations of oestradiol and progesterone drop (Fig. 4.3a), in parallel with a reduction in circulating corticosterone levels. In rats, a rapid postpartum ovulation occurs on the day of parturition or within 24 h of parturition, which increases progesterone levels. Progesterone levels are maintained during lactation in the mother because of the large suckling-induced increase in prolactin secretion, in contrast to plasma concentrations of oestradiol, which remain low throughout lactation.

4.4.1 Parturition and Activity of the HPA Axis

4.4.1.1 Rats

For obvious experimental considerations, very few studies have examined HPA axis reactivity during parturition in rodents. The general consensus is that the HPA axis hyporesponsiveness to stress observed in late pregnancy is maintained during parturition, despite enhanced basal corticosterone levels at the end of pregnancy. Expression of CRH mRNA is reduced in the hypothalamic PVN at parturition, together with a smaller number of PVN neurons expressing CRH (Douglas and Russell 1994). Basal secretion of ACTH and corticosterone decrease in the 4–5 h immediately before birth and remain low during and after delivery of the entire litter (Wigger et al. 1999).

4.4.1.2 Humans

In human mothers, basal plasma concentrations of CRH, ACTH, and cortisol are all increased at the time of labour onset (Mastorakos and Ilias 2003), but these increases do not necessarily indicate a greater activation of the maternal HPA axis. Indeed, as mentioned above, the human placenta and **endometrium** secrete CRH, which increases in the circulation as pregnancy progresses (Vannuccini et al. 2016), peaking at term (McLean et al. 1995). The increased levels of CRH likely maintains increased concentrations of ACTH and cortisol in the circulation until delivery. During the second stage of labour, plasma CRH concentrations decrease and in the third stage, plasma ACTH secretion is modestly increased, returning to control levels 2 days after delivery (Ochedalski et al. 2001). Although there are no studies investigating responses to psychological stressors during delivery, the mode of delivery in itself can be considered as a stressor and might provide some information on the state of activation of the maternal HPA axis. When compared with vaginal delivery (with or without epidural), planned Caesarean section is associated with lower circulating levels of cortisol in the mother (Okamoto et al. 1989; Vogl et al. 2006). Whereas, administration of oxytocin to stimulate contractions during labour is associated with a larger increase in ACTH and cortisol compared to vaginal delivery without oxytocin. These data suggest that this commonly used treatment might stimulate the maternal HPA axis during delivery (Ochedalski and Lachowicz 2004), or that the need for an intervention is in itself stressful. Interestingly, the metabolic state of the mother might also influence HPA axis activity during delivery, because maternal concentrations of cortisol are reduced in obese pregnancies and during labour, compared to mothers with a normal body mass index (BMI) (Stirrat et al. 2017). However, umbilical cord blood cortisol and corticosterone levels are not changed in obese pregnancies, perhaps mitigating the impact on the foetus of increased risk for preterm delivery in obese pregnancies. Immediately after parturition, the maternal HPA axis appears to be mildly suppressed due to a transient reduction in hypothalamic CRH secretion over the first 12 weeks postpartum (Magiakou et al. 1996a). Reduction in HPA axis activity during the postpartum period depends on several factors related to hormonal changes, the suckling stimulus and psychological factors, as discussed in Sect. 4.5.

4.4.2 Role for Endogenous Opioids and Oxytocin in HPA Axis Suppression at Parturition

Among the various peptides implicated in parturition, elevated endogenous opioids and oxytocin might actively suppress HPA axis activity during this specific period. Plasma levels of endogenous opioids increase during parturition (Petraglia et al. 1985) and serve, in fact, to inhibit premature secretion by oxytocin cells during parturition (Leng et al. 1987). When **parturient** rats are injected with naltrexone, an opioid receptor antagonist, plasma levels of ACTH and corticosterone are enhanced for an hour before returning to baseline levels, suggesting that endogenous opioids actively suppress ACTH and corticosterone secretion during the stimulus of parturition (Wigger et al. 1999). In response to air puff stress in rats (Neumann et al. 2003) or swim stress in mice (Douglas et al. 2003), parturient females display blunted ACTH and corticosterone responses that cannot be restored by prior treatment with an oxytocin receptor antagonist, suggesting that the increased activity of the brain oxytocin system at parturition is not responsible for dampened HPA axis responses during delivery in rodents.

4.5 Lactation

Like pregnancy, the postpartum period is a dynamic state, characterized by continuous hormonal and behavioural adaptations that serve the progressive development of the young. Thus, in early lactation in rats (around postpartum day, PPD1–3), regulation of the HPA axis might be quite different than that observed in mid (around PPD10–12) or late lactation (around PPD16–18), when the offspring have acquired sufficient independence from their mother to prepare for the weaning period around PPD21–22.

4.5.1 Basal Activity of the HPA Axis in Lactation

In both rodents and humans, the postpartum period is characterized by elevated basal levels of corticosteroid secretion that is necessary for adequate milk production. However, many studies report reduced stress responsiveness in lactating rodents and humans, as will be discussed below.

4.5.1.1 Rodents

Basal expression of CRH in the PVN has been shown to fluctuate in lactating rats. It is modestly increased over virgin levels in the early postpartum (PPD3) period (da Costa et al. 2001), followed by a longer lasting suppression of CRH expression in mid and late lactation (Lightman and Young 1989; Naik and de Jong 2017). The mechanisms maintaining low CRH expression during the postpartum period are still unclear, but increased GABAergic tone (Bosch et al. 2007b) and activity of the K^+/Cl^- cotransporter KCC2 in CRH neurons of the PVN (Melón et al. 2018) might be important. Down-regulation of KCC2 expression in the PVN in response to stress plays a role in the activation of the HPA axis (Sarkar et al. 2011). However, in contrast to virgin females, stress-induced down-regulation of KCC2 expression in the PVN is not observed in late pregnant or postpartum mice (Melón et al. 2018).

In parvocellular CRH neurons of the PVN, lactating females (PPD14) exhibit increased production of AVP that is likely accompanied by increased axonal transport to the median eminence (Walker et al. 2001a). In the anterior pituitary gland, corticotrophs become more sensitive to AVP in lactation (Toufexis et al. 1999b), allowing for the maintenance of normal to modestly increased basal levels of ACTH secretion in the face of reduced hypophysial CRH release.

Rat dams in mid-lactation (PPD10) maintain a diurnal rhythm in HPA axis activity, although basal concentrations of ACTH and corticosterone are higher during the trough of the cycle compared to virgin females (Walker et al. 1992). Diurnal and **ultradian** variations in ACTH and corticosterone secretion are observed in mid-lactation (Windle et al. 2013), although the peak concentration of corticosterone is lower in lactation compared to that in virgins, producing a flattening of the diurnal rhythm. The ACTH diurnal rhythm is not affected by the caloric drain of lactation, because supplying additional calories in the form of sucrose availability does not modify the diurnal ACTH rhythm (Walker et al. 1992). High morning basal ACTH secretion in lactating females is reduced if the litter size, and thus the intensity of the suckling stimulus, is decreased. Within 4 h of separation from the pups, morning concentrations of ACTH are restored to those seen in virgin female rats (Walker et al. 1992), indicating that increased ACTH secretion is maintained by the suckling stimulus. This rapid adaptation of ACTH secretion after removal of the suckling stimulus is somewhat mirrored by rapid changes observed in CRH mRNA expression that increases to levels seen in virgin mice 24 h after pup removal on PPD6 (Gustafson et al. 2017a) or 2 days after pup removal in rats on PPD7 (Lightman and Young 1989).

Glucocorticoid secretion during lactation is generally higher than that anticipated from the modest increase in basal ACTH secretion observed in both rodents and humans. Diurnal changes in adrenal sensitivity to ACTH are maintained in lactating rats in mid-lactation (Walker et al. 1992), but corticosterone secretion appears somewhat "uncoupled" from plasma ACTH concentrations in lactating mice in early lactation (PPD1). Thus, maintaining elevated glucocorticoid production relies on adrenal gland adaptations that increase circulating cholesterol bioavailability in PPD1 mice and in rats by mid-lactation (PPD8) (Perani et al. 2017). The net effect of elevated total corticosterone concentrations during lactation might be enhanced by the large reduction in **CBG** levels after parturition (Pawluski et al. 2009), leading to higher levels of free, biologically active glucocorticoids. Long-term glucocorticoid negative feedback on ACTH secretion is unaltered in mid-lactation (Walker et al. 1992), although rapid suppression of corticosterone secretion by intravenous methylprednisolone (a GR agonist) appears to be blunted in PPD9 lactating female rats (Windle et al. 2013).

4.5.1.2 Humans

Studies in humans have shown that postpartum HPA axis activity is mildly suppressed and that hypothalamic CRH secretion is reduced at 3–6 weeks postpartum before normalizing around 12 weeks postpartum (Magiakou et al. 1996a). When comparing diurnal cortisol secretion at 3 months postpartum, breastfeeding mothers exhibit a small suppression compared to non-postpartum mothers, but infant feeding mode significantly affects diurnal cortisol secretion, producing higher cortisol levels throughout the day in bottle-feeding mothers (Tu et al. 2006). Studies comparing healthy controls and women affected by either postpartum "blues" (O'Keane et al. 2011) or depression (de Rezende et al. 2016) have established that postpartum mothers exhibit a reduced **cortisol awakening response (CAR)** compared to non-postpartum women and that this CAR is further reduced in women with postpartum depression at 6 months. However, in a group of postpartum mothers (at 2 and 6 months postpartum) having experienced early adversity, levels of awakening cortisol were higher than in control mothers (Gonzalez et al. 2009), suggesting that early-life experiences have lasting effects on the regulation of the postpartum HPA axis that is distinct from the acute condition of postpartum depression.

4.5.2 Stress-Induced Activity of the HPA Axis in Lactation

Both the end of pregnancy and lactation are periods associated with blunted neuroendocrine responses to stress in rodents (Douglas et al. 2003; Hill et al. 2003; Slattery and Neumann 2008), as well as in breastfeeding mothers (Altemus et al. 2001; Heinrichs et al. 2001; Groer et al. 2002; Meinlschmidt et al. 2010). However, some of the mechanisms maintaining reduced responsiveness of the HPA axis differ between the two conditions. In rodents, down-regulation of the HPA axis response has been generalized to a wide array of **systemic** (e.g., immune challenge) and **psychogenic** (e.g., noise, restraint, predator odour) stressors. Although the stress hyporesponsiveness is well documented, the degree of suppression in ACTH and corticosterone responses is quite variable and depends on several factors, such as the stage of lactation (early versus late lactation), the intensity of the suckling stimulus (small versus large litters), as well as the condition in which the stressor occurs (e.g., in the presence or absence of the pups) and maternal traits, like anxiety. Exposure to prenatal stress has also been found to increase ACTH and corticosterone responses to stress in mid-lactation (PPD10) when these rats become mothers (Bosch et al. 2007a), but this effect might be dependent on genetically determined vulnerability to stress, as exhibited by high-anxiety compared to low-anxiety rats (Neumann et al. 2005). Of interest is the observation in rodents, as well as in humans, of a reduction in anxiety behaviour and emotional reactivity (Fleming et al. 1989; Toufexis et al. 1999a) in the early postpartum period, presumably to promote optimal maternal care. However, in contrast to HPA axis responses, anxiety responses return to basal levels during the last 2 weeks of lactation in rodents. These data suggest that overlapping pathways implicated in suppressing both anxiety behaviour and stress responsivity are primarily active early after birth, but that later in lactation these pathways become dissociated (Agrati et al. 2008) and/or other mechanisms play a predominant role in suppressing the neuroendocrine responses to stress.

4.5.2.1 Stage of Lactation and Suckling Stimulation

Many studies documenting stress hyporesponsiveness in lactating rodents have used females in mid-late lactation, once circulating levels of oestrogen and corticosteroids have stabilized and prolactin secretion is solidly established. By mid-lactation (PPD10) and persisting into late lactation (PPD15–16), CRH mRNA levels under basal and stress conditions are reduced. Stress responses are blunted and the HPA axis fails to respond to stressful conditions that would elicit responses in virgin females (Deschamps et al. 2003; Hillerer et al. 2018), and to direct PVN stimulation by an alpha-1 agonist (Windle et al. 1997a).

It is well known that suckling represents an important stimulus for prolactin secretion and that in the first days of lactation, the increase in prolactin secretion induced by suckling is already maximal (Mattheij et al. 1979). Early studies revealed that the size of the litter influences stress-induced secretion of both prolactin and corticosterone in late lactation (PPD21). Stress-induced prolactin secretion is increased with increased litter size, while stress-induced corticosterone secretion is inversely correlated with litter size (Myers et al. 1975). In mid-lactation (PPD10), baseline peak concentrations of corticosterone in the afternoon are higher in mothers nursing five pups compared to those nursing ten pups, but still reduced compared to virgins (Walker et al. 1992). This suggests that litter size and the intensity of the suckling stimulus is able to modulate the adaptations of the HPA axis to stress in lactation. As rat dams typically spend less time nursing large litters than small ones, it is also possible that the increased frequency of nursing small litters compensates for the reduction of the suckling stimulus intensity observed under these conditions and maintains some suppression on the HPA axis. Restoration of acute stress responses in mid-lactation are observed within 24 h of removal of the litter, even though circulating levels of prolactin drop to pre-pregnancy levels within 1 h of pup removal (Kalyani et al. 2017). It is suggested that diminished expression of the prolactin receptor (PRL-R) allows the return to normal HPA axis responsivity in mothers separated from their pups.

Human studies of stress responsiveness have for the most part examined mothers during the first 6–18 weeks postpartum, a time when breastfeeding is predominant and suckling stimulation provided by the infant is elevated. HPA axis responses to treadmill exercise are reduced in breastfeeding mothers versus those bottle-feeding their infants (Altemus et al. 1995), and the acute effect of suckling stimulation in reducing plasma ACTH and cortisol secretion is also observed following exposure to a psychogenic stressor, the **Trier Social Stress Test** (TSST) (Heinrichs et al. 2001).

4.5.2.2 Stressor Salience and HPA Axis Responses in Lactation

As maternal behaviour is oriented towards ensuring offspring survival, it may appear counterintuitive that hormonal responses to stress be blunted during this critical period and in particular in early lactation, when the pups are entirely dependent on their mother for survival. However, in early lactation, the nature of the stressor to which the mother is exposed evidently influences stress responsiveness, indicating that stressor **salience** is important at this time. Indeed, there is evidence that under certain conditions, early lactating females (PPD3–5), but not late lactating females (PPD15), can respond to stressors if they represent a direct threat to the pups (Deschamps et al. 2003). In early lactation when their litter is present, maternal ACTH and corticosterone responses to a male intruder in the home cage or a predator odour (both of which pose a potential real threat to her pups) are comparable to those of virgin females. Moreover, in this scenario, levels of CRH mRNA in the PVN of lactating dams do not significantly differ from virgins (da Costa et al. 2001; Deschamps et al. 2003). Hence, the presence of the litter at the time of exposure to a stressor that directly threatens the pups is sufficient to induce normal stress responses in early, but not late, postpartum females (Deschamps et al. 2003; Hillerer et al. 2018). In contrast, exposing early lactating mothers to tail pinch, even in the presence of their litter, does not restore HPA axis responsiveness. These results suggest that in early lactation, rats have the capacity to produce a hormonal response to stress that is similar to that of virgin rats when the situation demands it, whereas rats in late lactation do not show a similar capability. These observations also demonstrate that the presence of the litter is important to modulate stress responses in early, but not in late, lactating mothers.

These different capacities arise because the mechanisms promoting hyporesponsivity of the neuroendocrine stress system vary as a function of the stage of lactation. In early lactation the system has the capacity to respond, but in situations where there is no immediate or real threat to the young, the inputs that stimulate this system are reduced. By contrast, later in lactation the capacity of the system and notably of the CRH neurons in the pPVN are reduced so that a "normal" neuroendocrine response to stress cannot be induced, regardless of the stimulatory input. Functional MRI studies of females during mid-lactation support the notion that the presence or absence of pups changes maternal brain activity in situations of threat (Nephew et al. 2009b). In this study, the presence of the pups resulted in differential activation of the temporal cortex, nucleus accumbens, periaqueductal grey, and basolateral amygdala (BLA) in response to a male intruder.

To address the role of stressor context in mediating the amplitude of stress responses in postpartum women, two types of stressors, an emotionally relevant stressor (a film evoking threats to a child) and a non-child related stressor (**TSST**), were compared in primiparous and multiparous mothers 5–20 weeks postpartum (Tu et al. 2006). Interestingly, only multiparous, breastfeeding mothers showed blunted cortisol responses to the viewing of the emotional film and the **TSST** compared to bottle-feeding mothers. Thus, parity and infant feeding mode interacted to determine cortisol responses in postpartum mothers, but there were no significant differences attributed to the nature of the stressor.

4.5.2.3 Impact of Parity

In addition to significant longitudinal changes as a function of lactation progression, multiparity and the experience of rearing multiple litters represents another important factor that may modulate HPA axis activity in the long term. Indeed, morphological (Featherstone et al. 2000; Hatton and Zhao Yang 2002) and functional plasticity have been reported in pregnancy and lactation in rodents and humans (Kinsley and Lambert 2008; Kim et al. 2010) in structures that directly or indirectly regulate the activity of PVN CRH neurons (Slattery and Hillerer 2016; Workman et al. 2016). Some of these changes are enhanced with multiple reproductive episodes and may persist between reproductive episodes (Barha et al. 2015; Opala et al. 2019). For example, **multiparous** females exhibit reduced fear of novelty (Rima et al. 2009) and *c-fos* activation after restraint stress (Wartella et al. 2003), compared to **primiparous** or **nulliparous** females, even when both primiparous and multiparous females are tested between reproductive episodes (Fig. 4.5). Multiparous female rats also display enhanced maternal aggression (Nephew et al. 2009a), which might reflect reduced stress responses since maternal latency to

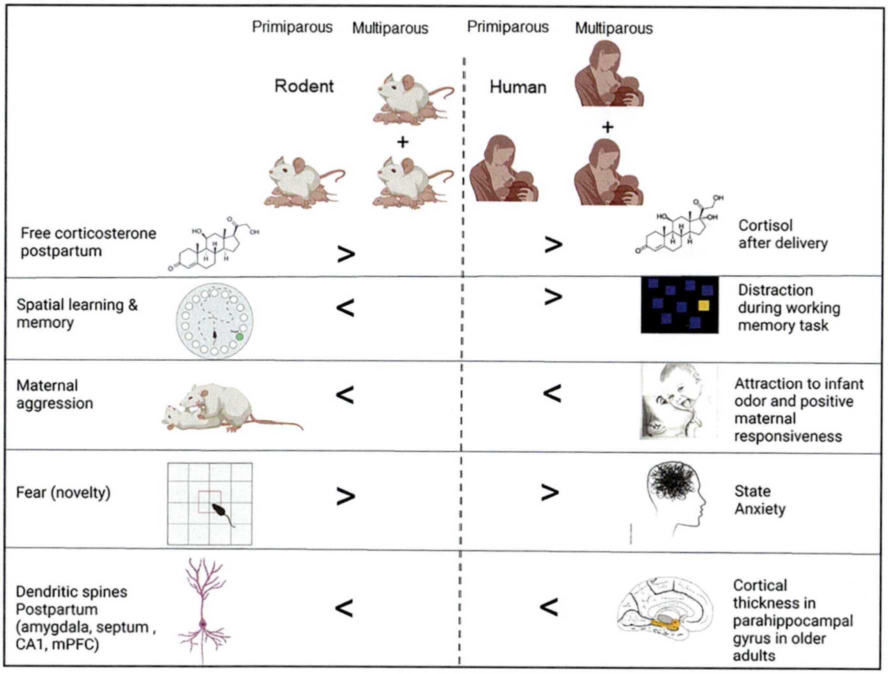

Fig. 4.5 The effects of parity on morphological, neuroendocrine, and behavioural outcomes in rodents and humans in the postpartum period. Endpoints illustrated here emphasize similarities found between the rodent and human studies, although it has to be emphasized that rodent studies comparing primiparous and multiparous mothers postpartum differ significantly from human studies because the context in which mothers are tested is fundamentally different. Rodent mothers are not tested in the presence of multiple sequential litters while the increased environmental complexity of parenthood continues for years—even decades—and is increased by each subsequent child, requiring simultaneous care across different stages of dependency. Note that the reported results for parity effects on human cortical thickness were not acquired during the postpartum period, but in older adults. Reported results are from rodents: fear responses (Rima et al. 2009), spatial learning (Kinsley and Lambert 2008), maternal aggression (Nephew et al. 2009a), free corticosterone (Pawluski et al. 2009), morphological changes (Leuner and Gould 2010; Hillerer et al. 2014; Opala et al. 2019); and from humans: cortisol after delivery (Tu et al. 2006), distraction during working memory task (Pérez-Hernández et al. 2021), attraction to infant odour (Fleming et al. 1997), anxiety state (Nakamura et al. 2020; Cena et al. 2021), cortical thickness in older non-postpartum adults (Orchard et al. 2020)

aggression in early lactating females is inversely correlated with ACTH responses to a male intruder (Deschamps et al. 2003).

Changes in stress responsiveness as a function of parity have also been documented in humans, where multiparous breastfeeding mothers have reduced responses to a psychological stressor, the **TSST**, compared to primiparous mothers (Tu et al. 2006) (Fig. 4.5).

4.5.3 Mechanisms of Suppressed HPA Axis Responses in Lactation

4.5.3.1 Central Inputs to PVN Neurons and Stress Processing Circuits

The ability of stressors to trigger a robust neuroendocrine response depends on the responsiveness of the CRH neurons in the pPVN, which constitute the final common pathway leading to neuroendocrine activation. Lactation is associated with significant reductions in basal and stress-induced CRH expression in the PVN and CeA (Windle et al. 1997b; Lightman et al. 2001; Deschamps et al. 2003). This suggests that these neurons become progressively desensitized or inhibited as lactation progresses. In support of this hypothesis, da Costa et al. (1996, 1997) showed that in early-mid lactating females, neither restraint stress nor CRH injection induced *c-fos* mRNA expression in input pathways to the pPVN neurons, including those originating in the MeA, lateral septum (LS), and cingulate cortex. However, virgin females showed a robust response, suggesting that these regions may be suppressed in lactation. In contrast to the reduction in CRH expression in pPVN and CeA in mid-lactation, the dorsolateral portion of the bed nucleus of the stria terminalis (BNST) exhibits a significant increase in CRH expression during the same period (Walker et al. 2001b). The BNST constitutes an important relay for the influence of the medial prefrontal cortex (mPFC), ventral hippocampus, and amygdala on pPVN neurons and various portions of the BNST can differentially affect CRH expression in the pPVN (Herman et al. 1994, 2020). In particular, anteroventral regions of the BNST integrate excitatory signals from limbic structures and translate them into GABAergic inhibitory outputs to the pPVN (Radley et al. 2009; Johnson et al. 2019). Thus, increased inhibition of pPVN neurons in lactation can be achieved through increased activity of neurons in the BNST. Finally, the influence of the mPFC on the activity of pPVN neurons needs to be considered, since this brain region is critical for evaluation of stressor salience (Figueiredo et al. 2003), a process that is critical, at least in early lactation. The prelimbic (PL) region of the mPFC exerts inhibitory actions over pPVN neurons via projections to the peri-PVN and BNST (Herman et al. 2020). Significant activity of the PL mPFC area was found during electroencephalographic (EEG) recordings in lactating rats after presentation of pup-associated odours in early- to mid-lactation (Hernandez-Gonzalez et al. 2005), suggesting that pup-associated cues can significantly modify mPFC activity and downstream pPVN neurons. Similar activation patterns are observed in response to a variety of visual and auditory infant stimuli in fMRI studies in humans (Kim et al. 2011). In contrast to the potential role of the PL mPFC on PVN neurons, local inhibition of the infralimbic (IL) mPFC does not modify stress responses in early lactating females exposed to a salient (predator odour) stressor (Hillerer et al. 2018).

Parvocellular CRH neurons also receive direct stimulatory noradrenergic inputs from A2 catecholaminergic cell groups in the NTS, the activity of which appear to be reduced in lactating rats (Toufexis and Walker 1996), as in late pregnant rats (Brunton et al. 2005). Indeed, concentrations of norepinephrine and alpha-2 adrenoreceptor density are decreased in the PVN of lactating females compared to virgin females (Toufexis et al. 1998), in parallel with a reduced responsiveness of CRH neurons to adrenergic alpha1 agonist stimulation (Windle et al. 1997a).

In summary, the precise mechanisms reducing responsiveness of pPVN neurons to stress in lactation remain unclear, although both a reduction of stimulatory inputs from the brainstem and CeA, as well as significant activation of inhibitory inputs from the mPFC and BNST, are plausible mechanisms. Alteration in these neuronal circuits might be a direct consequence of the combined effects of elevated prolactin, oxytocin, and corticosterone secretion in the face of low oestrogen levels on the activity of PVN neurons.

4.5.3.2 Pituitary Responsiveness to CRH and AVP

As is the case for late pregnant females, pituitary responsiveness to the main ACTH secretagogues, CRH and AVP, is altered during mid-lactation and might participate in reduced stress-induced HPA activation (Toufexis et al. 1999b). Interestingly, *in vivo* ACTH responses to CRH administration are blunted, but the pituitary corticotrophs respond to AVP stimulation in PPD10–12 lactating females. Following removal of the pups, partial restoration of pituitary responses to CRH are observed after 48 h. In contrast to pregnant females that display a reduction in pituitary Avpr1b, lactating females do not exhibit significant changes in corticotroph Avpr1b or Crhr1 expression that could explain the reduced pituitary sensitivity to these secretagogues (Toufexis et al. 1999b). It is interesting to note however, that the maintenance of responsivity to AVP, but not CRH, parallels the increased expression of AVP in pPVN CRH neurons in lactating females (Walker et al. 2001a). Because Avpr1b are generally reduced by increased circulating glucocorticoids in male rats, the lack of changes in pituitary Avpr1b in lactation suggests that the sustained sensory stimulation of suckling and/or milk ejection might increase AVP release and maintain Avpr1b density even in the face of **hypercorticosteronemia**. As a result, the shift in greater pituitary responsiveness to AVP during lactation might represent a mechanism that enables stress responses to be mounted to specific and salient threats to the infant or self, despite reduced CRH secretion. In the Brattleboro rat, which is deficient in AVP production, stress responses remain blunted in lactating females and oxytocin expression in the PVN is modestly increased to compensate for the AVP deficit (Fodor et al. 2013).

4.5.4 Hormonal Mechanisms Inducing Altered HPA Axis Output in Lactation

Comparable to the situation in late pregnancy, several of the same hormones (i.e., oxytocin and prolactin) might be implicated in the maintenance of reduced HPA axis responsiveness during lactation, although the mechanisms might be quite different because of the particular gonadal and glucocorticoid steroid milieu of each condition (Brunton et al. 2008).

4.5.4.1 Oxytocin

Oxytocin attenuates basal and stress-induced activity of the HPA axis in male and virgin females (Neumann et al. 2000b). Therefore, increased central and/or peripheral oxytocin secretion in lactation would also be expected to significantly reduce HPA axis activation. In males and cycling females, endogenous brain oxytocin regulates the activity of the HPA axis because central infusion of a selective oxytocin receptor antagonist is able to enhance basal and stress-induced ACTH and corticosterone secretion under several different stressor conditions (Neumann et al. 2000b). However, when tested in lactating females, the same antagonist is ineffective in disinhibiting HPA axis responses to different stressful stimuli (Neumann et al. 2000a), indicating a negligible role of central oxytocin in the hyporesponsiveness of lactation. Similar findings are observed at parturition when stress hyporesponsiveness is observed even after blocking central oxytocin receptors and despite maximal stimulation of the central oxytocin system and secretion of large pulses of oxytocin into the peripheral circulation (Douglas et al. 2003; Neumann et al. 2003). The lack of effect of oxytocin on neuroendocrine stress responses should not, however, preclude consideration of the important role of oxytocin in other aspects of emotional regulation and maternal behaviour during lactation (see Chap. 8).

4.5.4.2 Prolactin

Suckling-induced prolactin secretion and its lactogenic actions is necessary for successfully rearing the offspring in all mammalian species (see Chap. 7). Circulating prolactin concentrations increase dramatically in early lactation in rodents peaking in mid-lactation and then slowly declining as weaning approaches (see Fig. 3.5, Chap. 3). In the second half of lactation, the rise in plasma prolactin concentrations induced by suckling decreases gradually without decreases in the intensity of the suckling stimulus in older pups. Prolactin plays fundamental roles in the brain to regulate maternal mood, motivation, and behaviour (Georgescu et al. 2021), and also actively participates in the metabolic adaptations of lactation (see Chap. 3), as well as contributing to the stress hyporesponsiveness of lactation. This was initially indicated by observations that exogenous prolactin administration to cycling females reduced ACTH responses to stress (Donner et al. 2007) and that administration of antisense oligonucleotides to down-regulate expression of the prolactin receptor (PRL-R) had the opposite effect (i.e., increases stress responses) in cycling, as well as in lactating female rats (Torner et al. 2001).

Prolactin's action to reduce stress responses and anxiety in lactation is not mediated directly by pPVN CRH neurons, as these do not express the PRL-R and do not exhibit enhanced phosphorylation of the transcription factor STAT5 after prolactin administration (Gustafson et al. 2017b). Within the PVN, oxytocin neurons express PRL-R (Kokay et al. 2006) and these neurons are known to inhibit CRH gene transcription in male rats and mice (Jurek et al. 2015). The increased activation of these oxytocin neurons in lactation could provide one possible mechanism to relay prolactin-mediated inhibition of CRH neurons. The distribution of PRL-R in several brain areas indirectly involved in regulating the pPVN CRH neurons, such as the

BNST, MeA, and dorsomedial hypothalamic nucleus (DMH) (Herman et al. 2020), could also represent potential sites of prolactin-mediated suppression of CRH expression in lactation. Further studies are needed to clearly identify the primary region(s) implicated. One study has demonstrated that when the pups are removed from lactating mice, circulating prolactin levels decline, due to the loss of suckling stimulus. The reduction in prolactin levels is associated with up-regulation of CRH gene expression in the PVN, to levels similar to those in virgin mice (Gustafson et al. 2017a). Maintenance of elevated prolactin concentrations with exogenous administration of prolactin to lactating females in the absence of their pups reduces CRH gene expression in the PVN, suggesting that prolactin suppresses CRH gene transcription during lactation, even without the suckling stimulus (Gustafson et al. 2017a). Interestingly, however, the same study reported that the ongoing suckling stimulus in the absence of endogenous prolactin (achieved with bromocriptine treatment) was sufficient to maintain low CRH expression in lactating females (Gustafson et al. 2017a), suggesting that endogenous prolactin is not critical for the maintenance of low basal levels of CRH expression in the PVN during lactation.

4.5.4.3 Progesterone and Glucocorticoids

Circulating levels of progesterone and glucocorticoids are also high during lactation (see Fig. 3.5, Chap. 3), but it is unlikely that high progesterone is a causal factor in the stress hyporesponsiveness of lactation since **ovariectomy** of lactating females (and thus removal of progesterone) does not restore responsiveness to forced swim stress in mid-lactation (Walker et al. 1995).

Suckling-induced elevated levels of glucocorticoids during lactation might be expected to provide a more efficient suppression of the HPA axis. However, long-term glucocorticoid negative feedback sensitivity to corticosterone is not enhanced in mid-lactation in rats (Walker et al. 1992). Moreover, in experiments testing acute suppression of the HPA axis by the synthetic glucocorticoid methylprednisolone, efficient short-term feedback of endogenous corticosterone secretion is actually reduced in lactating females, concomitant with a reduction in hippocampal GR and MR expression (Windle et al. 2013). Taken together, these results suggest that elevated basal glucocorticoid secretion observed in lactating females probably does not contribute significantly to the stress hyporesponsiveness observed during this period.

4.5.4.4 Human Studies

In humans, suckling-related factors, possibly including activation of the brain oxytocin and prolactin systems, have been found to contribute to the reduced stress response (Heinrichs et al. 2001; Tu et al. 2006) in postpartum mothers, as well as to their positive mood state. There is a good consensus in the human literature to propose that disruption of mechanisms leading to stress hyporesponsiveness during lactation might lead to an increased vulnerability to postpartum depression and other mood disorders (See Chap. 10). Although still incompletely demonstrated, the role of both oxytocin and prolactin in the human postpartum period seems key to prevent the development of mood disorders in vulnerable women. In this respect, promoting breastfeeding might be beneficial to increase the activity of both the central oxytocin and prolactin systems and reduce stress responsiveness (Walker et al. 2004; Tu et al. 2005).

4.6 Perspectives

In summary, changes in basal and stimulated HPA axis activity during late pregnancy and lactation result in a general stress hyporesponsiveness that can be explained by mechanisms that are both similar and specific to each reproductive period (Fig. 4.6). Basal HPA axis activity in rodents and humans during pregnancy and lactation is characterized by elevated glucocorticoid concentrations. In rodents, there are only modest increases in ACTH secretion, while in humans high ACTH secretion is maintained by increased placental CRH stimulating the pituitary gland. During both pregnancy and lactation, changes in ACTH secretion might be predominantly dictated by a shift in the importance of AVP over that of CRH as an ACTH secretagogue. Stress hyporesponsiveness during pregnancy is due at least in part to an increase in endogenous opioid inhibition of noradrenergic stimulation of CRH neurons. Such an increased opioid tone is induced and maintained by increased allopregnanolone production in late gestation (Figs. 4.4 and 4.6). While opioid inhibition maintains hyporesponsiveness of the HPA axis in pregnant and parturient rats, increased prolactin secretion serves as the major driver suppressing the HPA axis during lactation. Interestingly, in the first few days of lactation in the rat, the HPA axis exhibits normal stress responses to stressors representing a direct threat to the offspring, but this situation is only transient and stress hyporesponsiveness is firmly established with increasing levels of circulating prolactin as lactation progresses. In humans, a modest suppression of stress-induced HPA axis activity is observed in late pregnancy and lactation, but divergent states of basal ACTH and cortisol secretion are found in pregnancy and lactation.

The reduction in HPA axis reactivity in late gestation and lactation has multiple benefits for both mother and offspring in many species. It limits exposure of the developing organism to excessively high levels of glucocorticoids that can increase the risk for multiple pathologies. Blunted maternal stress responses also limit catabolic expenses for the mother and promote energy storage in gestation or energy transfer to the offspring during lactation. The combination of these effects is beneficial to ensure optimal offspring nourishment and development, favours maternal care, and, ultimately, promotes the survival of the species.

As the perinatal period is associated with a broad array of novel stimuli and unique behavioural and physiological demands, all experienced during a period of hormonal and metabolic upheaval, it is perhaps not surprising that the peripartum period is one in which mothers are especially vulnerable to depressive and anxiety disorders. Hyporesponsiveness to stressors at this time presents a means of arming females, and hence their offspring, against the demands of this period and

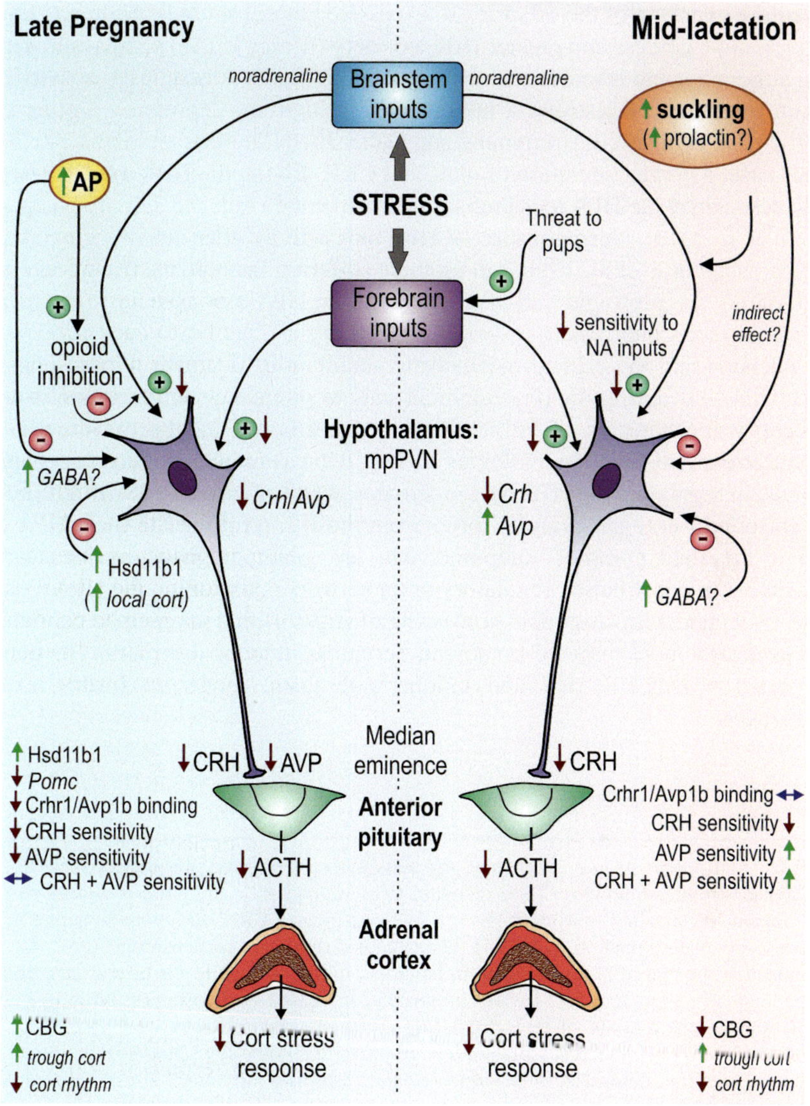

Fig. 4.6 Mechanisms involved in maintaining stress hyporesponsiveness during pregnancy and lactation in rats. Basal corticotropin-releasing hormone (*Crh*) and arginine vasopressin (*Avp*) gene expression in the medial parvocellular division of the paraventricular nucleus (mpPVN), CRH content in the median eminence, and pro-opiomelanocortin (*Pomc*) expression in the anterior pituitary gland are reduced in late pregnancy. Basal *Crh* expression in the mpPVN is also significantly lower in lactation, compared with virgins, however there is a compensatory increase in *Avp* expression. Activity of 11b-hydroxysteroid dehydrogenase type 1 (Hsd11b1) is increased in the PVN and anterior pituitary gland in late pregnancy, which is expected to increase local corticosterone concentrations and enhance glucocorticoid-mediated feedback of the HPA axis. CRH-receptor 1 (Crhr1) and arginine vasopressin receptor-1b (Avpr1b) binding in the anterior pituitary is reduced in late pregnancy, though in lactation is similar to that seen in virgin females.

preventing exposure of the offspring to excessive glucocorticoids. Failure to engage this protective process and greater HPA axis activity may lead to postpartum depression, although some researchers have argued that reduced basal HPA activity postpartum might also constitute a hallmark of postpartum depression (Jolley et al. 2007; Taylor et al. 2009; Brummelte and Galea 2010; Dickens and Pawluski 2018). In this case, a greater secretion of placental CRH driving the HPA axis in gestation could desensitize the HPA axis prepartum and maintain reduced activity postpartum (Glynn et al. 2013). Normalization of HPA axis activity after delivery can take 2–3 months (Magiakou et al. 1996a) in mothers suffering from postpartum depression, emphasizing the profound and lasting changes in HPA axis activity in postpartum depression (see Chap. 10).

Understanding the critical mechanisms maintaining adaptive hyporesponsivity of the HPA axis during the physiological state of pregnancy and lactation is useful to identify mechanisms of vulnerability and resilience in general situations of chronic stress and psychopathologies such as depression and anxiety. Mechanisms involved in regulating the HPA axis in females, whether associated with a reproductive episode or not, are likely to differ from those controlling the male HPA axis, even if the final common endpoints that are generally measured are similar. Refinement in constructing regulatory brain networks controlling the HPA axis and those associated with emotional processing of stressors and salience perception will help us design more targeted treatments for stress-induced disorders. The peripartum brain provides a rich and unique naturalistic model to further explore these issues.

Fig. 4.6 (continued) In late pregnancy, the sensitivity of pituitary corticotrophs to CRH or AVP alone is reduced, however their response to combined AVP and CRH is similar to virgin rats; whereas in lactation, corticotrophs are less sensitive to CRH, but more sensitive to AVP alone or given in combination with CRH. The circadian rhythm in corticosterone (cort) secretion is flattened at the end of pregnancy and in lactation, though trough levels of corticosterone are greater than in virgin females. In late pregnancy, increased corticosterone binding globulin (CBG) reduces the amount of free circulating corticosterone, whereas after parturition CBG levels significantly decline. Stressors are less effective in stimulating CRH/AVP neurons in late pregnancy/lactation, compared with virgin females, resulting in reduced stress-induced ACTH and corticosterone secretion. In late pregnancy and lactation, afferent drive to the pPVN neurons from limbic forebrain regions that are involved in processing psychological stressors is reduced. There is also reduced afferent noradrenergic (NA) drive from the brainstem to the PVN, despite physical stressors activating these neurons to a similar extent as seen in virgin females. In late pregnancy, increased levels of allopregnanolone (AP) induce an inhibitory endogenous opioid mechanism that presynaptically inhibits noradrenergic inputs to the PVN. The mpPVN CRH neurons are also less sensitive to excitatory noradrenergic inputs in lactation. Enhanced GABAergic inhibition of CRH neurons may also contribute to HPA axis hyporesponsiveness in pregnancy and lactation. The suckling stimulus from the pups plays a crucial role in maintaining HPA axis hyporesponsiveness in lactation. While the specific mechanism(s) are unclear, prolactin is likely to be involved, though probably via indirect actions on CRH neurons. Importantly, in lactation stressor salience plays a key role, and the maternal HPA axis is able to respond to stressors that pose a direct threat to the pups

Key References (see main list for reference details)

Brunton PJ et al (2009): Demonstrates that allopregnanolone induces an inhibitory endogenous opioid mechanism that prevents noradrenaline exciting CRH neurons in the PVN in response to stimulation by stress.

da Costa AP et al (1996): Demonstrates that induction of immediate early gene expression by stress is reduced in limbic forebrain regions during pregnancy and lactation.

Deschamps S et al (2003): Demonstrates the critical role of the pups in modulating the maternal HPA axis response to stressors that represent a threat for the litter.

Douglas AJ et al (1998): Established the predominant role of opioids in inhibiting HPA axis responsiveness to stress during pregnancy.

Douglas AJ et al (2003): Demonstrates that hyporesponsiveness to stress in pregnancy is maintained during parturition in mice.

Gustafson P et al (2017b): Provides evidence for the role of prolactin to indirectly inhibit CRH neurons during pregnancy and lactation.

Heinrichs et al (2001): Demonstrates that nursing an infant is important to reduce stress responses in early postpartum mothers.

Hillerer KM et al (2018): Demonstrates changes in stress responsiveness and attentional capabilities as a function of stage of lactation and stressor salience.

Kim P et al (2010): Highlights morphological changes in human brain anatomy during the early postpartum period.

Neumann ID et al (2000a): Illustrates the differential effects of oxytocin on stress and emotional responses in females during different reproductive states.

Torner L et al (2002): Demonstrates the critical role of prolactin in maintaining stress hyporesponsiveness in lactation, potentially through changes in oxytocin tone.

Wartella J et al (2003): Demonstrates that reproductive experience has long lasting effects on the maternal brain that lasts after pup care has ended in rodents.

References

Agrati D, Zuluaga MJ, Fernández-Guasti A, Meikle A, Ferreira A (2008) Maternal condition reduces fear behaviors but not the endocrine response to an emotional threat in virgin female rats. Horm Behav 53:232–240

Altemus M, Deuster PA, Galliven E, Carter CS, Gold PW (1995) Suppression of hypothalmic-pituitary-adrenal axis responses to stress in lactating women. J Clin Endocrinol Metab 80:2954–2959

Altemus M, Redwine LS, Leong YM, Frye CA, Porges SW, Carter CS (2001) Responses to laboratory psychosocial stress in postpartum women. Psychosom Med 63:814–821

Atkinson HC, Waddell BJ (1995) The hypothalamic-pituitary-adrenal axis in rat pregnancy and lactation: circadian variation and interrelationship of plasma adrenocorticotropin and corticosterone. Endocrinology 136:512–520

Augustine RA, Kokay IC, Andrews ZB, Ladyman SR, Grattan DR (2003) Quantitation of prolactin receptor mRNA in the maternal rat brain during pregnancy and lactation. J Mol Endocrinol 31:221–232

Bales J, Brunton PJ, Russell JA (2006) Suppressed hypothalamo-pituitary-adrenal (HPA) responses to central neuropeptide Y (NPY) in late pregnancy and the role of endogenous opioids. Front Neuroendocrinol 27:55

Barha CK, Lieblich SE, Chow C, Galea LA (2015) Multiparity-induced enhancement of hippocampal neurogenesis and spatial memory depends on ovarian hormone status in middle age. Neurobiol Aging 36:2391–2405

Bosch OJ, Müsch W, Bredewold R, Slattery DA, Neumann ID (2007a) Prenatal stress increases HPA axis activity and impairs maternal care in lactating female offspring: implications for postpartum mood disorder. Psychoneuroendocrinology 32:267–278

Bosch OJ, Sartori SB, Singewald N, Neumann ID (2007b) Extracellular amino acid levels in the paraventricular nucleus and the central amygdala in high- and low-anxiety dams rats during maternal aggression: regulation by oxytocin. Stress 10:261–270

Bridges RS (1984) A quantitative analysis of the roles of dosage, sequence, and duration of estradiol and progesterone exposure in the regulation of maternal behavior in the rat. Endocrinology 114:930–940

Brummelte S, Galea LA (2010) Depression during pregnancy and postpartum: contribution of stress and ovarian hormones. Prog Neuropsychopharmacol Biol Psychiatry 34:766–776

Brunton PJ, Bales J, Russell JA (2006) Neuroendocrine stress but not feeding responses to centrally administered neuropeptide Y are suppressed in pregnant rats. Endocrinology 147:3737–3745

Brunton PJ, McKay AJ, Ochedalski T, Piastowska A, Rebas E, Lachowicz A, Russell JA (2009) Central opioid inhibition of neuroendocrine stress responses in pregnancy in the rat is induced by the neurosteroid allopregnanolone. J Neurosci 29:6449–6460

Brunton PJ, Meddle SL, Ma S, Ochedalski T, Douglas AJ, Russell JA (2005) Endogenous opioids and attenuated hypothalamic-pituitary-adrenal axis responses to immune challenge in pregnant rats. J Neurosci 25:5117–5126

Brunton PJ, Russell JA (2003) Hypothalamic-pituitary-adrenal responses to centrally administered orexin-A are suppressed in pregnant rats. J Neuroendocrinol 15:633–637

Brunton PJ, Russell JA (2010) Prenatal social stress in the rat programmes neuroendocrine and behavioural responses to stress in the adult offspring: sex specific effects. J Neuroendocrinol 22:258–271

Brunton PJ, Russell JA, Douglas AJ (2008) Adaptive responses of the maternal hypothalamic-pituitary-adrenal axis during pregnancy and lactation. J Neuroendocrinol 20:764–776

Callander GE, Bathgate RA (2010) Relaxin family peptide systems and the central nervous system. Cell Mol Life Sci 67:2327–2341

Cena L, Gigantesco A, Mirabella F, Palumbo G, Trainini A, Stefana A (2021) Prevalence of maternal postnatal anxiety and its association with demographic and socioeconomic factors: a multicentre study in Italy. Front Psychiatry 12:737666

Concas A, Mostallino MC, Porcu P, Follesa P, Barbaccia ML, Trabucchi M, Purdy RH, Grisenti P, Biggio G (1998) Role of brain allopregnanolone in the plasticity of gamma-aminobutyric acid type A receptor in rat brain during pregnancy and after delivery. Proc Nat Acad Sci USA 95:13284–13289

Cullinan WE, Ziegler DR, Herman JP (2008) Functional role of local GABAergic influences on the HPA axis. Brain Struct Funct 213:63–72

Cunningham ETJ, Sawchenko PE (1988) Anatomical specificity of noradrenergic inputs to the paraventricular nucleus and supraoptic nucleus of the rat hypothalamus. J Comp Neurol 274:60–76

da Costa AP, Kampa RJ, Windle RJ, Ingram CD, Lightman SL (1997) Region-specific immediate-early gene expression following the administration of corticotropin-releasing hormone in virgin and lactating rats. Brain Res 770:151–162

da Costa AP, Ma X, Ingram CD, Lightman SL, Aguilera G (2001) Hypothalamic and amygdaloid corticotropin-releasing hormone (CRH) and CRH receptor-1 mRNA expression in the stress-hyporesponsive late pregnant and early lactating rat. Brain Res Mol Brain Res 91:119–130

da Costa AP, Wood S, Ingram CD, Lightman SL (1996) Region-specific reduction in stress-induced c-fos mRNA expression during pregnancy and lactation. Brain Res 742:177–184

de Rezende MG, Garcia-Leal C, de Figueiredo FP, Cavalli Rde C, Spanghero MS, Barbieri MA, Bettiol H, de Castro M, Del-Ben CM (2016) Altered functioning of the HPA axis in depressed postpartum women. J Affect Disord 193:249–256

Deschamps S, Woodside B, Walker CD (2003) Pups presence eliminates the stress hyporesponsiveness of early lactating females to a psychological stress representing a threat to the pups. J Neuroendocrinol 15:486–497

Dickens MJ, Pawluski JL (2018) The HPA axis during the perinatal period: Implications for perinatal depression. Endocrinology 159:3737–3746

Donner N, Bredewold R, Maloumby R, Neumann ID (2007) Chronic intracerebral prolactin attenuates neuronal stress circuitries in virgin rats. Eur J Neurosci 25:1804–1814

Douglas A, Meddle S, Toschi N, Bosch O, Neumann I (2005) Reduced activity of the noradrenergic system in the paraventricular nucleus at the end of pregnancy: implications for stress hyporesponsiveness. J Neuroendocrinol 17:40–48

Douglas AJ, Brunton PJ, Bosch OJ, Russell JA, Neumann ID (2003) Neuroendocrine responses to stress in mice: hyporesponsiveness in pregnancy and parturition. Endocrinology 144:5268–5276

Douglas AJ, Johnstone H, Brunton P, Russell JA (2000) Sex-steroid induction of endogenous opioid inhibition on oxytocin secretory responses to stress. J Neuroendocrinol 12:343–350

Douglas AJ, Johnstone HA, Wigger A, Landgraf R, Russell JA, Neumann ID (1998) The role of endogenous opioids in neurohypophysial and hypothalamo-pituitary-adrenal axis hormone secretory responses to stress in pregnant rats. J Endocrinol 158:285–293

Douglas AJ, Russell JA (1994) Corticotrophin-releasing hormone, proenkephalin A and oxytocin mRNA's in the paraventricular nucleus during pregnancy and parturition in the rat. Gene Ther 1(Suppl 1):S85

Featherstone RE, Fleming AS, Ivy GO (2000) Plasticity in the maternal circuit: effects of experience and partum condition on brain astrocyte number in female rats. Behav Neurosci 114:158–172

Figueiredo HF, Bruestle A, Bodie B, Dolgas CM, Herman JP (2003) The medial prefrontal cortex differentially regulates stress-induced c-fos expression in the forebrain depending on type of stressor. Eur J Neurosci 18:2357–2364

Fleming AS, Cheung U, Myhal N, Kessler Z (1989) Effects of maternal hormones on 'timidity' and attraction to pup-related odors in female rats. Physiol Behav 46:449–453

Fleming AS, Steiner M, Corter C (1997) Cortisol, hedonics, and maternal responsiveness in human mothers. Horm Behav 32:85–98

Fodor A, Pintér O, Domokos A, Langnaese K, Barna I, Engelmann M, Zelena D (2013) Blunted HPA axis response in lactating, vasopressin-deficient Brattleboro rats. J Endocrinol 219:89–100

Frim DM, Emanuel RL, Robinson BG, Smas CM, Adler GK, Majzoub JA (1988) Characterization and gestational regulation of corticotropin-releasing hormone messenger RNA in human placenta. J Clin Invest 82:287–292

Georgescu T, Swart JM, Grattan DR, Brown RSE (2021) The prolactin family of hormones as regulators of maternal mood and behavior. Front Glob Womens Health 2:767467

Glynn LM, Davis EP, Sandman CA (2013) New insights into the role of perinatal HPA-axis dysregulation in postpartum depression. Neuropeptides 47:363–370

Goland RS, Wardlaw SL, Blum M, Tropper PJ, Stark RI (1988) Biologically active corticotropin-releasing hormone in maternal and fetal plasma during pregnancy. Am J Obstet Gynecol 159:884–890

Gonzalez A, Jenkins JM, Steiner M, Fleming AS (2009) The relation between early life adversity, cortisol awakening response and diurnal salivary cortisol levels in postpartum women. Psychoneuroendocrinology 34:76–86

Grattan DR, Steyn FJ, Kokay IC, Anderson GM, Bunn SJ (2008) Pregnancy-induced adaptation in the neuroendocrine control of prolactin secretion. J Neuroendocrinol 20:497–507

Groer MW, Davis MW, Hemphill J (2002) Postpartum stress: current concepts and the possible protective role of breastfeeding. J Obstet Gynecol Neonatal Nurs 31:411–417

Gustafson P, Bunn SJ, Grattan DR (2017a) The role of prolactin in the suppression of Crh mRNA expression during pregnancy and lactation in the mouse. J Neuroendocrinol 29

Gustafson P, Kokay I, Sapsford T, Bunn S, Grattan D (2017b) Prolactin regulation of the HPA axis is not mediated by a direct action upon CRH neurons: evidence from the rat and mouse. Brain Struct Funct 222:3191–3204

Gustafson P, Ladyman SR, McFadden S, Larsen C, Khant Aung Z, Brown RSE, Bunn SJ, Grattan DR (2020) Prolactin receptor-mediated activation of pSTAT5 in the pregnant mouse brain. J Neuroendocrinol 32:e12901

Hartikainen-Sorri AL, Kirkinen P, Sorri M, Anttonen H, Tuimala R (1991) No effect of experimental noise exposure on human pregnancy. Obstet Gynecol 77:611–615

Hatton GI, Zhao YQ (2002) Peripartum interneuronal coupling in the supraoptic nucleus. Brain Res 932:120–123

Heiman ML, Ahima RS, Craft LS, Schoner B, Stephens TW, Flier JS (1997) Leptin inhibition of the hypothalamic-pituitary-adrenal axis in response to stress. Endocrinology 138:3859–3863

Heinrichs M, Meinlschmidt G, Neumann I, Wagner S, Kirschbaum C, Ehlert U, Hellhammer DH (2001) Effects of suckling on hypothalamic-pituitary-adrenal axis responses to psychosocial stress in postpartum lactating women. J Clin Endocrinol Metab 86:4798–4804

Herbison AE (2001) Physiological roles for the neurosteroid allopregnanolone in the modulation of brain function during pregnancy and parturition. Prog Brain Res 133:39–47

Herman JP, Cullinan WE, Watson SJ (1994) Involvement of the bed nucleus of the stria terminalis in tonic regulation of paraventricular hypothalamic CRH and AVP mRNA expression. J Neuroendocrinol 6:433–442

Herman JP, Nawreen N, Smail MA, Cotella EM (2020) Brain mechanisms of HPA axis regulation: neurocircuitry and feedback in context Richard Kvetnansky lecture. Stress 23:617–632

Hernandez-Gonzalez M, Prieto-Beracoechea C, Navarro-Meza M, Ramos-Guevara JP, Reyes-Cortes R, Guevara MA (2005) Prefrontal and tegmental electrical activity during olfactory stimulation in virgin and lactating rats. Physiol Behav 83:749–758

Hill PD, Chatterton RT Jr, Aldag JC (2003) Neuroendocrine responses to stressors in lactating and nonlactating mammals: a literature review. Biol Res Nurs 5:79–86

Hillerer KM, Jacobs VR, Fischer T, Aigner L (2014) The maternal brain: an organ with peripartal plasticity. Neural Plast 2014:574159

Hillerer KM, Woodside B, Parkinson E, Long H, Verlezza S, Walker CD (2018) Gating of the neuroendocrine stress responses by stressor salience in early lactating female rats is independent of infralimbic cortex activation and plasticity. Stress 21:217–228

Johnson SB, Emmons EB, Lingg RT, Anderson RM, Romig-Martin SA, LaLumiere RT, Narayanan NS, Viau V, Radley JJ (2019) Prefrontal-bed nucleus circuit modulation of a passive coping response set. J Neurosci 39:1405–1419

Johnstone HA, Wigger A, Douglas AJ, Neumann ID, Landgraf R, Seckl JR, Russell JA (2000) Attenuation of hypothalamic-pituitary-adrenal axis stress responses in late pregnancy: changes in feedforward and feedback mechanisms. J Neuroendocrinol 12:811–822

Jolley SN, Elmore S, Barnard KE, Carr DB (2007) Dysregulation of the hypothalamic-pituitary-adrenal axis in postpartum depression. Biol Res Nurs 8:210–222

Jurek B, Slattery DA, Hiraoka Y, Liu Y, Nishimori K, Aguilera G, Neumann ID, van den Burg EH (2015) Oxytocin regulates stress-induced Crf gene transcription through CREB-regulated transcription coactivator 3. J Neurosci 35:12248–12260

Kalyani M, Callahan P, Janik JM, Shi H (2017) Effects of pup separation on stress response in postpartum female rats. Int J Mol Sci 18

Kammerer M, Adams D, Castelberg Bv B, Glover V (2002) Pregnant women become insensitive to cold stress. BMC Pregnancy Childbirth 2:8

Kawai M, Yamaguchi M, Murakami T, Shima K, Murata Y, Kishi K (1997) The placenta is not the main source of leptin production in pregnant rat: gestational profile of leptin in plasma and adipose tissues. Biochem Biophys Res Commun 240:798–802

Kim P, Feldman R, Mayes LC, Eicher V, Thompson N, Leckman JF, Swain JE (2011) Breastfeeding, brain activation to own infant cry, and maternal sensitivity. J Child Psychol Psychiatry 52:907–915

Kim P, Leckman JF, Mayes LC, Feldman R, Wang X, Swain JE (2010) The plasticity of human maternal brain: longitudinal changes in brain anatomy during the early postpartum period. Behav Neurosci 124:695–700

Kinsley CH, Lambert KG (2008) Reproduction-induced neuroplasticity: natural behavioural and neuronal alterations associated with the production and care of offspring. J Neuroendocrinol 20:515–525

Kiss A, Palkovits M, Aguilera G (1996) Neural regulation of corticotropin releasing hormone (CRH) and CRH receptor mRNA in the hypothalamic paraventricular nucleus in the rat. J Neuroendocrinol 8:103–112

Kitay JI, Coyne MD, Newsom W, Nelson R (1965) Relation of the ovary to adrenal corticosterone production and adrenal enzyme activity in the rat. Endocrinology 77:902–908

Kokay IC, Bull PM, Davis RL, Ludwig M, Grattan DR (2006) Expression of the long form of the prolactin receptor in magnocellular oxytocin neurons is associated with specific prolactin regulation of oxytocin neurons. Am J Physiol Regul Integr Comp Physiol 290:R1216–R1225

Leng G, Mansfield S, Bicknell RJ, Brown D, Chapman C, Hollingsworth S, Ingram CD, Marsh MI, Yates JO, Dyer RG (1987) Stress-induced disruption of parturition in the rat may be mediated by endogenous opioids. J Endocrinol 114:247–252

Leuner B, Gould E (2010) Dendritic growth in medial prefrontal cortex and cognitive flexibility are enhanced during the postpartum period. J Neurosci 30:13499–13503

Liggins GC (1994) The role of cortisol in preparing the fetus for birth. Reprod Fertil Dev 6:141–150

Lightman SL, Windle RJ, Wood SA, Kershaw YM, Shanks N, Ingram CD (2001) Peripartum plasticity within the hypothalamo-pituitary-adrenal axis. Prog Brain Res 133:111–129

Lightman SL, Young WS III (1989) Lactation inhibits stress-mediated secretion of corticosterone and oxytocin and hypothalamic accumulation of corticotropin-releasing factor and enkephalin messenger ribonucleic acids. Endocrinology 124:2358–2364

Lindsay JR, Nieman LK (2005) The hypothalamic-pituitary-adrenal axis in pregnancy: challenges in disease detection and treatment. Endocr Rev 26:775–799

Lowrance SA, Ionadi A, McKay E, Douglas X, Johnson JD (2016) Sympathetic nervous system contributes to enhanced corticosterone levels following chronic stress. Psychoneuroendocrinology 68:163–170

Ma S, Shipston MJ, Morilak D, Russell JA (2005) Reduced hypothalamic vasopressin secretion underlies attenuated adrenocorticotropin stress responses in pregnant rats. Endocrinology 146:1626–1637

Magiakou MA, Mastorakos G, Rabin D, Dubbert B, Gold PW, Chrousos GP (1996a) Hypothalamic corticotropin-releasing hormone suppression during the postpartum period: implications for the increase in psychiatric manifestations at this time. J Clin Endocrinol Metab 81:1912–1917

Magiakou MA, Mastorakos G, Rabin D, Margioris AN, Dubbert B, Calogero AE, Tsigos C, Munson PJ, Chrousos GP (1996b) The maternal hypothalamic-pituitary-adrenal axis in the third trimester of human pregnancy. Clin Endocrinol (Oxf) 44:419–428

Majewska MD (1990) Steroid regulation of the GABAA receptor: ligand binding, chloride transport and behaviour. Ciba Found Symp 153:83–97

Mann PE, Bridges RS (2001) Lactogenic hormone regulation of maternal behavior. Prog Brain Res 133:251–262

Mastorakos G, Ilias I (2003) Maternal and fetal hypothalamic-pituitary-adrenal axes during pregnancy and postpartum. Ann N Y Acad Sci 997:136–149

Mattheij JA, Gruisen EF, Swarts JJ (1979) The suckling-induced rise of plasma prolactin in lactating rats: its dependence on stage of lactation and litter size. Horm Res 11:325–336

McLean M, Bisits A, Davies J, Woods R, Lowry P, Smith R (1995) A placental clock controlling the length of human pregnancy. Nat Med 1:460–463

Meinlschmidt G, Martin C, Neumann ID, Heinrichs M (2010) Maternal cortisol in late pregnancy and hypothalamic-pituitary-adrenal reactivity to psychosocial stress postpartum in women. Stress 13:163–171

Melón LC, Hooper A, Yang X, Moss SJ, Maguire J (2018) Inability to suppress the stress-induced activation of the HPA axis during the peripartum period engenders deficits in postpartum behaviors in mice. Psychoneuroendocrinology 90:182–193
Myers MM, Denenberg VH, Thoman E, Holloway WR, Bowerman DR (1975) The effects of litter size on plasma corticosterone and prolactin response to ether stress in the lactating rat. Neuroendocrinology 19:54–58
Naik RR, de Jong TR (2017) Transient and persistent behavioral and molecular changes in primiparous female Wistar rats. Eur J Neurosci 45:797–804
Nakamura Y, Okada T, Morikawa M, Yamauchi A, Sato M, Ando M, Ozaki N (2020) Perinatal depression and anxiety of primipara is higher than that of multipara in Japanese women. Sci Rep 10:17060
Nephew BC, Bridges RS, Lovelock DF, Byrnes EM (2009a) Enhanced maternal aggression and associated changes in neuropeptide gene expression in multiparous rats. Behav Neurosci 123:949–957
Nephew BC, Caffrey MK, Felix-Ortiz AC, Ferris CF, Febo M (2009b) Blood oxygen level-dependent signal responses in corticolimbic 'emotions' circuitry of lactating rats facing intruder threat to pups. Eur J Neurosci 30:934–945
Neumann ID, Bosch OJ, Toschi N, Torner L, Douglas AJ (2003) No stress response of the hypothalamo-pituitary-adrenal axis in parturient rats: lack of involvement of brain oxytocin. Endocrinology 144:2473–2479
Neumann ID, Johnstone HA, Hatzinger M, Liebsch G, Shipston M, Russell JA, Landgraf R, Douglas AJ (1998) Attenuated neuroendocrine responses to emotional and physical stressors in pregnant rats involve adenohypophysial changes. J Physiol 508:289–300
Neumann ID, Krömer SA, Bosch OJ (2005) Effects of psycho-social stress during pregnancy on neuroendocrine and behavioural parameters in lactation depend on the genetically determined stress vulnerability. Psychoneuroendocrinology 30:791–806
Neumann ID, Torner L, Wigger A (2000a) Brain oxytocin: differential inhibition of neuroendocrine stress responses and anxiety-related behaviour in virgin, pregnant and lactating rats. Neuroscience 95:567–575
Neumann ID, Toschi N, Ohl F, Torner L, Kromer SA (2001) Maternal defence as an emotional stressor in female rats: correlation of neuroendocrine and behavioural parameters and involvement of brain oxytocin. Eur J Neurosci 13:1016–1024
Neumann ID, Wigger A, Torner L, Holsboer F, Landgraf R (2000b) Brain oxytocin inhibits basal and stress-induced activity of the hypothalamo-pituitary-adrenal axis in male and female rats: partial action within the paraventricular nucleus. J Neuroendocrinol 12:235–243
O'Keane V, Lightman S, Patrick K, Marsh M, Papadopoulos AS, Pawlby S, Seneviratne G, Taylor A, Moore R (2011) Changes in the maternal hypothalamic-pituitary-adrenal axis during the early puerperium may be related to the postpartum 'blues'. J Neuroendocrinol 23:1149–1155
Oates M, Woodside B, Walker CD (2000) Chronic leptin administration in developing rats reduces stress responsiveness partly through changes in maternal behavior. Horm Behav 37:366–376
Obel C, Hedegaard M, Henriksen TB, Secher NJ, Olsen J, Levine S (2005) Stress and salivary cortisol during pregnancy. Psychoneuroendocrinology 30:647–656
Ochedalski T, Lachowicz A (2004) Maternal and fetal hypothalamo-pituitary-adrenal axis: different response depends upon the mode of parturition. Neuro Endocrinol Lett 25:278–282
Ochedalski T, Zylińska K, Laudański T, Lachowicz A (2001) Corticotrophin-releasing hormone and ACTH levels in maternal and fetal blood during spontaneous and oxytocin-induced labour. Eur J Endocrinol 144:117–121
Okamoto E, Takagi T, Makino T, Sata H, Iwata I, Nishino E, Mitsuda N, Sugita N, Otsuki Y, Tanizawa O (1989) Immunoreactive corticotropin-releasing hormone, adrenocorticotropin and cortisol in human plasma during pregnancy and delivery and postpartum. Horm Metab Res 21:566–572
Opala EA, Verlezza S, Long H, Rusu D, Woodside B, Walker CD (2019) Experience of adversity during a first lactation modifies prefrontal cortex morphology in primiparous female rats: lack of long term effects on a subsequent lactation. Neuroscience 417:95–106

Orchard ER, Ward PGD, Sforazzini F, Storey E, Egan GF, Jamadar SD (2020) Relationship between parenthood and cortical thickness in late adulthood. PLoS One 15:e0236031
Owens PC, Smith R, Brinsmead MW, Hall C, Rowley M, Hurt D, Lovelock M, Chan EC, Cubis J, Lewin T (1987) Postnatal disappearance of the pregnancy-associated reduced sensitivity of plasma cortisol to feedback inhibition. Life Sci 41:1745–1750
Patchev VK, Hassan AHS, Holsboer F, Almeida OFX (1996) The neurosteroid tetrahydroprogesterone attenuates the endocrine response to stress and exerts glucocorticoid-like effects on vasopressin gene transcription in the rat hypothalamus. Neuropsychopharmacology 15:533–540
Pawluski JL, Charlier TD, Lieblich SE, Hammond GL, Galea LA (2009) Reproductive experience alters corticosterone and CBG levels in the rat dam. Physiol Behav 96:108–114
Perani CV, Langgartner D, Uschold-Schmidt N, Füchsl AM, Neumann ID, Reber SO, Slattery DA (2017) Adrenal gland plasticity in lactating rats and mice is sufficient to maintain basal hypersecretion of corticosterone. Stress 20:303–311
Pérez-Hernández M, Hernández-González M, Hidalgo-Aguirre RM, Guevara MA, Amezcua-Gutiérrez C, Sandoval-Carrillo IK (2021) Multiparity decreases the effect of distractor stimuli on a working memory task: an EEG study. Soc Neurosci 16:277–288
Petraglia F, Baraldi M, Giarrè G, Facchinetti F, Santi M, Volpe A, Genazzani AR (1985) Opioid peptides of the pituitary and hypothalamus: changes in pregnant and lactating rats. J Endocrinol 105:239–245
Radley JJ, Gosselink KL, Sawchenko PE (2009) A discrete GABAergic relay mediates medial prefrontal cortical inhibition of the neuroendocrine stress response. J Neurosci 29:7330–7340
Rima BN, Bardi M, Friedenberg JM, Christon LM, Karelina KE, Lambert KG, Kinsley CH (2009) Reproductive experience and the response of female Sprague-Dawley rats to fear and stress. Comp Med 59:437–443
Robertson MC, Friesen HG (1981) Two forms of rat placental lactogen revealed by radioimmunoassay. Endocrinology 108:2388–2390
Robertson MC, Gillespie B, Friesen HG (1982) Characterization of the two forms of rat placental lactogen (rPL): rPL-I and rPL-II. Endocrinology 111:1862–1866
Robinson BG, Emanuel RL, Frim DM, Majzoub JA (1988) Glucocorticoid stimulates expression of corticotropin-releasing hormone gene in human placenta. Proc Natl Acad Sci USA 85:5244–5248
Ross HE, Young LJ (2009) Oxytocin and the neural mechanisms regulating social cognition and affiliative behavior. Front Neuroendocrinol 30:534–547
Sarkar J, Wakefield S, MacKenzie G, Moss SJ, Maguire J (2011) Neurosteroidogenesis is required for the physiological response to stress: role of neurosteroid-sensitive GABAA receptors. J Neurosci 31:18198–18210
Sasaki A, Shinkawa O, Yoshinaga K (1989) Placental corticotropin-releasing hormone may be a stimulator of maternal pituitary adrenocorticotropic hormone secretion in humans. J Clin Invest 84:1997–2001
Sawchenko PE, Li H-Y, Ericsson A (2000) Circuits and mechanisms governing hypothalamic responses to stress: a tale of two paradigms. Prog Brain Res 122:61–78
Schulte HM, Weisner D, Allolio B (1990) The corticotrophin releasing hormone test in late pregnancy: lack of adrenocorticotrophin and cortisol response. Clin Endocrinol 33:99–106
Sherwood OD, Crnekovic VE, Gordon WL, Rutherford JE (1980) Radioimmunoassay of relaxin throughout pregnancy and during parturition in the rat. Endocrinology 107:691–698
Slattery DA, Hillerer KM (2016) The maternal brain under stress: consequences for adaptive peripartum plasticity and its potential functional implications. Front Neuroendocrinol 41:114–128
Slattery DA, Neumann ID (2008) No stress please! Mechanisms of stress hyporesponsiveness of the maternal brain. J Physiol 586:377–385
Smith MS, Neill JD (1976) Termination at midpregnancy of the two daily surges of plasma prolactin initiated by mating in the rat. Endocrinology 98:696–701
Stirrat LI, Just G, Homer NZM, Andrew R, Norman JE, Reynolds RM (2017) Glucocorticoids are lower at delivery in maternal, but not cord blood of obese pregnancies. Sci Rep 7:10263

Taylor A, Glover V, Marks M, Kammerer M (2009) Diurnal pattern of cortisol output in postnatal depression. Psychoneuroendocrinology 34:1184–1188
Torner L, Toschi N, Nava G, Clapp C, Neumann ID (2002) Increased hypothalamic expression of prolactin in lactation: involvement in behavioural and neuroendocrine stress responses. Eur J Neurosci 15:1381–1389
Torner L, Toschi N, Pohlinger A, Landgraf R, Neumann ID (2001) Anxiolytic and anti-stress effects of brain prolactin: improved efficacy of antisense targeting of the prolactin receptor by molecular modeling. J Neurosci 21:3207–3214
Toufexis DJ, Rochford J, Walker CD (1999a) Lactation-induced reduction in rats' acoustic startle is associated with changes in noradrenergic neurotransmission. Behav Neurosci 113:176–184
Toufexis DJ, Tesolin S, Huang N, Walker C (1999b) Altered pituitary sensitivity to corticotropin-releasing factor and arginine vasopressin participates in the stress hyporesponsiveness of lactation in the rat. J Neuroendocrinol 11:757–764
Toufexis DJ, Thrivikraman KV, Plotsky PM, Morilak DA, Huang N, Walker CD (1998) Reduced noradrenergic tone to the hypothalamic paraventricular nucleus contributes to the stress hyporesponsiveness of lactation. J Neuroendocrinol 10:417–427
Toufexis DJ, Walker CD (1996) Noradrenergic facilitation of the adrenocorticotropin response to stress is absent during lactation in the rat. Brain Res 737:71–77
Tu MT, Lupien SJ, Walker CD (2005) Measuring stress responses in postpartum mothers: perspectives from studies in human and animal populations. Stress 8:19–34
Tu MT, Lupien SJ, Walker CD (2006) Multiparity reveals the blunting effect of breastfeeding on physiological reactivity to psychological stress. J Neuroendocrinol 18:494–503
Vannuccini S, Bocchi C, Severi FM, Challis JR, Petraglia F (2016) Endocrinology of human parturition. Ann Endocrinol (Paris) 77:105–113
Vogl SE, Worda C, Egarter C, Bieglmayer C, Szekeres T, Huber J, Husslein P (2006) Mode of delivery is associated with maternal and fetal endocrine stress response. BJOG 113:441–445
Waddell BJ, Atkinson HC (1994) Production rate, metabolic clearance rate and uterine extraction of corticosterone during rat pregnancy. J Endocrinol 143:183–190
Walker CD, Deschamps S, Proulx K, Tu M, Salzman C, Woodside B, Lupien S, Gallo-Payet N, Richard D (2004) Mother to infant or infant to mother? Reciprocal regulation of responsiveness to stress in rodents and the implications for humans. J Psychiatry Neurosci 29:364–382
Walker CD, Lightman SL, Steele MK, Dallman MF (1992) Suckling is a persistent stimulus to the adrenocortical system of the rat. Endocrinology 130:115–125
Walker CD, Tilders FJ, Burlet A (2001a) Increased colocalization of corticotropin-releasing factor and arginine vasopressin in paraventricular neurones of the hypothalamus in lactating rats: evidence from immunotargeted lesions and immunohistochemistry. J Neuroendocrinol 13:74–85
Walker CD, Toufexis DJ, Burlet A (2001b) Hypothalamic and limbic expression of CRF and vasopressin during lactation: implications for the control of ACTH secretion and stress hyporesponsiveness. Prog Brain Res 133:99–110
Walker CD, Trottier G, Rochford J, Lavallée D (1995) Dissociation between behavioral and hormonal responses to the forced swim stress in lactating rats. J Neuroendocrinol 7:615–622
Wartella J, Amory E, Lomas LM, Macbeth A, McNamara I, Stevens L, Lambert KG, Kinsley CH (2003) Single or multiple reproductive experiences attenuate neurobehavioral stress and fear responses in the female rat. Physiol Behav 79:373–381
Wigger A, Lörscher P, Oehler I, Keck ME, Naruo T, Neumann ID (1999) Nonresponsiveness of the rat hypothalamo-pituitary-adrenocortical axis to parturition-related events: inhibitory action of endogenous opioids. Endocrinology 140:2843–2849
Windle RJ, Brady MM, Kunanandam T, Da Costa AP, Wilson BC, Harbuz M, Lightman SL, Ingram CD (1997a) Reduced response of the hypothalamo-pituitary-adrenal axis to alpha1-agonist stimulation during lactation. Endocrinology 138:3741–3748
Windle RJ, Wood S, Shanks N, Perks P, Conde GL, da Costa APC, Ingram CD, Lightman SL (1997b) Endocrine and behavioural responses to noise stress: comparison of virgin and lactating female rats during non-disrupted maternal activity. J Neuroendocrinol 9:407–414

Windle RJ, Wood SA, Kershaw YM, Lightman SL, Ingram CD (2010) Reduced stress responsiveness in pregnancy: relationship with pattern of forebrain c-fos mRNA expression. Brain Res 1358:102–109

Windle RJ, Wood SA, Kershaw YM, Lightman SL, Ingram CD (2013) Adaptive changes in basal and stress-induced HPA activity in lactating and post-lactating female rats. Endocrinology 154:749–761

Workman JL, Gobinath AR, Kitay NF, Chow C, Brummelte S, Galea LAM (2016) Parity modifies the effects of fluoxetine and corticosterone on behavior, stress reactivity, and hippocampal neurogenesis. Neuropharmacology 105:443–453

Young LJ, Muns S, Wang Z, Insel TR (1997) Changes in oxytocin receptor mRNA in rat brain during pregnancy and the effects of estrogen and interleukin-6. J Neuroendocrinol 9:859–865

Yuen BS, Owens PC, Symonds ME, Keisler DH, McFarlane JR, Kauter KG, McMillen IC (2004) Effects of leptin on fetal plasma adrenocorticotropic hormone and cortisol concentrations and the timing of parturition in the sheep. Biol Reprod 70:1650–1657

5 Neuropeptide Pathways Controlling the Timing of Birth

Jonathan J. Hirst, Hannah K. Palliser, Julia C. Shaw, and Tamas Zakar

Abstract

The mechanisms regulating gestation length and birth differ markedly between species. In many mammals, corticotropin releasing hormone (CRH) plays a central role in the processes that trigger the progression from uterine quiescence to active labor. This results in processes being well coordinated with the maturation of the fetus and ensures sufficient maturation of fetal organ systems for successful extra-uterine life. In many mammalian species, CRH release and stimulation of the fetal hypothalamic-pituitary-adrenal (HPA) axis, with a consequent rise in cortisol levels, triggers the initiation of labor and birth. The onset of labor in women differs markedly, with initiation processes begun by rising levels of CRH produced by the syncytiotrophoblast of the placenta. This "placental clock" mechanism leads to increased production of estrogens toward term and a rise in the expression of contraction-associated proteins. A premature rise in CRH pro-

https://www.newcastle.edu.au/profile/jon-hirst
https://www.newcastle.edu.au/profile/hannah-palliser
https://www.newcastle.edu.au/profile/julia-c-shaw

J. J. Hirst (✉) · H. K. Palliser · J. C. Shaw
School of Biomedical Sciences and Pharmacy, University of Newcastle & Mothers and Babies Research Program, Hunter Medical Research Institute, Newcastle, Australia
e-mail: Jon.Hirst@Newcastle.edu.au; Hannah.Palliser@newcastle.edu.au; Julia.C.Shaw@newcastle.edu.au

T. Zakar
School of Medicine and Public Health, University of Newcastle & Mothers and Babies Research Program, Hunter Medical Research Institute, Newcastle, Australia
e-mail: Tamas.Zakar@newcastle.edu.au

P. J. Brunton, D. R. Grattan (eds.), *Neuroendocrine Regulation of Mammalian Pregnancy and Lactation*, Masterclass in Neuroendocrinology 15,
https://doi.org/10.1007/978-3-031-51138-7_5

duction leads to preterm labor in some instances; however, many cases of early preterm labor are started independently of these mechanisms and involve intrauterine infection and inflammation. Oxytocin is a maternal signal triggering labor once the uterine tissues are adequately prepared, and has a role in setting the time of day that labor is initiated. Myometrial stimulation by oxytocin is also important in the final stages of birth for expulsion of the fetus, separation of the placenta, and for reducing postpartum bleeding. Synthetic oxytocin may be used to promote uterine contractions for the augmentation of labor. Recent observations therefore suggest that CRH and oxytocin have key roles in controlling the onset of labor in women, whereas early preterm birth is controlled differently, involving poorly understood inflammatory mechanisms.

Keywords

Corticotropin-releasing hormone · Oxytocin · Placenta · Parturition · Labor · Preterm birth

5.1 Introduction

The processes involved in the transition from the quiescent state of late pregnancy to the onset of labor show extraordinary variation between species, none more so than the relative contributions of neuropeptide hormones. Short gestation species, including mice and rats, show markedly different mechanisms controlling the onset of labor compared to large-size and long-gestation species, such as sheep and humans (Mitchell and Taggart 2009; Smith et al. 2013). Even within these groups of species there are marked differences in the relative contribution of fetal and maternal processes in regulating the onset and progression of labor. Part of this difference may result from selective pressures on the ability to deliver offspring that are appropriately developed for their environments. Hence, ungulates (hoofed species) and grassland animals require the newborn to walk and run immediately after birth, whereas species where the offspring are maintained in a nest or with their mother after birth can afford more postnatal development. Therefore, ungulates, through both fetal and maternal signaling, tend to show a much tighter control over gestation length compared to humans.

Coordinated maturation of organ systems is required for newborn survival and much of this is accomplished by rising fetal cortisol levels. These maturational events have been best studied in sheep where maturation of the fetal **hypothalamic-pituitary-adrenal (HPA) axis** and the production of corticotropin-releasing hormone (CRH) coordinates development with labor onset. Rising CRH release stimulates adrenocorticotropic hormone (ACTH) secretion the from the corticotrophs of the anterior pituitary gland, leading to stimulation of the fetal adrenal glands to increase cortisol production. Cortisol has a major role in the maturation of organ systems, particularly the fetal lungs, in preparation for air breathing. Cortisol action includes stimulation of the development of type 2 alveolar epithetical cells (pneumocytes), which are critical for the production of surfactants that aid maintenance of the stability of the liquid/air interface. The continued rise in cortisol levels then initiates changes in placental hormone expression followed by downstream uterine stimulation, which leads to the onset and progression of labor in these species (Vannuccini et al. 2016) (Fig. 5.1). In contrast, developmental changes in the placenta, including the synthesis and release of local CRH, have a greater role in the onset of human labor at term (Smith et al. 2012). Importantly, in women, the mechanisms leading to spontaneous labor at term appear to differ considerably from those involved in preterm birth (Herrera et al. 2021). This is clinically important, as the incidence of preterm labor approaches 8–10% globally, and prematurity is responsible for the majority of adverse neonatal outcomes. Most evidence suggests that early (28–32 weeks of gestation) and extremely early preterm (<28 weeks of gestation) birth involve uteroplacental inflammation with relatively little input of maternal and fetal maturational factors (Challis et al. 2001; Phung et al. 2022).

Oxytocin is thought of as an essential hormone involved in labor; however, oxytocin has a relatively limited role in the initiation of labor in primates (Hirst et al. 1993; Arthur et al. 2007). In contrast, oxytocin is involved in mediating maternal influences over the onset of labor and is important in species that deliver their young within a narrow timeframe, including ungulates and species which deliver at night. Furthermore, despite the limited role of oxytocin in initiating labor in women, once labor is established, oxytocin strongly stimulates the progression of uterine contractions when adequate receptors are present (Arthur et al. 2007). Synthetic oxytocin preparations are used clinically to augment contractile activity, aid placental expulsion, and reduce excessive blood loss. This chapter will describe the regulatory roles of CRH and oxytocin in the timing of labor and highlight the major differences in these processes between species, particularly the mechanisms that are unique to higher primates. The chapter will also discuss differences between the control of the onset of term and preterm labor and the key role of inflammation and infection in triggering early preterm birth.

5.2 CRH and the HPA Axis Triggers the Onset of Labor at Term

Increasing levels of CRH and stimulation of the fetal HPA axis leads to the onset of labor in many mammalian species, and rising placental CRH is a contributor to labor onset in women (Whittle et al. 2001; Herrera et al. 2021). The most conclusive studies have been performed in sheep, in which hypothalamic **paraventricular nucleus (PVN)** and neurosecretory cell maturation increase fetal CRH output before labor onset (Challis and Brooks 1989). Maturation of input pathways to the fetal HPA axis, although unclear, may further contribute to increased CRH secretion (Fig. 5.1). This may involve reduced sensitivity to cortisol feedback through changes in glucocorticoid receptor levels, which could be programmed by earlier events in pregnancy. This ultimately leads to CRH mRNA levels in the fetal hypothalamus

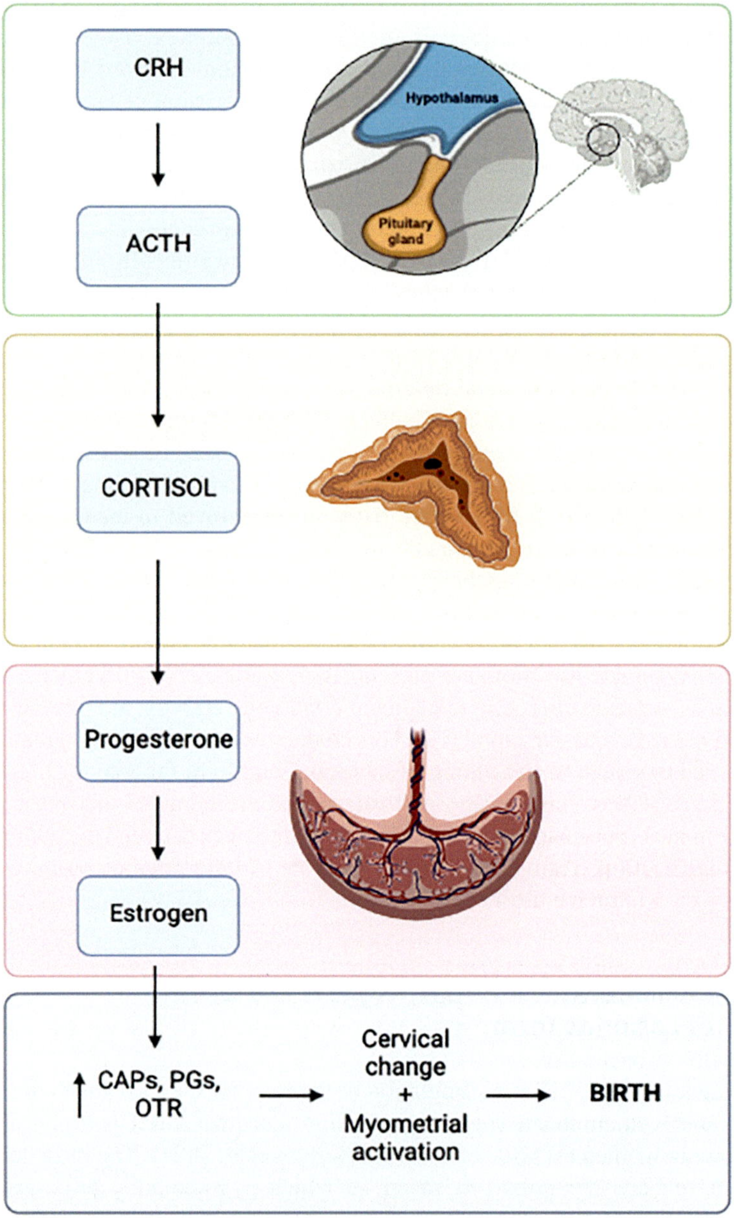

Fig. 5.1 Fetal development-driven processes leading to the initiation of labor and birth in the sheep. Maturation of fetal hypothalamic pathways and increased CRH release activate the HPA axis. Increased cortisol levels stimulate the expression of placental enzymes that convert progesterone to estrogens. The increased estrogen level leads to the onset of intrauterine processes that result in birth. Estrogens—largely estradiol-17β; *CAPs* contraction-associated proteins, *PGs* prostaglandins, *OTR* oxytocin receptors

rising rapidly prior to the onset of labor. The relative importance of these developmental processes in triggering labor depends on the contribution of maternal and fetal influences over the timing of labor. In species such as the sheep and other ungulates that deliver an advanced fetus that must be capable of near-independent life after delivery, the rise in activity of the fetal HPA axis exquisitely controls pathways leading to labor onset (Challis and Brooks 1989). The sheep was one of the first species in which increased fetal HPA axis activity, and the resulting rise in fetal plasma cortisol levels, was shown to control the onset of labor (Liggins 1969). As in the human, in ewes there is a shift from corpus luteum to placental production of progesterone early in pregnancy. The rising level of cortisol reaching the placenta toward term stimulates the synthesis of 17α-hydroxylase/17,20-lyase, a key rate-limiting enzyme in the pathway responsible for the conversion of progesterone to estradiol-17β (Thorburn and Challis 1979; Whittle et al. 2001). This leads to rising levels of estrogens, largely estradiol-17β. Several positive feedback processes amplify these changes once started, to ensure the rapid transition from uterine quiescence to active labor. These include cortisol and estrogens raising levels of the CRH receptor type-1 (CRHR-1) in the corticotrophs of the anterior pituitary gland and direct effects of cortisol on **prostaglandin** production by intrauterine tissues (Fig. 5.1) (Schwartz and McMillen 2001).

In the sheep, increasing levels of cortisol and estrogens enhance the synthesis of uterine stimulant prostaglandin levels by raising the expression of **prostaglandin-endoperoxide synthase** (PTGS) type-2 (cyclooxygenase [COX]-2) in the intrauterine issues, including the **amnion**, decidua, and **myometrium** and the cervical connective tissue (Olson et al. 2001). This action is potentiated by the associated decline in progesterone levels that would otherwise depress PTGS-2 expression. The prostaglandins increase uterine contractions and cause connective tissue changes in the cervix, including the upregulation of metalloproteases that contribute to **cervical dilation**. The prostaglandins produced include prostaglandin E_2 and $F_2\alpha$, which act locally on the uterus and the cervix (Olson et al. 1983). Importantly, prostaglandin E_2 is also released into the fetal circulation and may stimulate HPA axis development, further enhancing positive feedback (Young et al. 1996). The rising estrogen-dominated environment also increases the expression of other contraction-supporting agents (contraction-associated proteins), including the **connexins** that increase the coordination of more forceful contractions, and the expression of oxytocin receptors in the myometrium (Zakar and Mitchell 1996; Phung et al. 2022). This allows oxytocin to augment contractive activity (see below). In sheep, oxytocin release from the maternal posterior pituitary gland has a fine-tuning role that further boosts uterine contraction and may trigger labor onset at particular times during the diurnal cycle. Oxytocin is secreted in a pulsatile manner (see Chap. 6), increasing uterine activity and stimulating further prostaglandin output (see below). This leads to the final tipping point for the onset of labor when uterine receptors and prostaglandin-synthetic mechanisms are appropriately elevated.

5.3 Role of CRH in the Timing of Human Labor

The role of CRH in the control of labor onset in women and some higher primates is more complex than in many other species and involves the fetal and maternal HPA axis, as well as the placenta (Smith et al. 2012). Intrauterine inflammation, changes in progesterone action, and local actions of CRH (Chan et al. 1990; McLean et al. 1995) also contribute. Furthermore, the mechanisms triggering early preterm labor differ markedly from those that lead to the onset of labor close to normal term. The finding that increased circulating CRH levels precede some cases of preterm birth have resulted in investigation into CRH as a predictor of preterm birth risk (McLean et al. 1995). In contrast, most extremely preterm births are triggered by infection-induced inflammation, which seems relatively independent of the processes involved in the timing of normal term labor processes and CRH action. In these cases, bacteria move from the cervix to the intrauterine membranes, with resultant **choriodecidual** infection. This localized infection is often subclinical and not detected until the onset of preterm birth (Gravett et al. 1996; Adams Waldorf et al. 2011). The bacteria involved include *ureaplasma* species, which are not easily cultured, and until recently their role has been under-recognized (Viscardi 2010). This mechanism involves ascending infection passing through the cervix from the vagina. Bacteria may reach the **chorionic villi** and lead to preterm birth before the infection is detected and can be treated. The presence of certain bacteria in the vaginal microflora is a risk factor, but why this occurs in certain pregnancies is unclear. Infection can also trigger **chorioamnionitis** that may progress to the amniotic fluid, fetal infection, and sepsis, which has a major negative effect on pregnancy outcome (Hassan et al. 2011; Kelleher et al. 2020). Therefore, attempts to suppress preterm birth under this circumstance could lead to advanced fetal infection and is clinically regarded as a significant risk of adverse neonatal outcome (Kelleher et al. 2020). Assessing the triggering roles is further complicated as the effects of ascending bacteria may raise maternal CRH and cortisol levels as a consequence, and not a causative process, of the onset of preterm labor.

In women, CRH is synthesized in the placenta in addition to the fetal and maternal hypothalamus. Landmark studies have shown that placental CRH synthesis determines both local and circulating concentrations that ultimately promote and control the initiation of labor at normal term and in cases of moderately preterm labor (in the absence of infection) (McGrath et al. 2002; McLean and Smith 2001). As the placenta develops and pregnancy advances, the placental **cytotrophoblasts** undergo **syncytialization** to form the **syncytial layer**, which is the interface between the maternal and fetal circulations (Fig. 5.2). The **syncytiotrophoblast** is a major endocrine organ during pregnancy, producing progesterone, estrogens, and their metabolites. The metabolites of progesterone include key **neurosteroids**, such as allopregnanolone, and their precursors. Allopregnanolone concentrations increase as the placenta develops and has a significant role in fetal neurodevelopment, controlling levels of excitability in the fetal brain and supporting myelination (Herrera et al. 2021; Hirst et al. 2014). In this way, the placenta has a key endocrine regulatory role for the mother and fetus, as well as supporting fetal development and

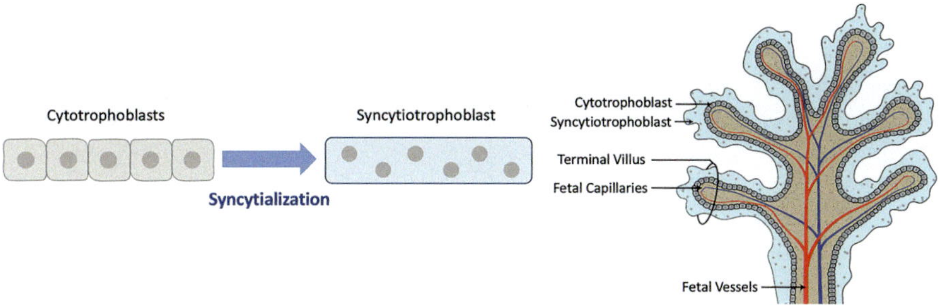

Fig. 5.2 Development of the placental villi outer syncytiotrophoblast layer. The cytotrophoblast undergoes syncytialization to form the outer syncytial layers that lie next to the maternal blood. The syncytiotrophoblast performs some of the major endocrine roles of the placenta secreting CRH, progesterone and estrogens. As the placenta grows the complexity of the villi increases and with advancing gestation the volume of the syncytiotrophoblast increases, which raises hormone output

nutrition. Indeed, rising placental steroidogenesis results in suppressive allopregnanolone concentrations that protect the developing fetal brain from excessive excitation. This contributes to the fetal brain being very resistant to episodes of hypoxia in late gestation and during delivery. Without this protection, hypoxia during birth can result in more frequent excitotoxic damage (Hirst et al. 2014, 2018). The syncytiotrophoblast is also the site of placental CRH production and significantly contributes to CRH levels in the fetal and maternal circulation (Leung et al. 1999). The growth of the placenta to support fetal development also results in an increasing volume of syncytial tissue and CRH production. This contributes to rising levels of CRH in the fetal and maternal circulation with advancing gestation (McGrath et al. 2002; McLean and Smith 2001). The regulation of CRH concentration is more complex, as fetal growth restriction, which typically results from a smaller placenta, and does not consistently lead to reduced CRH levels. Furthermore, fetal growth restriction is an associated risk factor for preterm birth triggered by a premature rise in CRH levels. Indeed, fetal and maternal stressors may elevate placental and central CRH expression (Mark et al. 2006).

5.4 Control of Placental CRH and Loss of Progesterone Action in the Primate

The rising levels of CRH production by the placenta toward term and the premature rise of circulating CRH levels associated with preterm birth suggest similar mechanisms of labor onset in the absence of infection (Fig. 5.3). These observations have led to the proposal that rising CRH levels form a "placental clock" that determines the time of labor onset (McGrath et al. 2002; McLean and Smith 2001). It was also suggested that CRH may be a biomarker for preterm birth (Herrera et al. 2021). However, as indicated above, most extremely preterm births involve intrauterine infection-induced inflammation. Therefore, many of the most concerning cases may

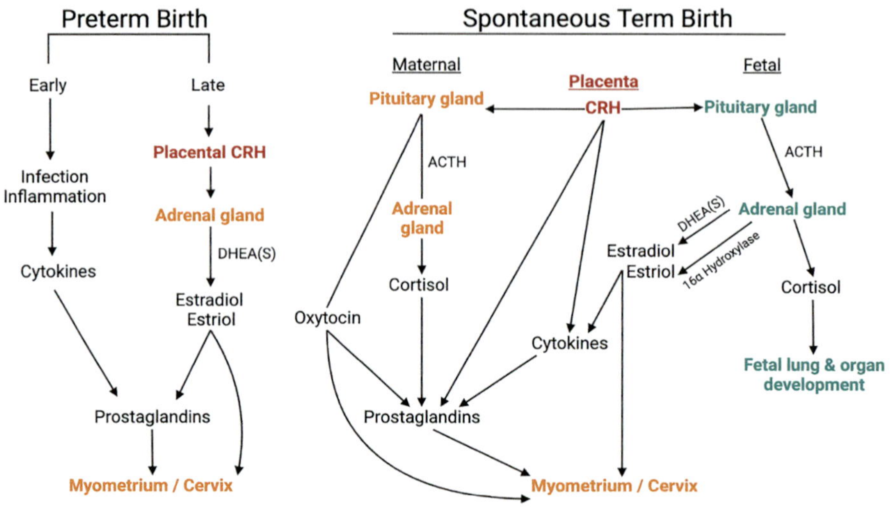

Fig. 5.3 Pathways to human preterm and spontaneous term birth showing the key roles of neuropeptides and the central role of CRH produced by the placenta. *CRH* corticotropin releasing hormone, *ACTH* adrenocorticotropic hormone, *DHEA(S)* dehydroepiandrosterone (sulfate)

not be detected by an early rise in CRH, lessening the value of CRH as a biomarker. The action of placental CRH in triggering labor involves interaction with the maternal and fetal anterior pituitary glands, leading to the release of ACTH into both circulations. During pregnancy, steadily rising CRH and ACTH levels in the fetus contribute to the development of the fetal zone of the fetal adrenal glands (Quinn et al. 2016). This structure, which is located at the junction of the adrenal cortex and medulla, is unique to higher primates and is critical to the production of steroids involved in maintaining human pregnancy and leading to birth (Ishimoto and Jaffe 2011; Ishimoto et al. 2008). The interaction between the fetal adrenal glands and placenta is termed the feto-placental unit. This interaction involves rising ACTH acting on the fetal zone to increase the production of the estrogen precursors, dehydroepiandrosterone (DHEA), and its sulfated conjugate DHEA sulfate (DHEAS) (Smith et al. 2012). These androgen steroids are converted to estradiol and estrone by aromatase in the placenta (Kaludjerovic and Ward 2012). In addition, some DHEA in the fetal circulation is converted to 16α–hydroxy-DHEA by 16α-hydroxylase in the fetal liver, where it then re-enters the fetal circulation and is converted to estriol by the placenta (Kaludjerovic and Ward 2012). Levels of these estrogens are markedly higher than in non-pregnant women, with urinary conjugate levels up to 1000 times higher than in non-pregnancy. The rising placental CRH with advancing gestation leads to increased ACTH secretion from both the fetal and maternal pituitary glands (Herrera et al. 2021). This in turn results in increased adrenal gland androgen output. The resulting increase in levels of estrogens in the maternal circulation toward term have significant actions on the myometrium and cervix. There is increased expression of contraction-associated proteins, including connexins that coordinate and strengthen contractions (Cook et al. 2000). As in non-pregnancy, higher levels of estrogen enhance the expression of estrogen receptors in the intrauterine tissues in a feed-forward manner, further increasing estrogen action. In the myometrium, estrogens also increase the expression of progesterone receptors (PR). This includes both progesterone receptor (PR) PRA and PRB (Zakar and Hertelendy 2007). The increased progesterone action has a suppressing effect on the uterus and reduces contraction-associated protein expression, promoting the continuance of uterine quiescence (Mitchell et al. 2011). As indicated above, a fall in maternal progesterone concentration is required to lift this suppression in many species, including the ungulates. In human pregnancy, a change in progesterone signaling has been suggested to fulfil this action (see below), reducing suppression. This allows increased expression of PTGS-2 and the associated production of prostaglandins, which activate the myometrium, soften the cervix, and weaken the fetal membranes (Hirst et al. 1995; Olson et al. 1993; Menon et al. 2021).

5.5 CRH and Uterine Stimulation

Rising maternal CRH, as well as fetal and maternal ACTH levels, lead to increased cortisol concentrations in the fetal circulation as term approaches. This has a major role in fetal organ development in preparation for extrauterine life and is common to most species (Young et al. 1996). As indicated above, cortisol-induced stimulation of pulmonary epithelial cells supports their production of surfactant which is essential for immediate neonatal life; other organs, including neuronal pathways, are also stimulated (Challis et al. 2001). In human pregnancy, the production of progesterone continues toward labor onset and circulating levels are maintained. Therefore, escape from the suppressive action of progesterone to allow a rise in contraction-associated proteins is required for labor onset (Welsh et al. 2014). This loss of progesterone suppression occurs due to the action of inflammatory mediators in infection-induced preterm labor. These mechanisms have been suggested to involve increased inflammatory regulator activation, which then reduces the influence of progesterone at normal term. This may involve effects on post-receptor pathways, as studies have shown that progesterone receptors are not downregulated and there is little change in the ratio of PRA to PRB expression. The mechanism of reduced progesterone action at term labor onset, in the absence of infection, remains unclear, but is likely also to involve a rise in the activity of inflammatory pathways and unique responses to this inflammation in uterine tissues. In most inflammatory processes, rising cortisol reduces the production of prostaglandin and other related endoperoxides by inhibiting phospholipase A_2 (PLA_2) in the upstream precursor pathway. In contrast, toward onset of labor there are parallel increases in cortisol and prostaglandin levels (Casey et al. 1985). This paradox is even more pronounced in the intrauterine tissues, particularly the human amnion (Sun et al. 2003), where glucocorticoids stimulate rather than reduce the expression of PTGS-2 expression (Zakar et al. 1994) (Fig. 5.3). This effect involves an action of glucocorticoids in the post-receptor cyclic AMP/protein kinase A (cAMP/PKA) pathway, resulting in stimulation of PTGS-2 expression (Guo et al. 2014; Wang et al. 2019; Lu et al.

2019). These and other studies suggest the amnion and decidual responses may be key to beginning the increased production of prostaglandins in the myometrium and hence the initiation process. In addition to cortisol, CRH has been shown to more directly influence amnion-decidual prostaglandin production and CRH can directly stimulate the myometrium by a nuclear factor kappa-B (NFκB)-mediated inflammatory process (Smith et al. 2007; Wang et al. 2020).

The expression of PTGS-2 is controlled by multiple factors, including the inflammatory **cytokines**, interleukin (IL)-1β, IL-6, and tumor necrosis factor-α (TNFα) (Ackerman et al. 2005). These mediators contribute to rising levels of prostaglandins following intra-uterine infection of the fetal membranes with the increase in levels starting in the amnion-decidua and progressing to the myometrium and cervix (Hertelendy and Zakar 2004). Studies show that these effects can be mimicked by intra-amniotic administration of immune stimulators such as lipopolysaccharide (a bacterial endotoxin from the cell wall of gram-negative bacteria) (Migale et al. 2016). This work suggests that the rising expression of PTGS-2 in these tissues following ascending infection of the fetal membranes progresses to elevated prostaglandins in the **chorion** and decidua. This leads finally to the weakening and rupture of the membranes, progression to elevated prostaglandin production in the myometrium, dilation of the cervix, and onset of infection-induced labor. In contrast, in the absence of infection, rising CRH levels toward the end of gestation directly stimulate inflammatory cytokine release (Migale et al. 2016). This further suggests that local synthesis of CRH in the placenta may trigger a rise in inflammatory cytokines and initiate labor in the absence of infection (Karalis et al. 1997). Furthermore, a premature rise in CRH may trigger preterm labor; however, the mechanisms leading to early placental CRH synthesis in the absence of infection are unclear. Collectively, these findings support the proposal that CRH induces the cascade of inflammatory mediators in human intrauterine tissues and, ultimately, the myometrium, leading to the stimulation of the uterus and the remodeling of the fetal membranes and dilation of the cervix (You et al. 2014).

The observation that CRH rises faster and earlier in gestation and long before the onset of preterm labor suggests changes in placental function may have a role in the relatively large range of timing in the initiation of late preterm-term births. Thus, genetic factors and sensitivity to stressors may be involved, increasing susceptibility to at least moderately preterm delivery. This may also account for a previous preterm labor being a major risk factor for preterm birth in subsequent pregnancies. These processes could be responsible for the higher incidences of preterm birth in women compared to many other species (Mitchell and Taggart 2009; Smith et al. 2013). Species in which rising HPA axis activation and cortisol production leads to rapid labor-inducing hormonal changes typically have a much narrower range of gestation length, such that fetal development is optimal for postnatal life. Importantly, we have reported methylation of the PTGS-2 gene at levels that would control transcription in late gestation (Mitchell et al. 2013). The addition of methyl groups to the DNA by **methyltransferases**, particularly in (**CpG**)-rich areas (islands), usually leads to repression of gene expression. These methylation processes may begin early in gestation and contribute to a transcriptionally repressed state of the PTGS-2 gene, reducing prostaglandin production. Differing early experiences or differences in placental development could lead to changes in the degree of PTGS-2 or inflammatory gene methylation and the capacity for transcriptional activation. This may lead to differences in patient susceptibility to early labor. Prenatal stress has been found to reduce methylation by a CRH-CRH receptor-1-mediated mechanism that suppresses DNA methyltransferase-1 (DNMT-1; the enzyme that catalyzes the transfer of methyl groups to CpG islands in the promoters of key genes) (Wang et al. 2014). Inhibition of DMNT-1 reduces the methylation rate and tends to shift the balance toward reduced overall methylation levels, and hence increased gene expression. However, to date, there is little evidence that CRH receptor pathways influence the methylation of PTGS2. Alternatively, we have found that the level of promoter methylation significantly contributes to control of CRH expression in trophoblastic cells (Pan et al. 2015). These observations suggest that changing conditions earlier in pregnancy may alter placental CRH expression with advancing gestation by influencing DNA methylation (Pan et al. 2015). DNA methylation may also contribute to the effects of adverse conditions such as fetal growth restriction and hypoxia, which are associated with an increased incidence of preterm labor. Collectively, these findings suggest that methylation of the PTGS2 and CRH genes may be altered by adverse conditions early in pregnancy and lead to increased susceptibility to preterm labor.

5.6 Oxytocin and Myometrial Activity

Oxytocin is a potent uterine smooth muscle stimulant that increases contractile activity when the uterine myometrium is primed by estrogen exposure, such that adequate levels of oxytocin receptors are present. Oxytocin acts to directly increase calcium entry into myometrial cells and to activate intra-cellular phospholipase C (PLC) pathways (Vannuccini et al. 2016). This leads to increased contraction of the myometrium. The activation of PLC pathways leads to activation of the inositol triphosphate signaling that increases prostaglandin production, which further increases contractile activity of the myometrium (Arthur et al. 2007). The human myometrium expresses relatively low levels of oxytocin receptors until near term, so oxytocin has little effect on contractility until the end of gestation (Ivell et al. 2001). Myometrial oxytocin receptor expression is at least partially controlled by estrogens and progesterone, with estrogen stimulating, and progesterone suppressing expression. As progesterone tends to suppress estrogen receptor expression, some of its effects may be mediated by decreased estrogen action (Arthur et al. 2007). Both estrogen and progesterone concentrations rise throughout human gestation, and (as indicated above) while there is a steady rise in estrogens levels, there is no dramatic decline in progesterone levels prior to labor onset as seen in other species, such as the sheep (Ivell et al. 2001). Oxytocin receptors, therefore, rise without a fall in progesterone levels in women, which indicates that oxytocin receptor expression becomes insensitive to progesterone, in a similar way to other contraction-associated genes. This may be due to the activation of inflammatory pathways, since several components of these pathways, including IL1β, enhance oxytocin receptor expression (Terzidou et al. 2011). The rise in oxytocin once labor has started enhances downstream uterine stimulatory pathways including the increased synthesis of PGE_2 that enhances cervical dilatation and $PGF_{2\alpha}$ that stimulates the myometrium.

5.7 Oxytocin and Maternal Control of Labor Onset

In human pregnancy, the fetal signaling that promotes labor remains undefined, and determining gestation length involves placental changes and inflammatory processes in the placenta and the amnion-decidua (as above) (Mitchell et al. 2008). These changes occur in the days before term and prime the myometrium for delivery. Delivery can occur without oxytocin stimulation when PGTS-2 levels are elevated and there is enhanced prostaglandin production. Maternal oxytocin signaling seems to be a fine-tuning mechanism for the timing of birth (Hirst et al. 1993). Accordingly, species that normally deliver at night rely on maternal oxytocin to prompt the final initiation process, once other activators are close to the tipping point. This triggering event involves oxytocin output that is increased by photoperiod leading to raised uterine activity at night (Hirst et al. 1993). The oxytocin-driven trigger very tightly controls the time of labor in some species, such as the alpaca and related South American camelids (Bravo 1994). These species generally live at high altitudes and routinely deliver in the early morning hours so that the offspring has most of the day to become mobile prior to the next cold night. Non-human primates and women also start labor with a greater frequency at night (Hirst et al. 1993). When carefully measured in primates, small increases in oxytocin-induced contractions occur in the preceding nights, but this is insufficient to initiate labor. Indeed in non-human primates, nocturnal oxytocin rises during the 3–4 nights prior to night-time labor onset (Hirst et al. 1993). These rises are associated with a minor increase in myometrial activity increasing on each preceding night until labor is triggered. Further priming results in the attainment of sufficient upregulation of the contractile mechanisms to lead to labor during a subsequent night. This may involve a combination of oxytocin-stimulated PTGS2 expression and rising myometrial activity, further enhancing prostaglandin production and the final and rapid induction of contractions (Fig. 5.3).

5.8 Oxytocin in Labor: Clinical Aspects

Some of the prelabor myometrial activity, referred to as Braxton-Hicks contractions, may be driven by oxytocin, with the stimulation of contractions sometimes called *practice labor*. Stress exposure around this time may lead to suppression of oxytocin release and a reduction in uterine activity (La Verde et al. 2022), with rises in oxytocin again triggering contractile activity, which progresses to labor during subsequent nights. This potential triggering in the period before labor is different from the reflex release of oxytocin following cervical dilation during the second and third stages of labor. The latter is termed the **Ferguson reflex** and involves neural signals following cervical stretch, which pass through spinal pathways and the brainstem and then the medial forebrain bundle, to increase oxytocin release from the maternal posterior pituitary gland (Komisaruk and Sansone 2003). This reflex is important in delivering the placenta and reducing the potential for postpartum bleeding, although in humans it is not essential for a successful delivery (Erickson et al. 2022). However, it may be important to reduce the duration of the period of fetal expulsion, given that this is a time when umbilical cord problems such as nuchal cord (the umbilical cord becomes wrapped around the baby's neck), various kinks, and cord prolapse (when the umbilical cord slips through the cervix ahead of the baby after the membranes have ruptured) may lead to cord compression, limiting blood flow to the fetus. The inhibition of oxytocin secretion may increase this period of delivery and result in fetal hypoxia. Importantly, as stretch-driven impulses pass through the lumbar spinal cord, neural conduction may be influenced by epidural analgesia (Flint et al. 1978). This may diminish this reflex release of oxytocin and, consequently, extend this phase of labor. This may be countered by reduced stress with epidural analgesia and hence stress-induced inhibition of oxytocin release. Reduced levels of oxytocin release may suggest that augmentation of contractions with oxytocin infusions could therefore be seen as a replacement treatment in this context.

5.9 Oxytocin for the Augmentation of Uterine Activity

The use of oxytocin infusions (Pitocin, Syntocinon, and other trade names—synthetically produced oxytocin) for inducing labor or augmenting contractile activity is well-established in clinical practice. There have been numerous trials of approaches and adjunct treatments, and actual treatment combinations vary between centers. The success of this treatment requires adequate uterine oxytocin receptor expression, which increases as term approaches and as cervical dilation occurs (Erickson et al. 2022). Oxytocin raises prostaglandin production and further contributes to uterine activity. This may not be sufficient to cause the necessary cervical dilation, at least without protracted uterine contractions. Therefore, most protocols for induction of labor involve the prior administration of PGE_2 (**Dinoprostone**), called "pre-induction." This is often administered as a pessary 12–24 h prior to induction with oxytocin. In some centers, prostaglandin alone is used for induction and oxytocin is employed for the augmentation of established labor. Oxytocin has the advantage of a short half-life when infused and so levels decline rapidly once the infusion is stopped (Son et al. 2023). Furthermore, oxytocin antagonists that can be injected intravenously are available to reduce contractile activity. This is important because the perfusion of the placenta is diminished during sustained contractions, which may cause marked fetal hypoxia. Synthetic oxytocin preparations are routinely used after Caesarean delivery to contract the myometrium to aid in placenta expulsion. Oxytocin-induced contractility also reduces bleeding and the risk of postpartum hemorrhage (Erickson et al. 2023). Postpartum hemorrhage contributes significantly to maternal mortality after delivery in remote locations in developing countries. The use of the more heat-stable oxytocin analogue, Carbetocin, to reduce bleeding in these settings requires further investigation.

5.10 Oxytocin Antagonists and Preterm Labor

The peptide oxytocin receptor antagonist (**Atosiban**) has been extensively examined in trials for reducing preterm labor contractile activity. Neither oxytocin nor related peptide antagonists cross the placenta significantly, and oxytocin augmentation and antagonist use are safe regarding direct fetal effects. Atosiban has been shown to have little or no benefit in delaying preterm delivery in a number of studies in women (Flenady et al. 2014). As indicated above, most early and extremely early preterm births, which have the greatest risk for neonatal morbidity and mortality, are triggered and maintained by intrauterine inflammation following ascending infection. Uterine oxytocin receptor expression is relatively low at this time, and elevated oxytocin levels are not involved in the infection-induced contractile progression. Therefore, antagonism with the low concentration of oxytocin receptors does not reduce the risk of delivery. In the late preterm period, oxytocin receptor expression may be higher and sufficient for oxytocin stimulation to contribute to uterine activity. However, preterm labor at these near-term gestational ages poses little risk to neonatal outcome, compared to allowing delivery, and therefore the use of oxytocin antagonists is not warranted. Since the development of peptide antagonists, a number of non-peptide antagonists (**Retosiban** and **Barusiban**) have been marketed for use to suppress preterm myometrial contractions (Helmer et al. 2020). While these antagonists may cross the placenta to a greater extent than Atosiban, their transfer is still slow (Helmer et al. 2020). These molecules, like the modified peptide antagonists, have little effect in reducing preterm myometrial activity, except close to normal term, as this activity is not driven by oxytocin. There is controversy surrounding the use of oral non-peptide antagonists for increasing embryo implantation in IVF procedures. This has been suggested to involve suppression of uterine activity after transfer; however, the results of trials have not been uniformly supportive of the treatment.

5.11 Oxytocin Receptors in the Myometrium During Pregnancy and at Term

Oxytocin receptors belong to the rhodopsin-type of G-protein coupled superfamily. Receptor levels are regulated by changes in expression and desensitization by oxytocin binding. Activation of oxytocin receptors in the myometrial cells leads to G-protein-mediated activation of voltage-activated Ca^{2+} channels and the influx of calcium ions. As indicated above, the activation of PLC and inositol triphosphate (IP3) formation leads to the release of calcium from intracellular stores, most importantly the sarcoplasmic reticulum, initiating uterine contractions (Ivell et al. 2001). Intrauterine oxytocin receptors are located mostly in the rat and human myometrial muscle; however, receptors are also present in other intrauterine tissues, including the amnion and the decidua. Oxytocin-mediated activation of receptors in these tissues leads to increased PTGS-2 activity, and prostaglandin (PGE_2 and $PGF_{2\alpha}$) release (Mitchell et al. 2008). This assists in driving myometrial contractions at term labor onset. Moreover, the rise in prostaglandin production is important, as oxytocin has a relatively weak action on the cervix, with dilation reliant on prostaglandin action. The lower level of oxytocin receptors in cervical tissue limits local prostaglandin production and is insufficient for successful delivery. Therefore, oxytocin induction is usually done after cervical dilation, occurring spontaneously or induced by local administration of PGE_2 (see above). In ungulates, estrogen dominance toward term increases the expression of a number of contraction-associated genes, including the oxytocin receptor, in preparation for birth (Arthur et al. 2007). These changes lead to an increase in the expression of oxytocin receptors prior to labor. Coordinated receptor expression and oxytocin availability allow the exact timing of labor in these species. Protracted labor with continued oxytocin release may lead to downregulation of oxytocin receptors. This may lessen the action of oxytocin on postpartum bleeding and contribute to postpartum hemorrhage, in particular in remote and developing world settings.

Relaxin and oxytocin have been suggested as an approach for augmentation of labor. Relaxin is a polypeptide hormone that has a structure similar to insulin-like growth factor (IGF) 1 and 2 and has growth-promoting actions in pregnancy. Relaxin has been found to increase metalloprotease levels in the cervical tissue (Goldsmith and Weiss 2009). There is support for a role of relaxin in labor in species such as the guinea pig, where the pubic ligament stretches markedly to allow delivery of the fetus, however the role of relaxin in human labor and dilation of the cervix has been found to be limited (Weiss 2013). The largest source of relaxin is the corpus luteum, which also maintains progesterone production for the first ~50 days of human pregnancy. The placenta takes over progesterone production (see above) after this time, but the corpus luteum of pregnancy remains and continues relaxin release. Pregnant women in whom the corpus luteum is absent, such as after embryo transfer, have markedly lower plasma relaxin levels but can go through labor and have a normal level of cervical dilation. The possible effects of the absence of the growth influencing roles of relaxin in uterine tissues are unclear.

5.12 Oxytocin in the Fetus and Newborn

There is minimal passage of oxytocin across the placenta from the maternal circulation to the fetus in the second half of gestation until term. This means that fetal and maternal oxytocin secretion and circulating levels are independent and levels in the fetus must be separately maintained by neurosecretion (Hirst et al. 1993). Oxytocin can control overall excitability in the late gestation fetal brain, which involves modulating effects on GABAergic action (Leonzino et al. 2016). The effect of stimulation of the $GABA_A$ receptor by GABA is controlled by the intracellular chloride concentration. This, in turn, is regulated by the concentration of sodium-chloride cotransporter-2 (NaCC2), which increases intracellular chloride and the potassium-chloride cotransporter-2 (KCC2) in the neuronal membrane, which decreases intracellular chloride levels. In early gestation, the low expression of KCC2 results in higher cellular chloride concentrations, so the opening of the $GABA_A$ receptor chloride channel decreases chloride levels and tends to depolarize the neuron, increasing excitability (Tyzio et al. 2008). This excitation in the early stages of pregnancy is important for neurodevelopment. In long gestation species, including humans, sheep, and guinea pigs, the rising expression of KCC2 with advancing gestation leads to changes in intracellular chloride levels that result in a switch in the response to $GABA_A$ receptor stimulation from excitatory to inhibitory (Coleman et al. 2013; Ben-Ari et al. 2012). Oxytocin receptor stimulation has been shown to have a role in regulating KCC2 expression and the onset of this switch (Leonzino et al. 2016). This suggests that in long-gestation species, maturation of hypothalamic oxytocin secretory neurons may have a role in controlling the switch to GABAergic stimulation becoming inhibitory (Khazipov et al. 2008). The production of neurosteroids, particularly allopregnanolone, by the placenta, combined with enhanced GABA action on $GABA_A$ receptors, may raise inhibition and protection from excitotoxicity (Hirst et al. 2009). Studies with oxytocin knockout animals suggest that oxytocin is needed to stimulate KCC2 expression and achieve the correct timing of the switch (Leonzino et al. 2016). In long gestation species, increases in GABAergic inhibition during the second half of gestation may be key to protecting the brain of the relatively advanced fetus from excitatory damage following episodes of hypoxia of asphyxia at birth (Bertoni et al. 2021). In rats, the rising levels of maternal oxytocin at the time of delivery have been suggested to transiently increase the inhibitory action of GABA during delivery and provide protection from excitotoxicity at birth (Tyzio et al. 2007). However, there is little information to indicate whether this mechanism applies in human pregnancy. Early preterm delivery, when maternal and possibly fetal oxytocin levels are lower than at term, may reduce the inhibitory action of $GABA_A$ receptor simulation, leaving the preterm neonate vulnerable to excessive excitation. However, further studies are required to determine the potential benefit of oxytocin therapy after preterm birth.

5.13 Conclusions

Neuropeptides play a major role in maintaining pregnancy and in the timing of birth. CRH and oxytocin are crucial in fetal and maternal signaling that leads to delivery at the most appropriate time for newborn survival. These mechanisms ensure that the fetus reaches an optimal developmental maturity to meet the challenges of extra-uterine life (Fig. 5.1). In most species, particularly the ungulates, the timing of labor onset is tightly controlled so that the newborn is well developed at birth and capable of near-independent life, essential for survival of the newborn. This is largely achieved by matching the development of the fetal HPA axis, which promotes development of fetal organ systems, with the labor initiation process. In the sheep, the model organism that has been most extensively studied, the maturation of input pathways to the CRH neurons in the fetal hypothalamus and the development of these neurons leads to increased CRH secretion. This activates the fetal HPA axis with a resultant rise in cortisol that drives key hormonal changes in the placenta, specifically the conversion of progesterone to estrogens. This frees the uterine tissues of the suppressive effects of progesterone on contraction-associated processes, and simultaneously, rising estrogens further stimulate the expression of these proteins. At the same time, the increasing fetal cortisol level leads to stimulation of the final stages of organ maturation.

The realization of the critical importance of rising cortisol in fetal maturation and labor onset in the sheep led to the development of corticosteroid therapies for hastening fetal lung development in human pregnancy around the onset of preterm labor. Corticosteroid treatment resulted in a tremendous improvement in neonatal survival, but also highlighted major differences in the mechanism of term and preterm labor in women.

In women and higher primates, the timing of term labor onset is unique, with a "placental clock" taking on the role of the HPA axis. The placental syncytiotrophoblast secretes CRH, with growth and development of the placenta determining CRH output and ultimately labor onset and gestation length (Fig. 5.3). The tying of placental development with the timing of labor may have evolved because of the importance of adequate support for maintaining late gestation brain development in higher primates. Rising CRH output leads to increased ACTH levels and increases the release of androgens from the fetal adrenal glands. These androgens are converted to estrogens, which stimulate the uterine changes that trigger labor in women. Cortisol levels also rise, supporting essential fetal organ growth. Premature increases in CRH can trigger moderately preterm labor, which may involve epigenetic mechanisms. However, prematurely elevated CRH is less important in early and extremely early preterm birth in women. The majority of these cases involve intrauterine infection and activation of inflammatory mediator-driven pathways. Thus, early preterm labor is likely to result from mechanisms that are quite different from those that lead to spontaneous delivery at term. Interestingly, increased activation of inflammatory pathways can trigger these preterm birth births in the absence of infection, in some cases with the initiation processes unclear.

Oxytocin is an important maternal trigger of the final stage of term labor onset, performing a fine-tuning role. The action of oxytocin requires a priming process in the myometrium so that the expression of oxytocin receptors is sufficient. Increased night-time oxytocin release leads to increased uterine contractile activity near term and is responsible for night-time delivery in many species, with labor initiated more often at night in women. Oxytocin release in response to cervical stretch contributes to more forceful contractions at the end of the second stage of labor and particularly aids in delivery of the placenta in the third stage of labor. This rise in oxytocin levels is important in stopping uterine blood flow and reducing the risk of postpartum hemorrhage. Oxytocin has important roles in obstetric practice and synthetic oxytocin is used clinically for augmenting myometrial activity and shortening labor, as well as aiding in placental separation and reducing postpartum hemorrhage (Erickson et al. 2023). Oxytocin has a limited role in the mechanism leading to preterm labor, and therefore antagonists have only marginal effectiveness in suppressing preterm myometrial activity (Erickson et al. 2022).

Perspectives

Preterm labor remains a major problem in obstetrics and continues to result in high newborn morbidity and mortality. Further studies are required to develop preventative therapies that are safe for the fetus, delay delivery, and reduce the mortality and morbidity associated with prematurity. The findings described in this chapter highlight the important differences between the mechanisms of human term and preterm labor. Future investigations into the effects of risk factors such as reduced fetal growth on epigenetic mechanisms controlling the expression of CRH and PTGS-2 and their receptors may provide targets that could be used to block non-infection related preterm birth. These studies may also explain the mechanism of increased risk for subsequent pregnancies and allow these risks to be prevented. Further research examining early markers of infection will allow earlier detection of infection-induced preterm labor. Moreover, studies of early antibiotic intervention may prevent infection-induced preterm labor while protecting the fetus (Kelleher et al. 2020). This work has the potential to markedly reduce brain injury following preterm delivery and improve newborn outcome and long-term health.

Acknowledgments JJH, HKP, and JCS are supported by the National Health and Medical Research Council of Australia and receive additional support from the Hunter Medical Research Institute (HMRI).

Key References (see main list for reference details)

Herrera, C.L., et al., (2021). This paper describes the most recent information on the role of rising CRH in the initiation of term birth and the "placental clock." The studies outline some of the problems with CRH measurement and give validated values for CRH changes during pregnancy and at term.

Mitchell, B.F., et al., (2013). The paper describes work outlining the key similarities and differences between animal models of term and preterm labor and the control of labor onset in women.

Mitchell, C.M., (2014). This study demonstrates the key role of DNA methylation in the expression of key genes involved in the shifting of uterine quiescence to an activated state. The work shows that differences in methylation may control the sensitivity of these genes to labor-triggering stimulation.

Vannuccini, S., et al., (2016). This paper provides a very good review of the pathways that lead to the onset of human labor, including the role of contraction-associated proteins and prostaglandins in the regulation of myometrial activity and in the breakdown of cervical connective tissue and cervical dilation.

Viscardi, R. M., et al., (2010). This study examined the role of intrauterine infection in preterm labor. The work particularly demonstrated the role of the relatively difficult to detect ureaplasma species in early preterm birth.

References

Ackerman WE, Zhang XL, Rovin BH, Kniss DA (2005) Modulation of cytokine-induced cyclo-oxygenase 2 expression by PPARG ligands through NFkappaB signal disruption in human WISH and amnion cells. Biol Reprod 73:527–535

Adams Waldorf KM, Gravett MG, McAdams RM, Paolella LJ, Gough GM, Carl DJ, Bansal A, Liggitt HD, Kapur RP, Reitz FB, Rubens CE (2011) Choriodecidual group B streptococcal inoculation induces fetal lung injury without intra-amniotic infection and preterm labor in Macaca nemestrina. PLoS One 6:e28972

Arthur P, Taggart MJ, Mitchell BF (2007) Oxytocin and parturition: a role for increased myometrial calcium and calcium sensitization? Front Biosci 12:619–633

Ben-Ari Y, Khalilov I, Kahle KT, Cherubini E (2012) The GABA excitatory/inhibitory shift in brain maturation and neurological disorders. Neuroscientist 18:467–486

Bertoni A, Schaller F, Tyzio R, Gaillard S, Santini F, Xolin M, Diabira D, Vaidyanathan R, Matarazzo V, Medina I, Hammock E, Zhang J, Chini B, Gaiarsa JL, Muscatelli F (2021) Oxytocin administration in neonates shapes hippocampal circuitry and restores social behavior in a mouse model of autism. Mol Psychiatry 26:7582–7595

Bravo PW (1994) Reproductive endocrinology of llamas and alpacas. Vet Clin North Am Food Anim Pract 10:265–279

Casey ML, MacDonald PC, Mitchell MD (1985) Despite a massive increase in cortisol secretion in women during parturition, there is an equally massive increase in prostaglandin synthesis. A paradox? J Clin Invest 75:1852–1857

Challis JR, Sloboda D, Matthews SG, Holloway A, Alfaidy N, Patel FA, Whittle W, Fraser M, Moss TJ, Newnham J (2001) The fetal placental hypothalamic-pituitary-adrenal (HPA) axis, parturition and post natal health. Mol Cell Endocrinol 185:135–144

Challis JRG, Brooks AN (1989) Maturation and activation of hypothalamic-pituitary-adrenal function in fetal sheep. Endocr Rev 10:182–203

Chan EC, Brinsmead MW, Chen SE, Nanra R, Simm B, McLean M, Smith R (1990) Urinary corticotropin-releasing hormone immunoreactivity is elevated during human pregnancy. Gynecol Endocrinol 4:233–244

Coleman H, Hirst JJ, Parkington HC (2013) The GABAA excitation-to-inhibition switch in the hippocampus of the perinatal guinea pig. Fetal Neonatal Physiol Soc Proc 40

Cook JL, Zaragoza DB, Sung DH, Olson DM (2000) Expression of myometrial activation and stimulation genes in a mouse model of preterm labor: myometrial activation, stimulation, and preterm labor. Endocrinology 141:1718–1728

Erickson EN, Krol KM, Perkeybile AM, Connelly JJ, Myatt L (2022) Oxytocin receptor single nucleotide polymorphism predicts atony-related postpartum hemorrhage. BMC Pregnancy Childbirth 22:884

Erickson EN, Myatt L, Danoff JS, Krol KM, Connelly JJ (2023) Oxytocin receptor DNA methylation is associated with exogenous oxytocin needs during parturition and postpartum hemorrhage. Commun Med (Lond) 3:11

Flenady V, Reinebrant HE, Liley HG, Tambimuttu EG, Papatsonis DN (2014) Oxytocin receptor antagonists for inhibiting preterm labour. Cochrane Database Syst Rev CD004452

Flint AP, Forsling ML, Mitchell MD (1978) Blockade of the Ferguson reflex by lumbar epidural anaesthesia in the parturient sheep: effects on oxytocin secretion and uterine venous prostaglandin F levels. Horm Metab Res 10:545–547

Goldsmith LT, Weiss G (2009) Relaxin in human pregnancy. Ann N Y Acad Sci 1160:130–135

Gravett MG, Haluska GJ, Cook MJ, Novy MJ (1996) Fetal and maternal endocrine responses to experimental intrauterine infection in rhesus monkeys. Am J Obstet Gynecol 174:1725–1731; discussion 31-3

Guo C, Wang W, Liu C, Myatt L, Sun K (2014) Induction of PGF2alpha synthesis by cortisol through GR dependent induction of CBR1 in human amnion fibroblasts. Endocrinology 155:3017–3024

Hassan HE, Othman AA, Eddington ND, Duffy L, Xiao L, Waites KB, Kaufman DA, Fairchild KD, Terrin ML, Viscardi RM (2011) Pharmacokinetics, safety, and biologic effects of azithromycin in extremely preterm infants at risk for ureaplasma colonization and bronchopulmonary dysplasia. J Clin Pharmacol 51:1264–1275

Helmer H, Saleh L, Petricevic L, Knofler M, Reinheimer TM (2020) Barusiban, a selective oxytocin receptor antagonist: placental transfer in rabbit, monkey, and humandagger. Biol Reprod 103:135–143

Herrera CL, Bowman ME, McIntire DD, Nelson DB, Smith R (2021) Revisiting the placental clock: early corticotrophin-releasing hormone rise in recurrent preterm birth. PLoS One 16:e0257422

Hertelendy F, Zakar T (2004) Prostaglandins and the myometrium and cervix. Prostaglandins Leukot Essent Fatty Acids 70:207–222

Hirst JJ, Haluska GJ, Cook MJ, Novy MJ (1993) Plasma oxytocin and nocturnal uterine activity: maternal but not fetal concentrations increase progressively during late pregnancy and delivery in rhesus monkeys. Am J Obstet Gynecol 169:415–422

Hirst JJ, Kelleher MA, Walker DW, Palliser HK (2014) Neuroactive steroids in pregnancy: key regulatory and protective roles in the foetal brain. J Steroid Biochem Mol Biol 139:144–153

Hirst JJ, Palliser HK, Shaw JC, Crombie G, Walker DW, Zakar T (2018) Birth and neonatal transition in the guinea pig: experimental approaches to prevent preterm birth and protect the premature fetus. Front Physiol 9:1802

Hirst JJ, Teixeira FJ, Zakar T, Olson DM (1995) Prostaglandin H synthase-2 expression increases in human gestational tissues with spontaneous labour onset. Reprod Fertil Dev 7:633–637

Hirst JJ, Walker DW, Yawno T, Palliser HK (2009) Stress in pregnancy: a role for neuroactive steroids in protecting the fetal and neonatal brain. Dev Neurosci 31:363–377

Ishimoto H, Jaffe RB (2011) Development and function of the human fetal adrenal cortex: a key component in the feto-placental unit. Endocr Rev 32:317–355

Ishimoto H, Minegishi K, Higuchi T, Furuya M, Asai S, Kim SH, Tanaka M, Yoshimura Y, Jaffe RB (2008) The periphery of the human fetal adrenal gland is a site of angiogenesis: zonal differential expression and regulation of angiogenic factors. J Clin Endocrinol Metab 93:2402–2408

Ivell R, Kimura T, Muller D, Augustin K, Abend N, Bathgate R, Telgmann R, Balvers M, Tillmann G, Fuchs AR (2001) The structure and regulation of the oxytocin receptor. Exp Physiol 86:289–296

Kaludjerovic J, Ward WE (2012) The interplay between estrogen and fetal adrenal cortex. J Nutr Metab 2012:837901

Karalis K, Muglia LJ, Bae D, Hilderbrand H, Majzoub JA (1997) CRH and the immune system. J Neuroimmunol 72:131–136

Kelleher MA, Lee JY, Roberts VHJ, Novak CM, Baschat AA, Morgan TK, Novy MJ, Rasanen JP, Frias AE, Burd I (2020) Maternal azithromycin therapy for Ureaplasma parvum intraamniotic infection improves fetal hemodynamics in a nonhuman primate model. Am J Obstet Gynecol 223:578 e1–78 e11

Khazipov R, Tyzio R, Ben-Ari Y (2008) Effects of oxytocin on GABA signalling in the foetal brain during delivery. Progress Brain Res 170:243–257

Komisaruk BR, Sansone G (2003) Neural pathways mediating vaginal function: the vagus nerves and spinal cord oxytocin. Scand J Psychol 44:241–250

La Verde M, Riemma G, Torella M, Torre C, Cianci S, Conte A, Capristo C, Morlando M, Colacurci N, De Franciscis P (2022) Impact of Braxton-Hicks contractions on fetal wellbeing; a prospective analysis through computerised cardiotocography. J Obstet Gynaecol 42:569–573

Leonzino M, Busnelli M, Antonucci F, Verderio C, Mazzanti M, Chini B (2016) The timing of the excitatory-to-inhibitory GABA switch is regulated by the oxytocin receptor via KCC2. Cell Rep 15:96–103

Leung TN, Chung TK, Madsen G, McLean M, Chang AM, Smith R (1999) Elevated mid-trimester maternal corticotrophin-releasing hormone levels in pregnancies that delivered before 34 weeks. Br J Obstet Gynaecol 106:1041–1046

Liggins GC (1969) Premature delivery of foetal lambs infused with glucocorticoids. J Endocrinol 45:515–523

Lu Y, Zhou Q, Lu JW, Wang WS, Sun K (2019) Involvement of STAT3 in the synergistic induction of 11beta-HSD1 by SAA1 and cortisol in human amnion fibroblasts. Am J Reprod Immunol 82:e13150

Mark PJ, Smith JT, Waddell BJ (2006) Placental and fetal growth retardation following partial progesterone withdrawal in rat pregnancy. Placenta 27:208–214

McGrath S, McLean M, Smith D, Bisits A, Giles W, Smith R (2002) Maternal plasma corticotropin-releasing hormone trajectories vary depending on the cause of preterm delivery. Am J Obstet Gynecol 186:257–260

McLean M, Bisits A, Davies J, Woods R, Lowry P, Smith R (1995) A placental clock controlling the length of human pregnancy. Nat Med 1:460–463

McLean M, Smith R (2001) Corticotrophin-releasing hormone and human parturition. Reproduction 121:493–501

Menon R, Lappas M, Zakar T (2021) Editorial: The role of the fetal membranes in pregnancy and birth. Front Physiol 12:653084

Migale R, MacIntyre DA, Cacciatore S, Lee YS, Hagberg H, Herbert BR, Johnson MR, Peebles D, Waddington SN, Bennett PR (2016) Modeling hormonal and inflammatory contributions to preterm and term labor using uterine temporal transcriptomics. BMC Med 14:86

Mitchell BF, Taggart MJ (2009) Are animal models relevant to key aspects of human parturition? Am J Physiol Regul Integr Comp Physiol 297:R525–R545

Mitchell C, Johnson R, Bisits A, Hirst J, Zakar T (2011) PTGS2 (prostaglandin endoperoxide synthase-2) expression in term human amnion in vivo involves rapid mRNA turnover, polymerase-II 5'-pausing, and glucocorticoid transrepression. Endocrinology 152:2113–2122

Mitchell CM, Johnson RF, Giles WB, Zakar T (2008) Prostaglandin H synthase-2 gene regulation in the amnion at labour: histone acetylation and nuclear factor kappa B binding to the promoter in vivo. Mol Hum Reprod 14:53–59

Mitchell CM, Sykes SD, Pan X, Pringle KG, Lumbers ER, Hirst JJ, Zakar T (2013) Inflammatory and steroid receptor gene methylation in the human amnion and decidua. J Mol Endocrinol 50:267–277

Olson DM, Mijovic JE, Zaragoza DB, Cook JL (2001) Prostaglandin endoperoxide H synthase type 1 and type 2 messenger ribonucleic acid in human fetal tissues throughout gestation and in the newborn infant. Am J Obstet Gynecol 184:169–174

Olson DM, Skinner K, Challis JR (1983) Estradiol-17 beta and 2-hydroxyestradiol-17 beta-induced differential production of prostaglandins by cells dispersed from human intrauterine tissues at parturition. Prostaglandins 25:639–651

Olson DM, Zakar T, Teixeira FJ, Hirst JJ, Guo F, Pankovich J (1993) Prostaglandins and parturition. In Morniex et al (ed) Progress in endocrinology 7:644–648

Pan X, Bowman M, Scott RJ, Fitter J, Nicholson RC, Smith R, Zakar T (2015) Methylation of the corticotropin releasing hormone gene promoter in BeWo cells: relationship to gene activity. Int J Endocrinol 2015:861302

Phung J, Wang C, Reeders J, Zakar T, Paul JW, Tyagi S, Pennell CE, Smith R (2022) Preterm labor with and without chorioamnionitis is associated with activation of myometrial inflammatory networks: a comprehensive transcriptomic analysis. Am J Obstet Gynecol

Quinn TA, Ratnayake U, Dickinson H, Castillo-Melendez M, Walker DW (2016) The feto-placental unit, and potential roles of dehydroepiandrosterone (DHEA) in prenatal and postnatal brain development: a re-examination using the spiny mouse. J Steroid Biochem Mol Biol 160:204–213

Schwartz J, McMillen IC (2001) Fetal hypothalamus-pituitary-adrenal axis on the road to parturition. Clin Exp Pharmacol Physiol 28:108–112

Smith R, Paul J, Maiti K, Tolosa J, Madsen G (2012) Recent advances in understanding the endocrinology of human birth. Trends Endocrinol Metab 23:516–523

Smith R, Zakar T, Madsen G (2013) Mammalian labor: variations on a theme by amniota. Endocrinology 154:584–588

Smith R, Van Helden D, Hirst J, Zakar T, Read M, Chan E-C, Palliser H, Grammatopoulos D, Nicholson R, Parkington HC (2007) Pathological interactions with the timing of birth and uterine activation. Austral New Zeal J Obstet Gynaecol 47:430–437

Son M, Roy A, Grobman WA, Miller ES, Dude A, Peaceman AM, Stetson B (2023) Maximum dose rate of intrapartum oxytocin infusion and associated obstetric and perinatal outcomes. Obstet Gynecol 141:379–386

Sun K, Ma R, Cui X, Campos B, Webster R, Brockman D, Myatt L (2003) Glucocorticoids induce cytosolic phospholipase A2 and prostaglandin H synthase type 2 but not microsomal prostaglandin E synthase (PGES) and cytosolic PGES expression in cultured primary human amnion cells. J Clin Endocrinol Metab 88:5564–5571

Terzidou V, Blanks AM, Kim SH, Thornton S, Bennett PR (2011) Labor and inflammation increase the expression of oxytocin receptor in human amnion. Biol Reprod 84:546–552

Thorburn GD, Challis JR (1979) Endocrine control of parturition. Physiol Rev 59:863–918

Tyzio R, Holmes GL, Ben-Ari Y, Khazipov R (2007) Timing of the developmental switch in GABA(A) mediated signaling from excitation to inhibition in CA3 rat hippocampus using gramicidin perforated patch and extracellular recordings. [Erratum appears in Epilepsia. 2007 Dec;48(12):2380]. Epilepsia 48(Suppl 5):96–105

Tyzio R, Minlebaev M, Rheims S, Ivanov A, Jorquera I, Holmes GL, Zilberter Y, Ben-Ari Y, Khazipov R (2008) Postnatal changes in somatic gamma-aminobutyric acid signalling in the rat hippocampus. Eur J Neurosci 27:2515–2528

Vannuccini S, Bocchi C, Severi FM, Challis JR, Petraglia F (2016) Endocrinology of human parturition. Annales d Endocrinologie 77:105–113

Viscardi RM (2010) Ureaplasma species: role in diseases of prematurity. Clin Perinatol 37:393–409

Wang A, Nie W, Li H, Hou Y, Yu Z, Fan Q, Sun R (2014) Epigenetic upregulation of corticotrophin-releasing hormone mediates postnatal maternal separation-induced memory deficiency. PLoS One 9:e94394

Wang LY, Wang WS, Wang YW, Lu JW, Lu Y, Zhang CY, Li WJ, Sun K, Ying H (2019) Drastic induction of MMP-7 by cortisol in the human amnion: implications for membrane rupture at parturition. Faseb J 33:2770–2781

Wang WS, Guo CM, Sun K (2020) Cortisol regeneration in the fetal membranes, a coincidental or requisite event in human parturition? Front Physiol 11:462

Weiss G (2013) Relaxin and the control of primate parturition. Ital J Anatomy Embryol 118:17–18

Welsh TN, Hirst JJ, Palliser H, Zakar T (2014) Progesterone receptor expression declines in the guinea pig uterus during functional progesterone withdrawal and in response to prostaglandins. PLoS One 9:e105253

Whittle WL, Patel FA, Alfaidy N, Holloway AC, Fraser M, Gyomorey S, Lye SJ, Gibb W, Challis JR (2001) Glucocorticoid regulation of human and ovine parturition: the relationship between fetal hypothalamic-pituitary-adrenal axis activation and intrauterine prostaglandin production. Biol Reprod 64:1019–1032

You X, Liu J, Xu C, Liu W, Zhu X, Li Y, Sun Q, Gu H, Ni X (2014) Corticotropin-releasing hormone (CRH) promotes inflammation in human pregnant myometrium: the evidence of CRH initiating parturition? J Clin Endocrinol Metab 99:E199–E208

Young IR, Deayton JM, Hollingworth SA, Thorburn GD (1996) Continuous intrafetal infusion of prostaglandin E2 prematurely activates the hypothalamo-pituitary-adrenal axis and induces parturition in sheep. Endocrinology 137:2424–2431

Zakar T, Mitchell BF (1996) The endocrinology of late pregnancy and parturition. In: Advances in organ biology. JAI Press

Zakar T, Hertelendy F (2007) Progesterone withdrawal: key to parturition. Am J Obstet Gynecol 196:289–296

Zakar T, Teixeira FJ, Hirst JJ, Guo F, MacLeod EA, Olson DM (1994) Regulation of prostaglandin endoperoxide H synthase by glucocorticoids and activators of protein kinase C in the human amnion. J Reprod Fertil 100:43–50

Oxytocin in Lactation and Parturition

6

Gareth Leng

Abstract

This chapter outlines the physiology of the oxytocin system as relevant to our understanding of the milk-ejection reflex, and of the control of parturition, as developed through experimental studies in many animal species. It is clear that oxytocin release is essential for milk let-down in lactation, and there has been substantial progress in understanding the milk-ejection reflex. However, considerable uncertainties remain in our understanding of the physiological role of oxytocin in parturition, and technical difficulties in the measurement of oxytocin in plasma are a significant impediment to progress.

Keywords

Milk-ejection · Supraoptic nucleus · Electrophysiology · Computational modelling · Pulsatility

6.1 Introduction

The first indication that a factor from the pituitary gland might be involved in parturition came in 1906 when Henry Dale observed that an extract of the pituitary could induce the uterus of an anesthetized cat to contract (Dale 1906). At first, Dale thought that this was a by-product of the known effects of pituitary extracts on blood pressure, but he changed his mind in response to evidence that the effects on blood pressure and uterine contractions involved two different principles. This stimulated

G. Leng (✉)
The University of Edinburgh, Edinburgh, UK
e-mail: Gareth.leng@ed.ac.uk

P. J. Brunton, D. R. Grattan (eds.), *Neuroendocrine Regulation of Mammalian Pregnancy and Lactation*, Masterclass in Neuroendocrinology 15,
https://doi.org/10.1007/978-3-031-51138-7_6

attempts to separate them, and, in 1928, Kamm and colleagues produced two extracts differentially enriched in these principles, which they named 'vasopressin' and 'oxytocin' (see Leng and Leng 2021).

In giving the name 'oxytocin' (from the Greek for "swift childbirth") to one of the principles, Kamm acknowledged the biological activity that he had sought to purify—the ability to induce uterine contractions at the end of pregnancy, and thereby speed the progress of parturition. When du Vigneau sequenced oxytocin and established a way to synthesize it, oxytocin soon became an indispensable part of managing labor in women. However, the physiological role that oxytocin plays in parturition is still unclear (see Chap. 5). Much better understood is its role in lactation: parturition can proceed near normally in the absence of oxytocin, but successful lactation depends absolutely on it (Russell and Leng 1998).

Oxytocin has many other effects, including on appetite, glucose homeostasis, sexual arousal, sodium balance and nociception, and on various behaviors, particularly some associated with stress and reproduction (Leng and Leng 2021). It is now thought that most of these effects reflect true physiological actions. These other actions and effects are in themselves outside the scope of this chapter, but how a single hormone can have many distinct physiological roles is an important question.

6.2 The Oxytocin System of the Hypothalamus and Pituitary

In mammals, oxytocin is mainly synthesized by large ('magnocellular') neurons in the hypothalamus that project to the posterior pituitary gland (Fig. 6.1). Most of these are in either the supraoptic nucleus, at the ventral surface of the brain rostrolateral to the optic chiasm, or in the paraventricular nucleus, adjacent to the dorsal limit of the third ventricle, but many are present in numerous small groupings between these major aggregations (Zhang et al. 2021). These neurons generate action potentials that govern the secretion of oxytocin from the pituitary.

The rate of generation of action potentials depends on the intrinsic excitability of the neurons, and on the synaptic input that they receive. The intrinsic excitability is affected by the external ionic composition and by neuroactive agents in the extracellular environment including hormones, neurosteroids, endocannabinoids and many neuropeptides. Some of these signals derive from oxytocin neurons themselves, and this 'autocrine' communication can 'bind' the activity of the oxytocin neurons together, as will be described later. These are slowly changing signals, distinct from the rapidly fluctuating barrage of excitatory and inhibitory synaptic potentials (EPSPs and IPSPs) imposed by synaptic activity. Most EPSPs appear to be mediated by synaptic release of glutamate, and most IPSPs by GABA, both of which derive from functionally heterogeneous afferent neurons in diverse locations. In addition, noradrenaline, dopamine, serotonin, acetylcholine and histamine are all involved in regulating the oxytocin neurons, but attributing specific functional roles to any of these factors has proved elusive. A noradrenergic input from the caudal

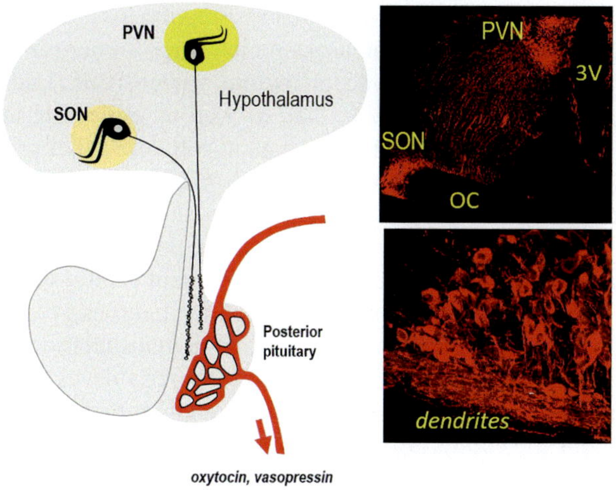

Fig. 6.1 The magnocellular oxytocin system. *Left:* Schematic of the system. Oxytocin is produced mainly by magnocellular neurons of the hypothalamus. These neurons are aggregated into two major nuclei—the supraoptic nucleus (SON) and the paraventricular nucleus (PVN), and a number of smaller 'accessory' nuclei between the two major nuclei. All magnocellular neurons project a main axon to the posterior pituitary, but many also send axon collaterals to various other forebrain regions. The posterior pituitary is filled with fenestrated capillaries, and oxytocin secreted at this site readily enters the blood. *Right:* Oxytocin immunoreactivity in the rat hypothalamus. The micrograph shows a coronal section through the SON and PVN; *OC*, optic chiasm; *3V*, third ventricle. The lower image is a higher magnification of the SON showing the dense layer of oxytocin-containing dendrites at the base of the nucleus

brainstem seems to be important in conveying signals from the contracting uterus to the oxytocin neurons, but noradrenergic inputs also appear to convey inputs from the gastrointestinal tract; whether these are from the same noradrenergic neurons or from different neurons is unclear.

There is little evidence of any substantive functional difference between the magnocellular oxytocin neurons at different locations, but neurons at different sites receive different proportions of their input from different locations, and this is likely to explain the minor differences in the responsiveness of oxytocin neurons at different locations: for example, compared to supraoptic oxytocin neurons, paraventricular oxytocin neurons are slightly more responsive to suckling and slightly less to osmotic challenge.

The rat supraoptic nucleus contains only magnocellular neurons, but the paraventricular nucleus contains some 'parvocellular' oxytocin neurons that do not project to the posterior pituitary; some of these project to the spinal cord, others to the dorsal vagal complex in the caudal brainstem, and some to the supraoptic nucleus. These neurons have different intrinsic membrane properties from the magnocellular neurons and receive different inputs, and thus they seem likely to have a functionally distinct role. It was once assumed that they mediated all of the behavioral effects of oxytocin, as magnocellular neurons were thought to project almost exclusively to the posterior pituitary, but that assumption was punctured first by the recognition that the magnocellular neurons secrete large amounts of oxytocin from their dendrites and from their abundant **axonal varicosities** (Ludwig and Leng 2006), and then by the demonstration that, as well as projecting to the posterior pituitary, some magnocellular neurons have axon collaterals that project to diverse sites in the forebrain (Zhang et al. 2021). At present it is unclear whether magnocellular oxytocin neurons that project centrally are in any other ways distinct in their intrinsic properties from those that project only to the posterior pituitary, and unclear whether magnocellular neurons that project to different central locations are regulated differently. In early development, magnocellular neurons give rise to multiple, branching axons that are subsequently pruned, so perhaps their central projections arise from a random outgrowth of collaterals during early development, collaterals that survive only when they happen to encounter a target that delivers a retrograde signal that sustains the connection. It is possible that evolution has delivered functionally specialized subpopulations of oxytocin neurons, but it is not obvious how a fully predetermined projection map might be implemented in the genetic code or what selection pressure could have delivered such a map.

In several species, oxytocin is also expressed at many peripheral locations, including the uterus, testes, ovaries, and gastrointestinal tract, and very large amounts of oxytocin mRNA are present at some of these sites. However, little oxytocin is produced at any of these sites, with the exception that, in cattle and sheep, large amounts of oxytocin are secreted from the **corpus luteum** during **luteolysis**. This exception aside, it seems likely that oxytocin expression at other sites is either superfluous expression of no physiological significance, or that oxytocin at these sites plays a local paracrine role. The oxytocin mRNA that is present in large amounts in the uterus has a much shorter poly-A tail than hypothalamic mRNA, with the apparent consequence that it is ineffectually translated, as it appears that the uterus contains very little endogenous oxytocin.

6.2.1 The Secreted Products of Oxytocin Neurons

The magnocellular neurons that project to the posterior pituitary generally synthesize large amounts of either oxytocin or vasopressin, but most appear to make small amounts of the 'other' peptide, and a few synthesize similar amounts of both (Xi et al. 1999). They also produce many other peptides in much lower quantities—including several factors with anorectic activity, including cholecystokinin, corticotrophin-releasing factor, nesfatin and secretin. Oxytocin is itself a potent anorectic and it appears to be involved in regulating appetite and glucose homeostasis (Leng and Sabatier 2017), so the presence of other anorectic factors in oxytocin neurons may have no physiological significance—there seems unlikely to be any strong selection pressure to suppress a low level of incidental production of factors whose biological activity is similar to that of oxytocin itself. However not all co-expressed peptides can be ignored; oxytocin neurons contain quite large amounts of dynorphin, which appears to have an important autocrine feedback role in regulating oxytocin secretion, as discussed later.

Oxytocin differs from vasopressin by just two amino acids, and the two genes are on the same chromosome, separated by just a few kilobases of sequence. The two peptides evolved from a common ancestral member of what is now a large family of nonapeptides, a family which includes vasotocin—the most likely common ancestor of vasopressin and oxytocin—in teleost fish and amphibia. But it is not just the peptides that are closely related—oxytocin neurons and vasopressin neurons are very similar, apparently diverging from a common ancestor population at about the time of evolution of vertebrates, probably at the same time as the gene duplication that gave rise to the oxytocin and vasopressin families. Thus, while we may think of oxytocin as being a quintessentially mammalian hormone because of its key role in lactation and parturition, oxytocin (or a close analog, such as mesotocin) is a neurohypophysial hormone in all vertebrates. Oxytocin has ancestors throughout the animal kingdom, including in annelids and insects that are commonly involved in egg-laying and other reproductive functions, ingestive behavior, and electrolyte homeostasis. Indeed, it appears that an ancestor hormone was expressed in *Urbilateria*, the hypothesized common ancestor of all bilateral animals. Comparative genetic analyses suggest that in this marine animal, the ancestral hormone was expressed in multisensory, multifunctional neurons that had properties of both endocrine cells and neurons, and which regulated reproductive and ingestive behavior and physiology in accordance with prevailing environmental conditions (Odekunle and Elphick 2020).

Like oxytocin, vasopressin is synthesized in magnocellular neurons of the supraoptic and paraventricular nuclei. Both peptides are packaged in large dense-cored neurosecretory vesicles, each of which contains ~80,000 molecules of oxytocin or vasopressin. Each neuron has one main axon, down which the vesicles are transported to be stored in axonal swellings and nerve endings in the posterior pituitary, from which they can be secreted in response to action potentials. The nerve terminals have none of the classic features of synapses—there is nothing analogous to a synaptic cleft, or a post-synaptic specialization; the contents of the vesicles are released into the extracellular space, from where they enter the capillaries that permeate the neurohypophysis via fenestrations in the capillary endothelium. Vesicle exocytosis can occur from any part of the axon—even from undilated segments of the axons—and it appears that the probability of a vesicle being exocytosed is mainly determined by its proximity to the axonal membrane.

The rat has about 5000 magnocellular oxytocin neurons, and the axon from each of these gives rise to about 2000 'release sites' (swellings and nerve terminals) in the pituitary. In all, the posterior pituitary of an adult virgin female rat contains 4–7 billion vesicles that contain 0.5–1 μg of oxytocin. These vesicles don't contain just oxytocin—oxytocin is synthesized as part of a large precursor molecule, and this entire precursor is packaged in the vesicles, along with the enzymes that process the precursor as it is conveyed to the posterior pituitary, and smaller amounts of various other coexisting peptides.

At basal release rates, a few thousand vesicles are released every second, so the pituitary content is equivalent to 20–30 days of secretion. This may seem like a large store, but if the secretion rate is increased tenfold, it can last for only 2–3 days. A sustained challenge will lead to an increase in the rate of synthesis—but this takes time, and it takes more time to transport newly synthesized vesicles to the pituitary. Accordingly, any sustained challenge to the oxytocin system will result in a depletion of the neural lobe content. In the rat, withholding drinking water or replacing drinking water with 2% saline solution leads to a fall of 30% in content of both oxytocin and vasopressin within a day, and within 4 days the content falls to about 10% of normal, despite a marked elevation in the rate of synthesis. Therefore, despite the apparent abundance of oxytocin in the pituitary, any sustained increase in demand must be accompanied by an increase in synthesis.

6.3 The Milk-Ejection Reflex

In eutherian mammals, milk production is governed by prolactin (see Chap. 7), but milk let-down in response to suckling depends on oxytocin (Russell and Leng 1998). In the rat, for about 21 days, a litter of pups depends on the mother's milk. When the young suckle, the stimulation is sensed by touch and pressure receptors and is conveyed to the brainstem by spinal nerves. These signals are conveyed by interneurons to the oxytocin cells in the hypothalamus, leading to secretion of oxytocin into the blood. The oxytocin acts at oxytocin receptors on the myoepithelial cells that surround the **alveolar cells**, which produce and store milk, and the contractions that oxytocin induces expel milk from the alveolar cells into the collecting ducts of the mammary gland. Continuous exposure to oxytocin is ineffective at maintaining contraction of the mammary gland, and will lead to the desensitization of hormone receptors on the target muscle. Instead, in response to suckling, oxytocin cells fire in sparse, synchronized bursts leading to intermittent, bolus secretion of the hormone from the pituitary (Fig. 6.2).

This 'pulsatile' secretion of oxytocin was first described from experiments in anesthetized lactating rats that involved recording the electrical activity of neurons in the hypothalamic supraoptic or paraventricular nucleus (Wakerley and Lincoln 1973; Lincoln and Wakerley 1974). Rat pups will suckle continuously at the nipples of an anesthetized mother, but every 5–15 min, they display a conspicuous 'stretch reflex' with their forelimbs rigidly extended, and this coincides with a sharp transient rise in intramammary pressure (attributable to the effect of a sharp pulse of oxytocin secretion). About 12 s before the intramammary pressure response, some of the neurons showed a brief, high-frequency burst of activity—a 'milk-ejection burst'—marking them out as the oxytocin cells. Subsequent studies in other species indicated that, while there are species differences in the behavioral patterning of suckling and milk let-down, milk let-down is commonly the result of pulsatile oxytocin secretion.

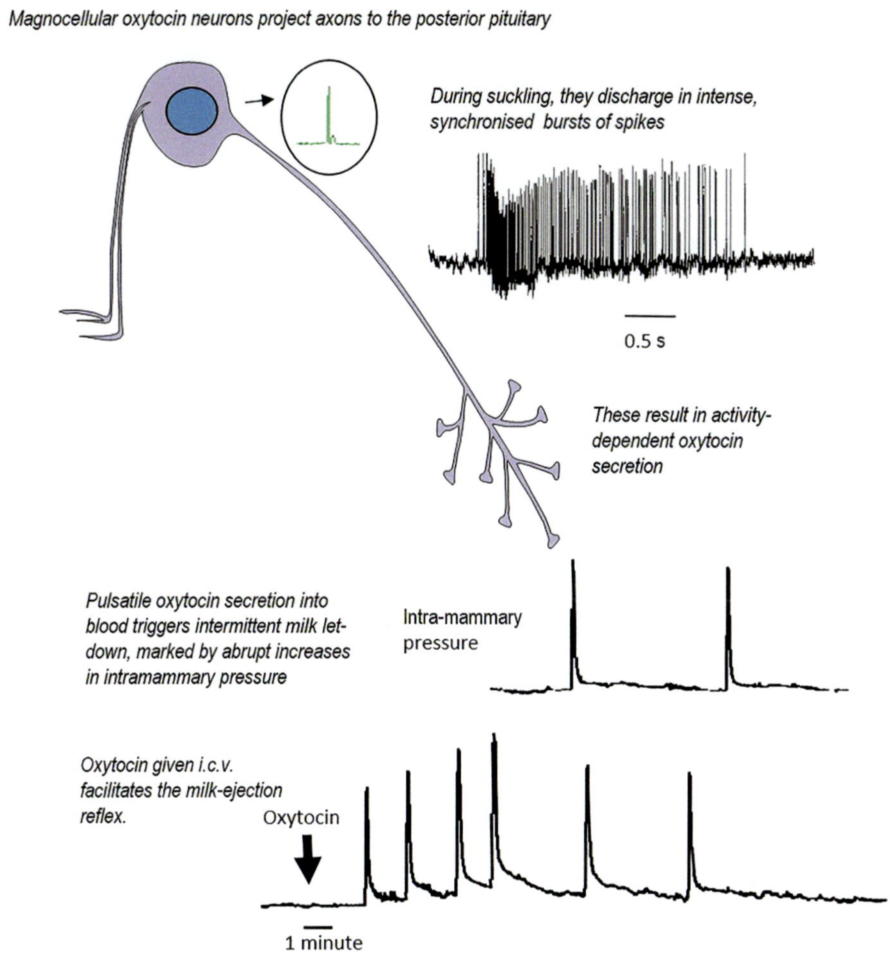

Fig. 6.2 The milk-ejection reflex. Schematic illustrating the pulsatile nature of suckling-induced oxytocin release. The extracellular recording of a milk ejection burst was recorded from an oxytocin neuron in the supraoptic nucleus in an anesthetized lactating rat; from Dyball and Leng (1986), and the recordings of intramammary pressure were also made in the course of those experiments. Although bursts in different cells can vary markedly in amplitude, the shape of bursts is very consistent between oxytocin cells, as well as from one burst to another in each cell

6.3.1 Physiological Importance of Bursting

The bursting pattern of oxytocin cells has an important consequence for the efficiency of hormone secretion: a given number of action potentials causes more secretion if they occur at a high frequency. This facilitation is caused by a broadening of individual action potentials when the neurons fire at high frequency, and this leads to a greater influx of calcium through the voltage-dependent membrane channels—a stronger stimulus for exocytosis. In addition, high-frequency firing increases the likelihood that all of the terminal endings will be invaded by the action potentials. In the rat, this burst in the entire population of magnocellular oxytocin cells releases 0.5–1 mU of oxytocin—just the right amount for a maximally effective contraction of the mammary gland; if less than 0.2 mU is released, there is scarcely any response from the mammary gland, while 2 mU is a supramaximal stimulus.

6.3.2 The Structure of Milk-Ejection Bursts

When suckling is established after a long break, the bursts are small at first, but become larger as the reflex becomes established. The amplitude of bursts is proportional to the number of pups that are suckling, but when eight or more pups suckling, the bursts typically last about 2 s and contain about 100 spikes (action potentials). All magnocellular oxytocin cells fire together during a burst, but the spikes are not strictly synchronized as they are in cells tightly coupled by electrical synapses. There is some evidence of occasional gap junction coupling between oxytocin neurons in the rat, but this seems too weak and sparse to contribute much to burst synchronization. Simultaneous recordings from pairs of oxytocin neurons—either adjacent cells recorded with a single electrode, or cells in two different nuclei—showed that burst onsets varied typically by around 200 ms, and that there were no consistent 'leaders' and 'followers'—one cell might begin a burst earlier at one reflex and later at the next. The bursts varied in intensity from one cell to another, but their shape was very consistent between cells as well as within cells (Belin and Moos 1986).

The association of synchronous burst discharge with a specific physiological state (lactation) led to speculation that some structural change in the network of oxytocin neurons might be involved. A case for an anatomical reorganization emerged from ultrastructural studies that showed that in lactation, the glial processes that normally intervened between adjacent oxytocin neurons were retracted, allowing direct dendro-dendritic appositions, and it was speculated that this might favor interactions between the neurons. However, it eventually emerged that the morphological reorganization in the rat hypothalamus could be blocked pharmacologically with no apparent effect on the milk-ejection reflex (Catheline et al. 2006). The functional significance of the glial reorganization in lactation is still unclear.

6.3.3 The Burst-Generating Mechanism

The stereotyped nature of the bursts suggested that some form of positive-feedback process was probably involved. Many different hypotheses were proposed, but a conundrum was that bursts were associated uniquely with suckling; when excited by other kinds of stimuli, the oxytocin neurons did not exhibit such bursts. It was natural therefore to conceive that something in the pattern of the suckling stimulus might be integral to the triggering of a burst. The stretch reflex was typically followed by frenzied activity before the pups settled down to quietly suck again, so perhaps this sequence entrained the pattern of secretion. However, this hypothesis was refuted by Sutherland et al. (1987), who showed that milk-ejection bursts could be induced in anesthetized lactating rats if the suckling of the pups was replaced by electrical stimulation of the nipples. Remarkably, continuous stimulation for up to 120 min resulted in milk ejections at intervals of approximately 8 min, apparently indistinguishable from the typical responses to suckling.

Clearly, milk-ejection bursts could not be explained by any behavioral feedback from the pups; it seemed that the 'burst generator' mechanism must lie within the brain, but might be anywhere on the afferent pathway. However, in the 1990s it was shown that oxytocin itself, when administered into the third ventricle of a suckled, lactating rat at a dose of about 1 ng, could increase the frequency and amplitude of milk-ejection bursts for about 15–30 min (Richard et al. 1991). Similar doses given to lactating rats while no pups were suckling had no effect. Moreover, a specific oxytocin antagonist injected centrally would block the milk-ejection reflex, demonstrating the involvement of endogenous oxytocin (Lambert et al. 1993). Oxytocin could facilitate the reflex when injected directly into the supraoptic nucleus at even lower amounts—and bursts occurred in both supraoptic nuclei even when only one site was injected. It thus appeared that a milk-ejection burst might begin in any of the four magnocellular nuclei as a result of oxytocin release in that nucleus, and that this would then trigger a burst in all other nuclei.

Oxytocin is indeed released in the supraoptic nucleus in response to suckling, and this release precedes the appearance of milk-ejection bursts (Moos et al. 1989). But where this oxytocin came from was a mystery, as there were very few oxytocin-containing synapses in the supraoptic nucleus and no evidence of any synaptic excitation directly between the oxytocin neurons.

6.3.4 Dendritic Oxytocin Release

The answer came with the demonstration that oxytocin can be released from the soma and dendrites of oxytocin cells (Morris and Pow 1991); these were known to be full of neurosecretory vesicles, but it had been assumed that this was a store awaiting transit to the pituitary. However, identifying oxytocin receptors on magnocellular neurons proved difficult. Autoradiographic methods of identifying binding sites revealed that oxytocin receptors were abundantly expressed in many parts of the brain—including in the amygdala and the ventromedial nucleus of the hypothalamus, where it was hard to find any innervation by oxytocin-containing fibers, but these techniques stubbornly failed to reveal any expression in magnocellular neurons. However, it became clear that oxytocin neurons express oxytocin receptor mRNA (Yoshimura et al. 1993), and that oxytocin would mobilize intracellular calcium in isolated magnocellular oxytocin neurons *in vitro* (Lambert et al. 1994). This could be blocked by selective oxytocin antagonists—and oxytocin release from the supraoptic nucleus *in vitro* could be evoked by application of oxytocin agonists.

It thus appeared that oxytocin neurons express functional oxytocin receptors, but that these were not visible to autoradiographic techniques perhaps because of the high levels of endogenous oxytocin in the supraoptic nucleus. Oxytocin receptors bound by oxytocin are rapidly internalized and must be returned to the plasma membrane before they can be re-activated by oxytocin, so perhaps most oxytocin receptors on the oxytocin neurons were either occupied or internalized. This hypothesis was supported by a striking experiment: if the supraoptic nucleus was first exposed to an oxytocin antagonist (to block endogenous oxytocin release), then oxytocin receptors on the dendrites of oxytocin neurons subsequently became visible to autoradiography (Freund-Mercier and Stoeckel 1995).

Thus, oxytocin cells synthesize oxytocin receptors, oxytocin can mobilize intracellular calcium stores in oxytocin neurons, and oxytocin can stimulate dendritic oxytocin release, presumably as a consequence of the intracellular calcium release. If dendritic release of oxytocin was indeed part of a positive-feedback loop, oxytocin release should, at the time of a burst, both stimulate electrical activity and be evoked by electrical activity. Experiments on oxytocin neurons *in vitro* showed that it could depolarize oxytocin neurons, and, under certain pharmacological conditions, oxytocin administered to supraoptic neurons *in vitro* could even trigger milk-ejection like bursts (Hatton and Wang 2008). But *was* oxytocin released in the hypothalamus in response to electrical activity? Increases in intracellular calcium evoked some dendritic oxytocin release, but this seemed to be too slow and late to be effective in contributing to the generation of a burst. And still there was the conundrum of why bursts were only seen in response to suckling.

In anesthetized rats, Ludwig et al. electrically stimulated the axons entering the pituitary; action potentials triggered in this way are propagated antidromically to the cell bodies as well as orthodromically to the axonal endings. They used a microdialysis probe to collect samples of extracellular fluid in the supraoptic nucleus, and measured oxytocin in these samples. However, while such stimulation could evoke massive secretion of oxytocin into the blood, it seemed that it was ineffective in triggering dendritic oxytocin release (Ludwig and Leng 2006).

6.3.5 Priming Dendritic Release

However, perhaps very particular circumstances were required to evoke activity-dependent dendritic release. The shape of an answer emerged from experiments inspired by studies of pituitary gonadotrophs. These cells secrete luteinizing hormone (LH) in response to gonadotrophin-releasing hormone (GnRH). This reflects calcium-dependent exocytosis of large dense-cored vesicles that are located close to the gonadotroph plasma membrane; vesicles deeper within the cell cannot be readily released. In steroid-primed pituitaries from female rats, GnRH not only evokes exocytosis of LH-containing vesicles, it also evokes a 'margination' of secretory vesicles. This margination involves a migration of vesicles from deep in the cell interior to the periphery, from where they can be readily released. As a consequence, repeated exposures to GnRH result in a progressively escalating secretion of LH, and this 'self-priming' is key to the generation of the preovulatory LH surge (Leng et al. 2008).

Ludwig et al. (2002) hypothesized that mobilization of intracellular calcium might have a similar 'priming' effect in oxytocin dendrites, and tested this by applying thapsigargin to the supraoptic nucleus via the microdialysis probe. Thapsigargin is a pharmacological blocker of the calcium pump that maintains high levels of calcium in the intracellular stores, and it causes an increase in free intracellular calcium as the stores empty. Thapsigargin itself, as expected, caused some oxytocin release from the dendrites. But, 30–90 min after applying thapsigargin, electrical stimulation of the pituitary axons could now evoke a very large release of oxytocin. It appeared that dendritic release could be 'primed' just as LH secretion from gonadotrophs could be. In subsequent experiments, it was established that neuropeptides that mobilized intracellular calcium stores could also prime dendritic oxytocin release, and that this involved margination of the vesicles as a result of a reorganization of the actin skeleton in the dendrites (Brown et al. 2020).

In 2008, Rossoni et al. published a mathematical model of the milk-ejection reflex (Rossoni et al. 2008) that represented these elements of the biology of oxytocin neurons as differential equations (Fig. 6.3). In its final form, it comprised 300

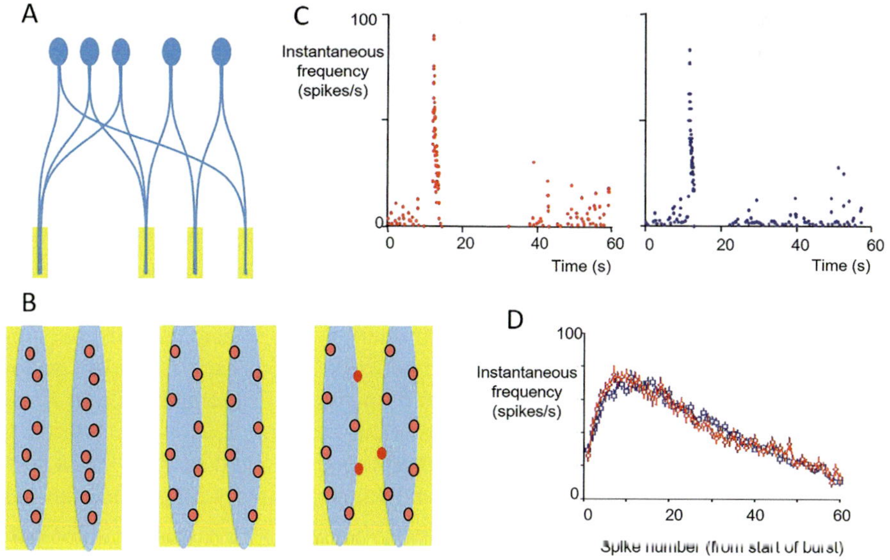

Fig. 6.3 Mathematical modelling of the milk-ejection reflex (see Rossoni et al. 2008). (**a**) The model comprises a population of up to 300 neurons, each with two dendrites, that are each bundled together with dendrites from several other oxytocin neurons; the two dendrites of each cell are randomly assigned to different bundles (indicated in yellow). (**b**) In the model, suckling progressively primes the dendritic stores of oxytocin; vesicles are translocated from the interior of the dendrites to the perimeter, where they are sensitive to the actions of calcium entering through membrane channels in response to spike activity. The oxytocin that is released in response to spiking depolarizes other neurons in dendro-dendritic contact, leading by positive feedback to a burst of electrical activity and secretion that depletes the store of readily releasable vesicles, terminating the burst. (**c**) Burst profiles in a model cell (red) and in an oxytocin cell recorded in vivo (blue; the cell shown in Fig. 6.1); the cell activity is plotted as instantaneous frequency (the reciprocals of the intervals between successive spikes). (**d**) Burst profiles in a model cell (red) and an oxytocin cell (blue) plotted as the mean instantaneous frequency (calculated from 13 successive bursts) of each interval from the first spike in each burst

independent model neurons, each of which was given intrinsic membrane properties based on published work, and an array of 'synaptic inputs' mimicking the frequent random arrival of small transient EPSCs and IPSCs. In response to these, the model neurons generated spike activity that closely matched the spontaneous activity recorded from oxytocin neurons *in vivo*. The model neurons were then each given two 'dendrites', each of which was bundled together with dendrites from several other model neurons, following published descriptions of the morphology of supraoptic neurons. Each dendrite was given a store of oxytocin, which could be released by electrical activity to affect the other dendrites with which it was bundled. Vesicles in this store could be progressively primed by a model suckling stimulus to become part of a 'readily-releasable store'. Release was assumed to be proportional to the size of this readily releasable store, and to be proportional to the calcium entry into the dendrites that accompanied action potentials, and also to be subject to 'frequency-facilitation', a phenomenon known from studies of stimulus-secretion coupling in the posterior pituitary (see below). Released oxytocin could excite the other cells that shared the same dendritic bundle, and was accompanied by the production of endocannabinoids which had the effect of suppressing afferent excitatory synaptic input by presynaptic actions.

The model would generate 'milk-ejection bursts' quantitatively indistinguishable from bursts observed in magnocellular oxytocin neurons *in vivo*; the bursts occurred approximately synchronously in all neurons of the network, and at intervals of 5–8 min. It could also reproduce a range of electrophysiological phenomena that had been reported in the literature and which were commonly considered paradoxical. Exactly how the model works is not easy to explain, because it was not designed to implement a deterministic burst-generating mechanism, but simply to reproduce the known electrophysiological properties of oxytocin neurons; bursting is thus an 'emergent property' of the model network, and its expression depends on the scale and complexity of the network and on stochastic elements within it. By analogy, we might say that the model oxytocin neurons behave like 'electrical capacitors', slowly charged by priming, and rapidly discharged as a result of a burst that arises semi-randomly in the network and which is rapidly propagated through the dendro-dendritic connections. The 'discharge' of oxytocin in a burst depletes the dendritic stores that must be rebuilt beyond a critical level before another burst can occur.

We thus seem to have a reasonable understanding of the generation of milk-ejection bursts, but there remain many important gaps. For example: the model of Rossoni et al. assumes that the suckling stimulus primes the dendritic stores of oxytocin, and that this is a consequence of the actions of a neuropeptide released from afferent neurons—but it is not known which, if any, peptides are released in the hypothalamus by suckling. One possible candidate is alpha-melanocyte-stimulating hormone (α-MSH), present in a prominent input to oxytocin neurons from neurons of the arcuate nucleus. In virgin rats α-MSH can prime dendritic oxytocin release but also inhibits oxytocin neuron electrical activity (Sabatier et al. 2003). However, this α-MSH inhibition of oxytocin neuron firing is attenuated in mid-pregnancy and in late pregnant and lactating rats it is excitatory to oxytocin neurons (Perkinson et al. 2022).

We also have a very incomplete knowledge of the pathway mediating the effects of suckling; it appears to pass through the lateral tegmentum, as lesions to this area block the reflex (Juss and Wakerley 1981), but the precise neuronal elements involved are still unknown. No neurons have been shown to respond to suckling other than neurons in the spinal cord that mediate the primary afferent stimuli, except for some neurons in the dorsomedial hypothalamus that behave so like magnocellular oxytocin neurons that they may be scattered members of the same population (Honda and Higuchi 2010).

It is also unclear how a burst that originates in one nucleus spreads to the other nuclei; it is possible that sparse long dendrites link the nuclei, but it has also been proposed that a subset of parvocellular oxytocin neurons in the paraventricular nucleus have the task of coordinating the nuclei (Tang et al. 2020). There is partial evidence of some involvement of the mammillary nucleus (Wang et al. 2013) and of a bilaterally projecting cell group in the medial medulla (Moos et al. 2004) in burst synchronization amongst the nuclei, but the evidence for these two possibilities appears to be mutually incompatible.

6.3.6 Synthesis of Oxytocin in Lactation

Lactation is obviously an extended time of high demand for oxytocin, and the pituitary content, as expected, is relatively depleted throughout lactation, while the expression of oxytocin mRNA is elevated. At present however we have a very incomplete understanding of the mechanisms involved in increasing oxytocin synthesis in lactation. There is no evidence that depletion itself is a stimulus for increased synthesis, instead it appears that the same signals that drive secretion also stimulate synthesis. A curious feature of the milk-ejection reflex is that the massively elevated secretion of oxytocin is achieved with a marked change in the pattern of the electrical activity of the oxytocin neurons, but hardly any increase in their overall electrical activity. Thus, an oxytocin neuron firing spontaneously at 2 spikes/s will discharge 1200 spikes in 10 min; during suckling, the background activity is unaltered, but every 10 min it may fire a 2-s burst of 100 spikes followed by about 30 s of silence; for a total of 1240 spikes every 10 min. As explained above, the change in pattern produces a massive facilitation in secretion, and presumably the same change in pattern must produce a similar facilitation of synthesis. How it does so remains unknown; suckling-induced reflex milk ejection is not accompanied by any measurable increase in the expression of the immediate-early gene c-*fos*, a long-standing candidate for linking activity to synthesis.

6.4 Oxytocin Secretion at Parturition

There is one other known circumstance in which oxytocin neurons fire in bursts. Recordings of the electrical activity of supraoptic neurons in conscious, freely moving rats generated several seminal observations. First, they showed that, in conscious lactating rats, the milk-ejection reflex discharge of oxytocin neurons is essentially indistinguishable from that seen in anesthetized rats (Summerlee and Lincoln 1981). Second, oxytocin neurons showed similar intense synchronized bursts during parturition, bursts that were contemporaneous with delivery of a fetus or of the afterbirth (Summerlee 1981). In parturition however the bursts were superimposed on a strongly elevated background activity.

6.4.1 The Prelude to Parturition

A rat is pregnant for about 21 days, and during this time, the store of oxytocin in the pituitary gland increases by about 50%. About 24 h before the pups are born, increased production of prostaglandin by the uterus induces luteolysis, and ovarian progesterone production falls dramatically. This fall is an essential prelude to parturition; if prevented, then the rat pups will remain unborn, and will die within the uterus. The fall in progesterone leads to increased production of prostaglandins by the **endometrium**, and to many other changes that prepare the uterus and birth canal for parturition. In the last few hours of pregnancy, oxytocin receptors appear at high concentrations in the uterus. In many mammalian species, including the sheep and nonhuman primates, there is a progressive increase in uterine activity toward the end of pregnancy that can be attenuated by infusion of a specific oxytocin antagonist, but this need not imply any increase in oxytocin secretion: oxytocin is the most powerful uterotonic agent known, and the marked increase in sensitivity to oxytocin may result in an unchanging basal level of oxytocin becoming effective. However once parturition is underway, oxytocin secretion from the pituitary is increased in all placental mammals so far studied, including the human. Strong uterine contractions, triggered by local prostaglandin production or by the increasing sensitivity of the uterus to basal levels of oxytocin, excite the oxytocin cells and the oxytocin secretion triggers further prostaglandin production and further uterine contraction. Then, typically in the afternoon of the 21st day, the rat will give birth, normally to between 8 and 18 pups (Fig. 6.4a, b). Over the 90 min or so that this process occupies, the accumulated excess in the pituitary store of oxytocin is secreted. If the actions of oxytocin are prevented by drugs, or if the secretion of oxytocin is interrupted for example as a result of stress, then the progress of parturition is impaired (Russell et al. 2003).

Pulsatile secretion of oxytocin in labor has been observed in rats, rabbits, pigs, and cattle, and in some but not all studies large, brief pulses of oxytocin have been measured in the plasma of women during labor, thus it is generally recognized that oxytocin secretion accompanies the expulsive phase of labor in mammals generally (Fig. 6.4c). It should be noted however that, in women, there is considerable

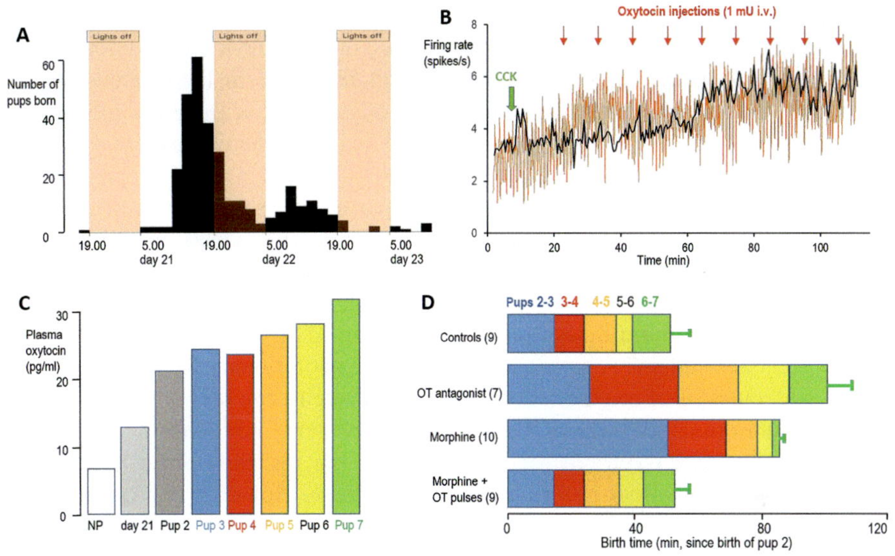

Fig. 6.4 Oxytocin secretion during parturition in the rat. (**a**) Diurnal rhythm in time of onset of parturition in the rat. Meta-analysis of observations on parturition in 293 rats, with lights off for 10 h each day. In most rats, the first pup is born in the afternoon of day 21; few are born at night, and rats that do not start parturition on day 21 mainly give birth to their first pup on the afternoon of day 22. The time of birth can be advanced by shifting the lighting cues in mid-pregnancy, so it appears that the timing of the first birth is determined by a maternal signal late in pregnancy (see Russell and Leng 1998). (**b**) Recording from a magnocellular neuron of the supraoptic nucleus in an anesthetized rat in the afternoon of day 21 of pregnancy. The rat was given an injection of CCK; the transient excitatory response is characteristic of oxytocin cells. The rat was then given an i.v. injections of 1 mU oxytocin every 10 min. These injections resulted in a gradual increase in the force of uterine contractions (orange) accompanied by a gradual increase in the firing activity of the oxytocin cell. Such experiments indicate that uterine activity feeds back to increase the background activity of oxytocin cells, leading to an increase in oxytocin secretion (see Douglas et al. 2001). (**c**) Mean levels of oxytocin in conscious rats. In the morning of day 21, oxytocin levels are higher than in nonpregnant (NP) rats, and rise further during parturition—samples taken at the time of birth of each pup (see Russell and Leng 1998). (**d**) Once parturition has begun injections of an oxytocin antagonist will interrupt parturition if given early—at around the time of birth of the second or third pups, but have no effect if given later. Oxytocin neurons in the rat can be very potently inhibited by morphine, and the third bar shows that parturition is profoundly interrupted by an i.v. injection of 1 mg/kg morphine at the time of birth of the second pup. Pulsatile i.v. injection of oxytocin (1 mU every 10 min) after a single injection of morphine given after the birth of pup 2 fully restored the progress of normal parturition; continuous infusion (not shown) of the same total dose of oxytocin was ineffective (modified from Russell and Leng 1998)

variation in reported measures of oxytocin in late pregnancy (Uvnas-Moberg et al. 2019). A particular difficulty is associated with the presence of very high levels of 'human pregnancy oxytocinase'. Oxytocinase, also known as PLAP (placental alkaline phosphatase) generally occurs as a membrane-bound enzyme, but in the pregnant human only, it is abundant as an enzyme that circulates at high levels in the blood. This necessitates stringent precautions to inactivate enzyme activity in blood samples from pregnant women. As well as making measurement of oxytocin concentrations technically difficult, the presence of this enzyme led to doubts that endogenous oxytocin was of any physiological relevance in labor. However, by shortening the half-life of oxytocin in plasma, oxytocinase will enhance the pulsatile patterning of the hormone in the circulation, and may thereby assist its biological effectiveness.

But while elevated oxytocin secretion in parturition seems to be universally evident in mammals, in some species, parturition can proceed apparently normally in its absence. For example, transgenic mice wholly deficient in either oxytocin (Nishimori et al. 1996) or in oxytocin receptors (Lee et al. 2008) will give birth at about the normal time. So, is the secretion of oxytocin from the pituitary at term an anachronism? A remnant perhaps of an evolutionary history in which a hormoneal ancestor of oxytocin had an essential role in egg laying?

In all mammals that have been studied, at term pregnancy, the uterus expresses abundant oxytocin receptors, and in the rat, the uterus is sensitive only in the last few hours before the young are born. However, the influence of endogenous oxytocin on the initiation of labor in the rat appears to be weak; high doses of oxytocin antagonists do not consistently delay the onset of birth in the rat, and the observed delays are relatively short. Indeed, low doses of oxytocin produce a longer delay to subsequent birth than high doses of antagonist, possibly reflecting desensitization of uterine oxytocin receptors, but perhaps also reflecting fatigue of other uterotonic control mechanisms within the uterus itself.

By contrast, systemic administration of oxytocin antagonists or pharmacological inhibition of oxytocin secretion is consistently effective in slowing the progress of parturition when given shortly after the onset of parturition (see Chap. 5). Interestingly, in rats, the opioid agonist morphine, which profoundly inhibits oxytocin secretion in this species, will cause a prolonged interruption of parturition if given just after the birth of the first pup, but this interruption does not occur if the rat is given an intravenous injection of 1 mU oxytocin every 10 min, indicating the functional role of pulsatile secretion of oxytocin at this stage (Luckman et al. 1993) (Fig. 6.4d).

However, why the pulses should be important is not clear; while pulses of oxytocin were shown to be highly effective in rescuing parturition when endogenous secretion was blocked, the delivery of a pulse of oxytocin did not seem to directly result in pup expulsion. The conundrum was made very apparent in studies of parturition in the pig: in this species, oxytocin secretion can be monitored during parturition by measuring intramammary pressure (Gilbert et al. 1994). These studies demonstrated that while a pulse of oxytocin secretion is associated with the delivery of each piglet, the pulse is triggered as the piglet is expelled—the pulse thus appears to be more a *consequence* of delivery than a *cause* of it. In the rat, bursts of oxytocin cell activity are triggered by cervical distension as the pup is expelled, and the pulse of oxytocin again occurs after delivery. It appears possible that the hemodynamic consequences of bolus secretion of oxytocin may be critically important. In the rat, fetuses are delivered alternately from each uterine horn, at about 10-min intervals.

Perhaps, in the rat, the bolus of oxytocin, by its hemostatic actions, precipitates the separation of the next most distal fetal sac from the uterine wall, initiating its transit from the site of implantation to the cervix. In women, the highest levels of oxytocin are generally observed during the third stage of labor, and at this stage the hemostatic actions of oxytocin are of considerable importance, in minimizing postpartum hemorrhage.

6.4.2 Oxytocin and the Timing of Parturition

Rats generally give birth in the light phase, in the afternoon of day 22 of pregnancy (where day 1 is the day on which sperm is detected in a vaginal smear; Fig. 6.4a). It seems that the day of birth is determined by the fetus, through the maturation of the fetal pituitary-adrenal-placental axis, but the mother may determine the precise hour of birth (see Chap. 5). This is affected not only by endogenous circadian rhythm but also by environmental factors—environmental stress will delay parturition in many species. The mechanisms of stress-induced disruption of parturition can involve inhibition of oxytocin secretion, as shown in the rat and pig, and suppression of uterine contractility by noradrenergic stimuli, as shown in the mouse.

In a seminal series of experiments, Lincoln and Porter (1976) showed that, rats, if the light cycle is advanced or delayed by up to 8 h from mid-pregnancy onwards, they will still give birth either earlier or later than normal so that births still mainly occur in the light phase. However, light itself did not seem to provide an acute cue for the onset of birth; when rats were subjected to continuous darkness on the last day of pregnancy, the distribution of the deliveries was not retarded. Thus, it appeared that some endogenous circadian rhythm determines the hour of birth.

Thirty years later, similar experiments were performed in mice, accompanied by measurements of oxytocin secretion (Roizen et al. 2007). Mice kept on a 12 h:12 h light dark cycle give birth in the dark phase, in the early morning of day 20, and serum measurements suggested that this timing might reflect a circadian rhythm in oxytocin secretion. When the light cycle was advanced or delayed by 6 h in early pregnancy the cycle of oxytocin secretion was unaffected, and the mice still give birth at the same absolute time regardless of the light cycle. By contrast, while oxytocin-deficient mice also give birth in the dark phase of day 20, the circadian gating of birth was lost following a shift in the light cycle—their birth times were distributed apparently randomly on day 20. The authors concluded that, in mice, oxytocin plays a critical role in minimizing the disruption to labor that follows a change in the light-dark cycle.

6.4.3 Oxytocin and Prostaglandins

As mentioned, transgenic mice that lack oxytocin nevertheless deliver viable pups at about the normal time. However, oxytocin is not the only agent involved in uterine contractions at term. In particular, prostaglandin F2α (PGF2α) is a potent uterotonic agent that is secreted from the uterus at term in many mammals, stimulated both directly by uterine distension and by the actions of oxytocin. In women, the increase in uterine oxytocin receptor expression at term occurs in both the myometrium and in the decidua/uterine epithelium. Oxytocin directly induces uterine contractions via its effects on the **myometrium**. On the human decidua/uterine epithelium (and on many other tissues including the cow and sheep endometrium and rabbit amnion) oxytocin can stimulate the production of PGF2α.

In mice, luteolysis in late pregnancy is triggered by a surge in ovarian PGF2α production; after the ensuing progesterone withdrawal, ovarian oxytocin receptor levels in the mouse decrease while uterine receptor levels increase dramatically as in all mammalian species tested so far. Thus, in transgenic mice that lack PGF2α, parturition fails because progesterone levels remain high, and the same is true of mice that lack cyclooxygenase-1 (COX-1), an enzyme essential for production of PGF2α. However surgical or pharmacological induction of luteolysis in these mice restores progesterone withdrawal and normal labor, thus neither oxytocin nor prostaglandins are essential for labor in the mouse once progesterone withdrawal has been initiated. Surprisingly, luteolysis occurs as normal in mice that lack both COX-1 and oxytocin. These mice thus initiate labor on the normal day of delivery, but labor is very prolonged. Accordingly, it seems that in mice, oxytocin has a luteotrophic action. The prolonged labor suggests that neither prostaglandin or oxytocin are essential for the normal progress of parturition, as either can compensate for the absence of the other (Blanks and Thornton 2003).

6.4.4 Preparation for Parturition

The increase in pituitary oxytocin content before parturition ensures that enough oxytocin is available for secretion during parturition without depleting the stores so much as to endanger lactation, and the question arises as to how exactly the increase in stores is achieved. An increase in pituitary content implies an excess of synthesis above secretion, so might arise either from more synthesis or from less secretion. However, there normally appears to be a tight entanglement between synthesis and secretion, apparently because the signals that stimulate secretion also stimulate synthesis. It has been suggested that the increase in pituitary oxytocin content in pregnancy might be a consequence of the increase in estrogen production. However, magnocellular oxytocin neurons do not express Erα (the estrogen receptor linked to oxytocin synthesis in other tissues) and there is no consistent evidence that oxytocin mRNA expression is increased in magnocellular neurons before parturition, although it is dramatically elevated in lactation.

Oxytocin secretion in rats is only modestly increased in pregnancy until close to the end—but this is despite substantial changes in the inputs to the oxytocin neurons. In pregnancy, blood volume expands to meet the additional demands of the growing placentae and fetus(es). As plasma volume is homeostatically regulated, how the volume can expand requires explanation. Normally an increase in plasma volume would elicit an inhibition of vasopressin secretion, entailing increased loss of water through the urine. However, in mammals pregnancy is accompanied by secretion of relaxin, produced in pregnancy by the ovaries or by the uterus itself in different species. Relaxin prepares the birth canal for parturition, by ripening the cervix, softening the epiphyseal ligaments, and it quietens the emergently excitable uterus while progressively sensitizing it to oxytocin, but it also acts on the brain, at sites, such as the subfornical organ, that lack an effective blood-brain barrier. By its actions at the subfornical organ, relaxin activates an excitatory pathway to the magnocellular neurons of the hypothalamus (Way and Leng 1992). The excitatory drive to vasopressin is balanced by inhibitory plasma volume feedback; as a consequence, the vasopressin system is re-set to new set-points for osmoregulation and volume homeostasis, and this is reflected in the sustained hyponatremia and hypervolemia of pregnancy.

The excitation of oxytocin neurons by relaxin is not opposed to the same extent by volume feedback, and oxytocin release is modestly but persistently elevated in pregnancy, at least in the rat. However, dynorphin is co-packaged with oxytocin in the secretory vesicles, and it acts as an autocrine negative feedback at the pituitary to restrain oxytocin secretion. This system is upregulated in pregnancy, providing an elegant solution to the problem of how to accumulate stores when the stimulus to synthesis is also the stimulus to release (Douglas et al. 1993; Leng et al. 1997). However, this would be inappropriate for parturition—the time of high demand for secretion, and as the end of pregnancy approaches, the κ-opioid receptors in the pituitary at which dynorphin acts are downregulated. Yet oxytocin secretion must still be restrained, as the last hours of pregnancy are when oxytocin receptors in the uterus are gathering apace. Premature oxytocin secretion will not necessarily trigger early delivery but will desensitize the uterine oxytocin receptors, delaying parturition while also depleting the pituitary store. At this time oxytocin neuronal activity is strongly inhibited, by an endogenous opioid system that acts at μ-opioid receptors on the cell bodies of oxytocin neurons. The presence of this inhibition can be inferred from the fact that oxytocin neurons at the end of pregnancy but not earlier can be activated by an opioid antagonist. The source of this inhibition is not established, but noradrenergic afferents to oxytocin neurons that are activated by uterine distension co-express a μ-opioid agonist, met-enkephalin, and the arcuate neurons that project to them co-express another μ-agonist, β-endorphin.

6.4.5 Concluding Remarks

Our understanding of the neuroendocrine control of lactation is substantially founded on electrophysiological studies of oxytocin neurons and measurements of plasma oxytocin. The key electrophysiological studies were mainly performed in anesthetized laboratory rats, and were supported by electrophysiological studies in conscious rats that demonstrated that moderate levels of anesthesia do not impair the reflex. By the beginning of the twenty-first century, electrophysiological studies of oxytocin neurons became focused on neuronal membrane properties through *in vitro* approaches, and the expertise in *in vivo* electrophysiology became depleted, as its main practitioners retired, and with pressure to reduce the use of live animals in experiments. The key measurements of oxytocin in plasma were enabled by the development of specific and sensitive radioimmunoassays, such as that developed by Higuchi et al. (1985, 1986a, b) which enabled measurement in frequent blood samples in the rat. In the 1970s and 1980s, many labs developed oxytocin assays and validated them carefully and systematically by comparison with bioassays to ensure that they reliably measured biologically active oxytocin, and to ensure that there was no interference with plasma matrix factors. That expertise and understanding has largely been lost over time (Box 6.1).

However, with new technical approaches using fiber photometry (Perkinson et al. 2021; Yukinaga et al. 2022), there is the beginning of a resurgence in *in vivo* electrophysiology, and with recent fresh interest in oxytocin there has been increasing awareness of the importance of sensitive and reliable assays, and a drive to develop new measurement techniques (Tabak et al. 2022). There is accordingly some reason to be optimistic that the current hiatus in our mechanistic understanding might be coming to an end.

Box 6.1 Problems with measuring oxytocin

Until the 1970s, oxytocin was measured by bioassays. The most common preparations measured the ability of samples to evoke contractions in strips of guinea pig uterus, but other more sensitive assays measured effects on milk let-down. These assays could be extremely sensitive, but required highly specialized expertise, and only a few samples could be measured in a day.

The advent of radioimmunoassays made it possible to measure oxytocin in hundreds of samples in a single assay. In the early days, developing a radioimmunoassay involved (1) assembling the necessary equipment and facilities, including centrifuges, counters and access to radiolabelled peptide or to the means for labelling peptide; (2) raising an antibody, usually in rabbits, goats or sheep by injecting the animals with exogenous antigen conjugated to an adjuvant to provoke an immune response; (3) testing the antibody for sensitivity and specificity; (4) establishing the optimal concentration of antibody to use in an assay; and (5) validating the assay, typically by comparing the assay measurements with those made by bioassays. This was a considerable undertaking, yet such was the demand for radioimmunoassays that many laboratories around the world devoted themselves to developing and running them.

A radioimmunoassay involves mixing three things: (1) a sample, containing an unknown amount of oxytocin; (2) a known amount of radiolabelled oxytocin; and (3) a fixed amount of a specific antibody. When these are mixed, together, the labelled oxytocin and the oxytocin in the sample will compete for binding to the antibody. The mixture is then centrifuged, and the amount of radioactivity in the pellet is measured: the more oxytocin there is in the sample, the less labelled oxytocin will bind, so a low amount of radioactivity

(continued)

Box 6.1 (continued)
in the pellet should indicate that the sample contains a high concentration of oxytocin (Fig. 6.5).

However, other things present in the sample can affect the measurements. These problems are grouped together collectively as 'sample matrix interference'.

One of the most troublesome issues arises from nonspecific interactions between the assay antibody and factors in plasma. Raw plasma contains extremely high concentrations of immunoglobulins—large, sticky molecules that can become entangled with the assay antibody. If such entanglement occludes the antigen recognition site, this will reduce the number of available sites at which oxytocin can bind: accordingly, the pellet will contain less labelled oxytocin than it should, and the assay will erroneously report a high amount of oxytocin in the sample. This issue often affected assays that sought to measure low concentrations of small peptides in plasma, and it affected almost every radioimmunoassay for oxytocin; a rare exception was the Higuchi assay (Higuchi et al. 1985, 1986a, b) which could measure oxytocin in rat plasma reliably—but not in human plasma. This problem, if present, cannot be corrected for, as its extent varies from sample to sample with individual differences in immunological status. However, it can be eliminated by removing high molecular weight components from the plasma with an 'extraction' protocol. Many different protocols have been used; in any given assay a protocol must be found that removes the interfering factor without also removing more than a small proportion of the endogenous oxytocin.

Another problem is that the sample may contain an enzyme that degrades oxytocin. In vivo, oxytocin in the blood is rapidly degraded in the course of the passage of blood through the liver and kidneys. However, once a blood sample is removed from the animal, its oxytocin content typically remains stable for tens of minutes even at room temperature, and remains very stable if placed on ice. However, there is an important exception: in pregnant women, very high levels of oxytocinase circulate in the blood. If this is not neutralized, labelled oxytocin mixed with the plasma will be degraded leaving the label free in solution. Thus, the pellet will contain low levels of bound label, which is then misread as implying a high concentration of endogenous oxytocin. Accordingly, when measuring oxytocin in women, it is essential to take rigorous precautions to rapidly block oxytocinase activity in blood samples (see Tabak et al. 2022).

Other issues can affect assay integrity, but are generally less problematic. The sample may contain other oxytocin-like molecules that bind to the antibody; generally, antibodies are checked to establish that there is no cross-reaction from the obvious candidates, such as vasopressin. The concentration of ions can affect antibody binding—this is only likely to become critical if samples have to be concentrated massively before measurement, but it can present problems for measurements in urine for example.

(continued)

Box 6.1 (continued)

The problems that affect radioimmunoassays affect all immunoassays equally. Since the start of the twenty-first century, the academic laboratories that developed radioimmunoassays closed as the demand for plasma measurements declined. Commercial ELISA kits became available to measure oxytocin; these were generally much less sensitive than radioimmunoassays, but were extremely easy to use. The lack of sensitivity went largely unnoticed because the assays typically reported very high levels of oxytocin in unextracted plasma. It appears that, while the vendors of the kits typically advised users that it was important to extract plasma samples, this advice was widely ignored. Importantly, the measurements of oxytocin in unextracted sample are not simply exaggerated estimates of the true amounts of oxytocin—typically there is no correlation at all between oxytocin levels measured in samples with and without extraction (see Leng and Sabatier 2016). For the moment, it must be assumed that all measurements of oxytocin by immunoassay in unextracted plasma must be assumed to be erroneous unless the assay has been properly validated in a way that excludes plasma matrix interference.

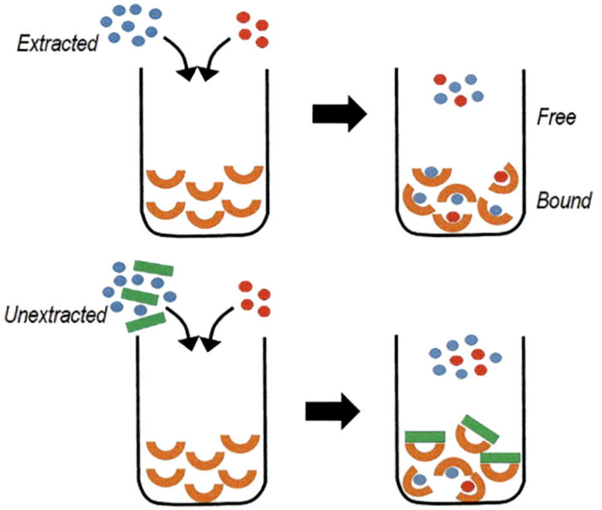

Fig. 6.5 Schematic of radioimmunoassay. *Top*: An unknown amount of oxytocin in an extracted plasma sample (blue) is combined with a known amount of radiolabelled oxytocin (red) and mixed in a solution containing a specific oxytocin antibody (brown). The labelled and unlabelled oxytocin compete for binding, so the more oxytocin there is in the sample the less labelled oxytocin binds. *Bottom*: Unextracted plasma contains abundant large molecules (immunoglobulins), some of which (green) can interfere nonspecifically with the antibody. This interference can occlude the antigen recognition site, meaning that less labelled oxytocin can bind—resulting in an erroneously high reading of oxytocin content

References

Belin V, Moos F (1986) Paired recordings from supraoptic and paraventricular oxytocin cells in suckled rats: recruitment and synchronization. J Physiol 377:369–390

Blanks AM, Thornton S (2003) The role of oxytocin in parturition. BJOG 110(Suppl 20):46–51

Catheline G, Touquet B, Lombard MC, Poulain DA, Theodosis DT (2006) A study of the role of neuro-glial remodeling in the oxytocin system at lactation. Neuroscience 137:309–316

Dale HH (1906) On some physiological actions of ergot. J Physiol 34:163–206

Douglas AJ, Dye S, Leng G, Russell JA, Bicknell RJ (1993) Endogenous opioid regulation of oxytocin secretion through pregnancy in the rat. J Neuroendocrinol 5(3):307–314

Douglas AJ, Scullion S, Antonijevic IA, Brown D, Russell JA, Leng G (2001) Uterine contractile activity stimulates supraoptic neurons in term pregnant rats via a noradrenergic pathway. Endocrinology 142:633–644

Dyball RE, Leng G (1986) Regulation of the milk ejection reflex in the rat. J Physiol 380:239–256

Freund-Mercier MJ, Stoeckel ME (1995) Somatodendritic autoreceptors on oxytocin neurones. Adv Exp Med Biol 395:185–194

Gilbert CL, Goode JA, McGrath TJ (1994) Pulsatile secretion of oxytocin during parturition in the pig: temporal relationship with fetal expulsion. J Physiol 475:129–137

Hatton GI, Wang YF (2008) Neural mechanisms underlying the milk ejection burst and reflex. Prog Brain Res 170:155–166

Higuchi T, Tadokoro Y, Honda K, Negoro H (1986a) Detailed analysis of blood oxytocin levels during suckling and parturition in the rat. J Endocrinol 110:251–256

Higuchi T, Uchide K, Honda K, Negoro H (1985) Functional development of the oxytocin release mechanism and its role in the initiation of parturition in the rat. J Endocrinol 106:311–316

Higuchi T, Uchide K, Honda K, Negoro H (1986b) Oxytocin release during parturition in the pelvic-neurectomized rat. J Endocrinol 109:149–154

Honda K, Higuchi T (2010) Electrical activities of neurones in the dorsomedial hypothalamic nucleus projecting to the supraoptic nucleus during milk-ejection reflex in the rat. J Reprod Dev 56:336–340

Juss TS, Wakerley JB (1981) Mesencephalic areas controlling pulsatile oxytocin release in the suckled rat. J Endocrinol 91:233–244

Lambert RC, Dayanithi G, Moos FC, Richard P (1994) A rise in the intracellular Ca2+ concentration of isolated rat supraoptic cells in response to oxytocin. J Physiol 478(Pt 2):275–287

Lambert RC, Moos FC, Richard P (1993) Action of endogenous oxytocin within the paraventricular or supraoptic nuclei: a powerful link in the regulation of the bursting pattern of oxytocin neurons during the milk-ejection reflex in rats. Neuroscience 57:1027–1038

Lee HJ, Caldwell HK, Macbeth AH, Tolu SG, Young WS (2008) A conditional knockout mouse line of the oxytocin receptor. Endocrinology 149:3256–3263

Leng G, Dye S, Bicknell RJ (1997) Kappa-opioid restraint of oxytocin secretion: plasticity through pregnancy. Neuroendocrinology 66:378–383

Leng G, Caquineau C, Ludwig M (2008) Priming in oxytocin cells and in gonadotrophs. Neurochem Res 33:668–677

Leng G, Leng RI (2021) Oxytocin: a citation network analysis of 10 000 papers. J Neuroendocrinol 33:e13014

Leng G, Sabatier N (2016) Measuring oxytocin and vasopressin: bioassays, immunoassays and random numbers. J Neuroendocrinol 28. *(Review analysing the problems in measuring oxytocin in plasma)*

Leng G, Sabatier N (2017) Oxytocin – the sweet hormone? Trends Endocrinol Metab 28:365–376

Lincoln DW, Porter DG (1976) Timing of the photoperiod and the hour of birth in rats. Nature 260:780–781

Lincoln DW, Wakerley JB (1974) Electrophysiological evidence for the activation of supraoptic neurones during the release of oxytocin. J Physiol 242:533–554

Luckman SM, Antonijevic I, Leng G, Dye S, Douglas AJ, Russell JA, Bicknell RJ (1993) The maintenance of normal parturition in the rat requires neurohypophysial oxytocin. J Neuroendocrinol 5:7–12

Ludwig M, Sabatier N, Bull PM, Landgraf R, Dayanithi G, Leng G (2002) Intracellular calcium stores regulate activity-dependent neuropeptide release from dendrites. Nature 418:85–89

Ludwig M, Leng G (2006) Dendritic peptide release and peptide-dependent behaviours. Nat Rev Neurosci 7:126–136

Moos F, Marganiec A, Fontanaud P, Guillou-Duvoid A, Alonso G (2004) Synchronization of oxytocin neurons in suckled rats: possible role of bilateral innervation of hypothalamic supraoptic nuclei by single medullary neurons. Eur J Neurosci 20:66–78

Moos F, Poulain DA, Rodriguez F, Guerne Y, Vincent JD, Richard P (1989) Release of oxytocin within the supraoptic nucleus during the milk ejection reflex in rats. Exp Brain Res 76:593–602

Morris JF, Pow DV (1991) Widespread release of peptides in the central nervous system: quantitation of tannic acid-captured exocytoses. Anat Rec 231:437–445

Nishimori K, Young LJ, Guo Q, Wang Z, Insel TR, Matzuk MM (1996) Oxytocin is required for nursing but is not essential for parturition or reproductive behavior. Proc Natl Acad Sci U S A 93:11699–11704

Odekunle EA, Elphick MR (2020) Comparative and evolutionary physiology of vasopressin/oxytocin-type neuropeptide signaling in invertebrates. Front Endocrinol (Lausanne) 11:225

Perkinson MR, Kirchner MK, Zhang M, Augustine RA, Stern JE, Brown CH (2022) alpha-Melanocyte-stimulating hormone inhibition of oxytocin neurons switches to excitation in late pregnancy and lactation. Physiol Rep 10:e15226

Perkinson MR, Kim JS, Iremonger KJ, Brown CH (2021) Visualising oxytocin neurone activity in vivo: the key to unlocking central regulation of parturition and lactation. J Neuroendocrinol 33(11):e13012

Richard P, Moos F, Freund-Mercier MJ (1991) Central effects of oxytocin. Physiol Rev 71:331–370. *(Seminal review on the physiological role of centrally released oxytocin)*

Roizen J, Luedke CE, Herzog ED, Muglia LJ (2007) Oxytocin in the circadian timing of birth. PLoS One 2(9):e922

Rossoni E, Feng J, Tirozzi B, Brown D, Leng G, Moos F (2008) Emergent synchronous bursting of oxytocin neuronal network. PLoS Comput Biol 4:e1000123. *(Mathematical model of the milk-ejection reflex)*

Russell JA, Leng G (1998) Sex, parturition and motherhood without oxytocin? J Endocrinol 157:343–359

Russell JA, Leng G, Douglas AJ (2003) The magnocellular oxytocin system, the fount of maternity: adaptations in pregnancy. Front Neuroendocrinol 24:27–61. *(Review of the plasticity of the oxytocin system in pregnancy)*

Sabatier N, Caquineau C, Dayanithi G, Bull P, Douglas AJ, Guan XM, Jiang M, Van der Ploeg L, Leng G (2003) Alpha-melanocyte-stimulating hormone stimulates oxytocin release from the dendrites of hypothalamic neurons while inhibiting oxytocin release from their terminals in the neurohypophysis. J Neurosci 23:10351–10358

Summerlee AJ (1981) Extracellular recordings from oxytocin neurones during the expulsive phase of birth in unanaesthetized rats. J Physiol 321:1–9. *(Key electrophysiological studies in conscious rats)*

Summerlee AJ, Lincoln DW (1981) Electrophysiological recordings from oxytocinergic neurones during suckling in the unanaesthetized lactating rat. J Endocrinol 90:255–265

Sutherland RC, Juss TS, Wakerley JB (1987) Prolonged electrical stimulation of the nipples evokes intermittent milk ejection in the anaesthetised lactating rat. Exp Brain Res 66:29–34

Tabak BA, Leng G, Szeto A, Parker KJ, Verbalis JG, Ziegler TE, Lee MR, Neumann ID, Mendez AJ (2022) Advances in human oxytocin measurement: challenges and proposed solutions. Mol Psychiatry. (Recommendations on measurement of oxytocin in human plasma)

Tang Y, Benusiglio D, Lefevre A, Hilfiger L, Althammer F, Bludau A, Hagiwara D, Baudon A, Darbon P, Schimmer J, Kirchner MK, Roy RK, Wang S, Eliava M, Wagner S, Oberhuber M, Conzelmann KK, Schwarz M, Stern JE, Leng G, Neumann ID, Charlet A, Grinevich V (2020)

Social touch promotes interfemale communication via activation of parvocellular oxytocin neurons. Nat Neurosci 23:1125–1137

Uvnas-Moberg K, Ekstrom-Bergstrom A, Berg M, Buckley S, Pajalic Z, Hadjigeorgiou E, Kotlowska A, Lengler L, Kielbratowska B, Leon-Larios F, Magistretti CM, Downe S, Lindstrom B, Dencker A (2019) Maternal plasma levels of oxytocin during physiological childbirth – a systematic review with implications for uterine contractions and central actions of oxytocin. BMC Pregnancy Childbirth 19:285

Wakerley JB, Lincoln DW (1973) The milk-ejection reflex of the rat: a 20- to 40-fold acceleration in the firing of paraventricular neurones during oxytocin release. J Endocrinol 57:477–493. *(Key paper reporting the first evidence of milk-ejection bursts)*

Wang YF, Negoro H, Higuchi T (2013) Lesions of hypothalamic mammillary body desynchronise milk-ejection bursts of rat bilateral supraoptic oxytocin neurones. J Neuroendocrinol 25:67–75

Way SA, Leng G (1992) Relaxin increases the firing rate of supraoptic neurones and increases oxytocin secretion in the rat. J Endocrinol 132:149–158

Xi D, Kusano K, Gainer H (1999) Quantitative analysis of oxytocin and vasopressin messenger ribonucleic acids in single magnocellular neurons isolated from supraoptic nucleus of rat hypothalamus. Endocrinology 140:4677–4682

Yoshimura R, Kiyama H, Kimura T, Araki T, Maeno H, Tanizawa O, Tohyama M (1993) Localization of oxytocin receptor messenger ribonucleic acid in the rat brain. Endocrinology 133:1239–1246

Yukinaga H, Hagihara M, Tsujimoto K, Chiang HL, Kato S, Kobayashi K, Miyamichi K (2022) Recording and manipulation of the maternal oxytocin neural activities in mice. Curr Biol 32:3821–3829

Zhang B, Qiu L, Xiao W, Ni H, Chen L, Wang F, Mai W, Wu J, Bao A, Hu H, Gong H, Duan S, Li A, Gao Z (2021) Reconstruction of the hypothalamo-neurohypophysial system and functional dissection of magnocellular oxytocin neurons in the brain. Neuron 109:331–346e337

Further Recommended Reading

Althammer F, Jirikowski G, Grinevich V (2018) The oxytocin system of mice and men-Similarities and discrepancies of oxytocinergic modulation in rodents and primates. Peptides 109:1–8

Arrowsmith S, Wray S (2014) Oxytocin: its mechanism of action and receptor signalling in the myometrium. J Neuroendocrinol 26:356–369

Brown CH (2016) Magnocellular neurons and posterior pituitary function. Compr Physiol 6:1701–1741

Brown CH, Ludwig M, Tasker JG, Stern JE (2020) Somato-dendritic vasopressin and oxytocin secretion in endocrine and autonomic regulation. J Neuroendocrinol 32:e12856

Gimpl G, Fahrenholz F (2001) The oxytocin receptor system: structure, function, and regulation. Physiol Rev 81:629–683

Grinevich V, Ludwig M (2021) The multiple faces of the oxytocin and vasopressin systems in the brain. J Neuroendocrinol 33:e13004

Tang Y, Benusiglio D, Lefevre A, Kuppers S, Lapies O, Kerspern D, Charlet A, Grinevich V (2022) Viral vectors for opto-electrode recording and photometry-based imaging of oxytocin neurons in anesthetized and socially interacting rats. STAR Protoc 3:101032

Lactation and the Control of the Prolactin Secretion

7

Raphael E. Szawka, Stephen J. Bunn, Paul Le Tissier, Siew Hoong Yip, and David R. Grattan

Abstract

Prolactin is a polypeptide hormone secreted from the lactotrophs in the anterior pituitary gland. It is recognized as a pleiotropic hormone with multiple biological actions, which seem to have evolved to provide physiological adaptations required for the success of reproduction. Among its numerous roles, lactation is one to which prolactin is essential, being the primary hormone responsible for the synthesis of milk. The neuroendocrine control of prolactin secretion is unique among the anterior pituitary hormones. The lactotrophs have a high spontaneous capacity for proliferation and secretion of prolactin that is suppressed by dopamine. Neuroendocrine dopaminergic (NEDA) neurons release dopamine into the hypothalamic-pituitary portal blood system to inhibit prolactin secretion. Prolactin exerts a short-loop negative feedback effect stimulating several aspects of NEDA neuronal activity, resulting in inhibition of its own secretion. The nipple suckling by the offspring during lactation is the most powerful stimulus to increase prolactin secretion. Thus, lactation is a hyperprolactinemic state, characterized by suckling-induced prolactin surges and chronically elevated basal

R. E. Szawka (✉)
Department of Physiology and Biophysics, Institute of Biological Sciences, Federal University of Minas Gerais (UFMG), Belo Horizonte, Brazil
e-mail: reszawka@icb.ufmg.br

S. J. Bunn · S. H. Yip · D. R. Grattan
Centre for Neuroendocrinology and Department of Anatomy, University of Otago, Dunedin, New Zealand
e-mail: joe.yip@otago.ac.nz; dave.grattan@otago.ac.nz

P. Le Tissier
Centre for Discovery Brain Sciences, University of Edinburgh, Edinburgh, UK
e-mail: Paul.LeTissier@ed.ac.uk

P. J. Brunton, D. R. Grattan (eds.), *Neuroendocrine Regulation of Mammalian Pregnancy and Lactation*, Masterclass in Neuroendocrinology 15,
https://doi.org/10.1007/978-3-031-51138-7_7

levels of prolactin. Phenotypic and functional modifications in the NEDA neurons associated with changes in the number and organization of lactotrophs are responsible for the remarkable increase in prolactin secretion during lactation. Hypothalamic prolactin-releasing factors seem to also stimulate the rise in prolactin secretion, but their identity remains to be determined. In this chapter, we review the classical concepts and recent advances in our understanding of the neuroendocrine control of prolactin secretion and how it changes to promote the physiological state of hyperprolactinemia during lactation.

Keywords

Dopamine · Hypothalamus · Lactation · Lactotroph · Prolactin-releasing factor · Prolactin

7.1 Introduction

Prolactin is a polypeptide hormone secreted from the lactotrophs in the anterior pituitary gland that is essential to lactation and milk production in mammals. Prolactin was given its name based on its effect of inducing enlargement and function of the crop sac in pigeons and milk secretion in rodents (Riddle et al. 1933), but it should be kept in mind that it is present in all vertebrates apart from jawless fishes, and therefore likely to have many more ancestral functions (Dobolyi et al. 2020). Prolactin belongs to a large family of homologous proteins, known as the group I of the helix bundle peptide hormones, which encompasses prolactin, growth hormone and placental lactogen (Horseman and Yu-Lee 1994). The genes encoding these hormones appear to have evolved from a single ancestral gene, and it is postulated that the chromosomal duplication segregating prolactin and growth hormone genes occurred about 392 million years ago (Cooke et al. 1981). Rodents express the prolactin gene on chromosome 13 (mice) or 17 (rats), which is expressed mainly in the anterior pituitary gland, although it can be expressed in other tissues such as the **decidua** and the mammary glands. In humans, the prolactin gene is located on chromosome 6. It is expressed in the pituitary gland but displays a more widespread expression in other tissues than in rodents due to the presence of an additional supra-distal promoter that drives extra-pituitary expression (Ben-Jonathan et al. 2008). Circulating prolactin is secreted mainly by the lactotrophs, which in rodents represent between 20% and 35% of the pituitary cells, with a female-biased sexual dimorphism (Tierney and Robinson 2002).

A variety of biological actions have been attributed to prolactin since its identification. Accordingly, prolactin has long been recognized as a **pleiotropic hormone** with more biological actions than all the other pituitary hormones combined (Nicoll 1980). Prolactin actions were thus distributed in different categories such as water and electrolyte balance, growth and development, endocrinology and metabolism, brain and behavior, reproduction, and immune-regulation and protection (Bole-Feysot et al. 1998). More recently, it has been proposed that many of the apparently diverse prolactin actions can be seen as directed towards the primary outcome of successful reproduction, promoting adaptive changes required by the organism to cope with the physiological demands of pregnancy and lactation (Grattan and Le Tissier 2015). Nevertheless, among its multiple homeostatic functions, lactation is one to which prolactin is essential. Although the hormonal role in mammary gland development and induction and maintenance of milk production varies across species, prolactin is the primary hormone responsible for the synthesis of milk. Thus, the powerful stimulus of nipple suckling by the young to increase prolactin secretion is common to all mammals. In this chapter, we aim to discuss the classical concepts as well as the recent advances in our understanding of the neuroendocrine control of prolactin secretion and how it changes to promote the physiological state of hyperprolactinemia during lactation.

7.2 The Prolactin Axis

The secretory output of the anterior pituitary gland is controlled by the hypothalamus through the release of hormones into the hypophyseal portal blood. In contrast to the other pituitary hormones, secretion of which demands stimulation by hypothalamic peptides, prolactin is released spontaneously by the lactotrophs. Thus, the control of prolactin secretion is unique because it is characterized by a predominant hypothalamic inhibition. Early studies showed that lactotrophs have a high intrinsic capacity to secrete prolactin, whenever the hypothalamic inhibition is interrupted. For instance, **hypophysectomized** rats, in which the pituitary gland was autotransplanted under the kidney capsule, were found to maintain functional corpora lutea for several months (Everett 1956). At the time, this suggested an increased secretion of prolactin, which is a powerful *luteotropic* hormone in rodents. Accordingly, ectopic pituitary grafts were also reported to be able to maintain lactation in hypophysectomized rats (Cowie et al. 1960). With the advent of hormonal measurements by radioimmunoassay in the 1970s, it was possible to demonstrate the presence of high prolactin levels after pituitary stalk section (Kanematsu and Sawyer 1973), consistent with the seminal concept of its tonic inhibition by a hypothalamic factor secreted into the portal blood.

Prolactin effects are exerted through the prolactin receptor, which belongs to class 1 of the cytokine receptor superfamily. It is a single-pass transmembrane protein, devoid of intrinsic tyrosine kinase activity, that is phosphorylated by receptor-associated proteins after formation of receptor-prolactin-receptor heterodimers. There are multiple different receptor isoforms that are formed by post-transcriptional processing of the prolactin receptor transcript, differing mostly in the intracellular domains. A number of shorter isoforms in different species can activate some of the cellular signaling pathways, but the full signal transduction requires the long form of the prolactin receptor. For example, the tyrosine residues present in the long form of the prolactin receptor are phosphorylated by the receptor-associated Janus kinase 2 (Jak2) to create docking sites of Src homology 2 (SH2) domain, of which the main substrates are the signal of transducer and activator of transcription (STAT)-5. The prolactin receptor has widespread and variable expression in the body. In the female rat, for instance, both short and long isoforms are highly expressed in the ovaries, choroid plexus, liver, adrenal cortex, anterior pituitary gland, kidney, mammary glands, hypothalamus and uterus (for references see Bole-Feysot et al. 1998). This has been further emphasized in mouse models using genetically targeted reporters and showing prolactin receptor expression in all of these tissues (Aoki et al. 2019).

7.2.1 Dopamine: The Hypothalamic Prolactin-Inhibiting Hormone

The catecholamine dopamine has long been established as the hypothalamic hormone responsible for inhibition of prolactin secretion (for references see Ben-Jonathan and Hnasko 2001). This is interesting, as all of the other anterior pituitary gland hormones are regulated by hypothalamic neuropeptides, again making the prolactin system stand out as distinct from other pituitary hormones. As expected for a hypothalamic prolactin-inhibiting hormone, dopamine is present in high concentrations in the median eminence of the hypothalamus and pituitary portal blood, and dopamine levels in the portal circulation are frequently inversely related to changes in pituitary prolactin secretion. Lactotrophs express the dopamine D2 receptor (D2R) subtype, which belongs to the D2-like class of inhibitory receptors that suppress adenylyl cyclase activity. Accordingly, dopamine can directly suppress prolactin release from lactotrophs *in vitro*, while reduction in the availability of dopamine markedly increases prolactin secretion. Moreover, as part of the inhibitory control, dopamine plays a role in restraining the spontaneous capacity of the lactotroph for proliferation. For instance, the lack of this anti-proliferative role in D2R-deficient mice results in hyperprolactinemia and lactotroph hyperplasia that develops into adenoma in aged females. It is accepted that dopamine inhibition of prolactin secretion is also present in humans. This is evidenced by the fact that prolactinomas, a prevalent benign pituitary tumor, are commonly treated with D2R agonists, such as bromocriptine and cabergoline, which are effective at reducing prolactin levels and tumor size in most patients. Despite the unquestionable role of dopamine in suppressing prolactin secretion, it has also been reported that very low concentrations of dopamine, a thousand-fold lower than those required for maximal inhibition, can stimulate prolactin release from pituitary cells *in vitro* (for references see Freeman et al. 2000). However, the use of dynamic *in vitro* systems has further demonstrated that dopamine suppression of prolactin is time-dependent and that a rebound increase in its secretion occurs after dopamine exposure (Denef et al. 1984; He et al. 2011). These findings may explain the earlier reports of a paradoxical stimulatory effect of dopamine. Thus, considering the physiological regulation of lactotrophs, dopamine produced by hypothalamic neurons and released into the hypophyseal portal blood vasculature tonically inhibits prolactin secretion, representing the main hypothalamic output in the control of the prolactin axis (Fig. 7.1). This characteristic of these hypothalamic dopamine neurons to release their "neurotransmitter" into the blood, as distinct from the typical neurotransmission at

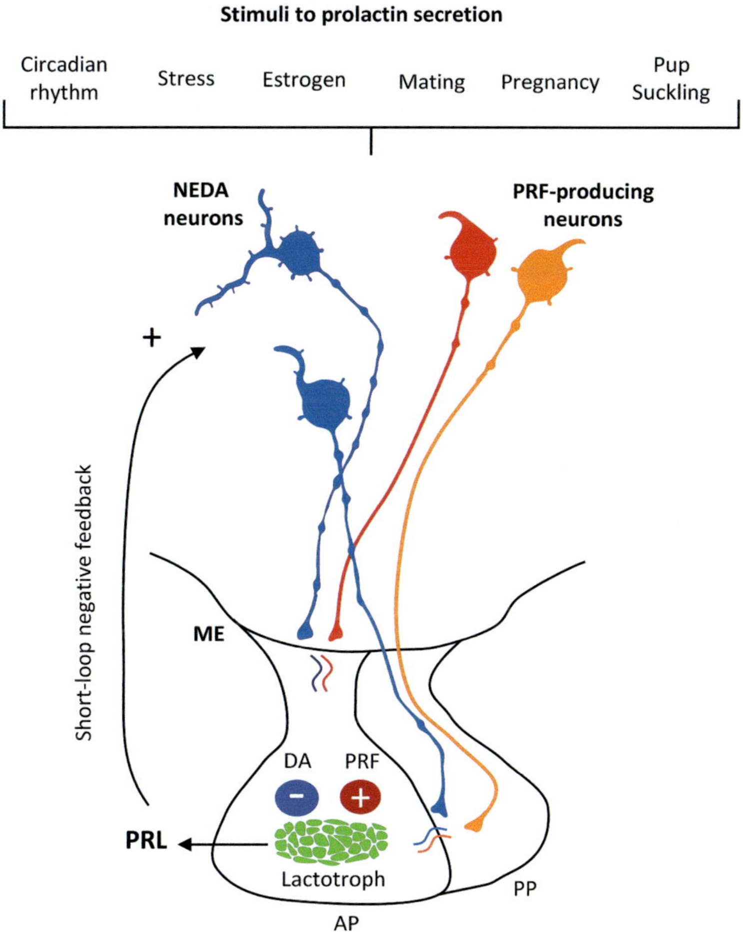

Fig. 7.1 Diagrammatic representation of the prolactin (PRL) axis. Neuroendocrine dopaminergic (NEDA) neurons release dopamine (DA) into the hypophyseal portal blood vasculature at the median eminence (ME) of the hypothalamus and the posterior pituitary (PP). DA acts on the lactotrophs of the anterior pituitary (AP) to tonically suppress PRL secretion. PRL auto-regulates its own secretion through a short-loop negative feedback to stimulate NEDA neuronal activity and DA release. Putative prolactin-releasing factors (PRFs) may exert stimulatory effects on the lactotrophs and increase PRL secretion. A PRF is produced by hypothalamic neurons and secreted into the pituitary portal blood at the ME or PP. PRL secretion responds to a variety of environmental and physiological stimuli, consistent with its diverse biological actions. These stimuli modulate the secretory activity of NEDA and PRF-producing neurons to increase PRL secretion

synaptic connections between neurons, enables us to broadly define them as the neuroendocrine dopaminergic (NEDA) neurons. It remains possible, however, that some of the NEDA neurons also use dopamine as a neurotransmitter in more classical neural networks, both within and outside of the hypothalamus.

At least three distinct populations of hypothalamic NEDA neurons are recognized to provide dopamine to the anterior pituitary gland (see Sect. 7.3.1). Among them, the tuberoinfundibular dopaminergic (TIDA) neurons, whose cell bodies are located in the arcuate nucleus of the hypothalamus and projections reach the capillaries of the portal blood at the median eminence, are the best characterized and possibly the main population responsible for prolactin regulation. Earlier functional studies estimate that the amount of dopamine released by the TIDA neurons into the pituitary stalk blood is responsible for approximately two-thirds of the tonic inhibition of prolactin secretion (Leong et al. 1983). The biosynthesis of dopamine is under the control of tyrosine hydroxylase (TH), which catalyzes the rate-limiting step in its production, the conversion of tyrosine to L-3,4-dihydroxyphenylalanine (L-Dopa). TH is activated by phosphorylation at the serine residues 8, 19, 31, and 40, with the latter having the greatest effect. Thus, the degree of TH phosphorylation in the median eminence reflects the state of enzymatic activity of TIDA neurons (Liu and Arbogast 2008). Once enzymatically converted from L-Dopa, dopamine is stored in secretory vesicles at nerve terminals. The fusion of the vesicles with the plasma membrane releases dopamine into the synaptic cleft or the extracellular space, as in the case of the NEDA neurons. After secretion, dopamine can be recaptured by the nerve terminals through the dopamine transporter (DAT) and metabolized by monoamine oxidase, present in the external membrane of the mitochondria, to the inactive metabolite dihydroxyphenylacetic acid (DOPAC). For this reason, neurochemical measurement of DOPAC levels in the median eminence reflects dopamine release by TIDA neurons (Lookingland et al. 1987), and the DOPAC/dopamine ratio can be used as an index of rapid changes in the activity of TIDA terminals (Aquino et al. 2019) (Fig. 7.2). Because TIDA neurons release dopamine into the blood instead of a synaptic cleft, the actual role of dopamine reuptake has been a long-standing question. It would be predicted that most dopamine would immediately diffuse into the blood and therefore not be available for recapture. Nevertheless, DAT is expressed in the arcuate-median eminence region, with experimental evidence indicating that TIDA neuron reuptake of dopamine may be functionally relevant to prolactin secretion (for references see Ben-Jonathan and Hnasko 2001). A detailed investigation in transgenic mice has demonstrated, however, that less than half of TIDA neurons express DAT (Yip et al. 2018), suggesting that only a subpopulation of these neurons are involved in the reuptake and metabolism of dopamine. This view is consistent with the existence of different subpopulations of hypothalamic dopaminergic neurons, as demonstrated by single-cell RNA sequencing of the mouse hypothalamus (Campbell et al. 2017; Romanov et al. 2017). Moreover, as will be described in Sect. 7.3.2, recent electrophysiological evaluations have provided new insights into our understanding of the functionality of TIDA neurons at the cellular level. Thus, future studies should determine how the electrical activity of functional subpopulations of TIDA neurons changes in different physiological states and what are the consequences for their secretory output and, therefore, prolactin secretion.

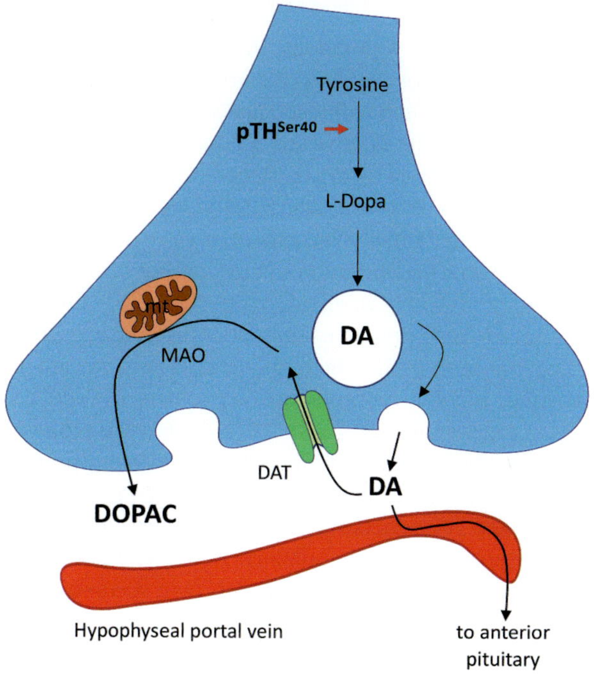

Fig. 7.2 Release and metabolism of dopamine (DA) from the terminals of neuroendocrine dopaminergic (NEDA) neurons. The biosynthesis of DA is under the control of tyrosine hydroxylase (TH), which catalyzes the conversion of tyrosine to L-Dopa. TH is activated by phosphorylation at the serine residues, especially serine 40 (pTH^{Ser40}). Upon conversion from L-Dopa, catalyzed by the enzyme aromatic L-amino acid decarboxylase, DA is stored in secretory vesicles. The fusion of the vesicles with the plasma membrane releases DA in the proximity of the vasculature of the hypophyseal portal blood system. After secretion, DA can be recaptured by the nerve terminals through the DA transporter (DAT) and metabolized by monoamine oxidase (MAO), present in the external membrane of the mitochondria, to the inactive metabolite dihydroxyphenylacetic acid (DOPAC). The measurement of DOPAC levels in the median eminence reflects DA release, and the DOPAC/DA ratio provides an index of rapid changes in the activity of NEDA terminals

7.2.2 Prolactin Autoregulation by Short-Loop Negative Feedback

Prolactin auto-regulates its own secretion through a short-loop negative feedback effect on NEDA neurons. Through this mechanism, prolactin directly stimulates the activity of NEDA neurons, which increases dopamine concentrations in the portal blood suppressing prolactin secretion. As a polypeptide hormone, prolactin would not be expected to diffuse across the **blood-brain barrier**. Nevertheless, circulating prolactin is known to enter the cerebrospinal fluid, and prolactin levels in this compartment reflect changes in its systemic concentration. The mechanism seems to involve a carrier-mediated, saturable transport that is not completely understood, but is not dependent on the prolactin receptor (Brown et al. 2016a). Using this mechanism, prolactin can rapidly cross the blood-brain barrier and activate its receptors in multiple neuronal populations, including the NEDA neurons, as indicated by using STAT5 phosphorylation as a marker of prolactin receptor activation. Furthermore, because the median eminence lacks a blood-brain barrier, circulating prolactin can readily access the neurosecretory terminals of TIDA neurons in this region. Finally, there is evidence of the existence of a retrograde flow in the portal blood vasculature (Oliver et al. 1977), which could locally transport prolactin from the pituitary to the hypothalamus. Thus, there are different ways through which secreted prolactin can reach the *perikarya* or terminals of NEDA neurons. However, which pathway is more prevalent *in vivo* remains to be fully elucidated. As further discussed later in this chapter, prolactin stimulates different aspects of TIDA neuron function, including gene expression, their neurochemistry, and electrophysiological activity. On the other hand, hypoprolactinemia results in reduced dopaminergic activity. Thus, there is a homeostatic feedback loop between prolactin and hypothalamic dopamine. Accordingly, the neuron-specific deletion of prolactin receptors in female mice results in hyperprolactinemia (Brown et al. 2016b), and a loss-of-function variation in the gene encoding the prolactin receptor has been associated with idiopathic hyperprolactinemia in humans (Kobayashi et al. 2018). Thus, under normal circumstances, prolactin auto-regulates its own secretion through a negative feedback loop with hypothalamic NEDA neurons. Importantly, however, as described later in this chapter, this homeostatic control changes markedly during lactation, and in doing so plays a crucial role is establishing a state of high prolactin secretion.

7.2.3 Role of Prolactin-Releasing Factors (PRFs)

While the inhibitory effect of hypothalamic dopamine on prolactin secretion was being characterized, parallel efforts in search of a hypothalamic PRF were also underway, driven by the expectation that such a releasing factor should exist, as it does for all other anterior pituitary hormones. As is the case for any hypothalamic **hypophysiotropic** hormone, a PRF should be produced in the hypothalamus and secreted into the pituitary portal blood at greater concentrations than in the general circulation. Moreover, a PRF would be expected to act directly on the lactotrophs to stimulate prolactin release and should fluctuate in the portal blood in direct association with physiological changes in prolactin secretion. After decades of ongoing research in the field, however, the question of whether there is one or more PRFs and their identity if so, has never been convincingly resolved. Dopamine antagonists, by themselves, are sufficient to markedly increase prolactin secretion, meaning that a physiological PRF may not be necessary to explain many situations of prolactin secretion. Nevertheless, it is safe to say that the hypothalamus contains multiple molecules capable of stimulating prolactin release from the pituitary and, importantly, the physiological increases in prolactin secretion are, in many cases, better explained by the involvement of a PRF, acting in concert with a reduction in inhibitory dopaminergic tone. Early studies observed that hypothalamic extracts could induce milk production in the mammary gland when injected into estrogen-treated rats, which was later shown to be due to the increase in prolactin secretion (Valverde et al. 1972). In this sense, thyrotropin-releasing hormone (TRH) was one of the first hypothalamic peptides identified as a potent prolactin stimulator. However, the prolactin-releasing activity of porcine hypothalamic extracts does not depend on the presence of TRH or vasopressin (Boyd et al. 1976). Likewise, neither TRH nor thyroid-stimulating hormone (TSH), its downstream anterior pituitary gland hormone, is increased in response to suckling, which excludes the role of TRH as a physiological PRF during lactation in mammals. While removal of dopaminergic inhibition could account for a rise in prolactin secretion, studies evaluating stress- or nipple suckling-induced prolactin secretion have provided evidence that it is not this simple. For example, stress-induced prolactin release occurs in the face of a pharmacological "clamp" of the dopamine D2R (Shin 1980), and the measured fall in dopamine levels in the portal blood was deemed insufficient to account for the scale of increase in prolactin secretion measured after mammary gland nerve stimulation (de Greef et al. 1981), supporting the existence of a physiological PRF. In line with this view, a putative PRF originating from the posterior pituitary gland was found to be involved in the suckling-induced release of prolactin in lactating rats (Murai and Ben-Jonathan 1987). This surprising observation, with its completely novel origin for a pituitary regulatory factor, provided a potential explanation for why so many studies looking for a hypothalamic releasing factor had not found it. However, despite many efforts over the years, the chemical identity of the PRF of the posterior pituitary and its possible physiological role remain elusive. Thus, this posterior pituitary PRF joins the many hypothalamic peptides, such as TRH, angiotensin 2, oxytocin, vasopressin, neurotensin and **enkephalin**, that have been found to exert prolactin-releasing activity (for references see Freeman et al. 2000), but none of these have accumulated enough evidence to be considered the physiological PRF. There was also much interest in the proposal that the dopamine metabolite, salsolinol, could be a PRF from the posterior pituitary gland involved in prolactin release induced by stress and lactation. Salsolinol stimulates prolactin secretion after intravenous injection in rats and increases prolactin release from pituitary cell cultures, although the latter effect is much less prominent (Toth et al. 2001). Subsequent investigation of its physiological effects, however, have revealed that salsolinol most likely acts via inhibition of dopamine release from the neuroendocrine dopaminergic terminals, which would indirectly cause the increase in prolactin secretion observed (Hashizume et al. 2017). This is a common complication with studies of this sort, in that the powerful role of dopamine and the prolactin-releasing consequences of dopamine withdrawal make interpretation complicated. Thus, although there is compelling evidence of the prolactin-releasing activity of the hypothalamus, the identity of the hypothalamic PRF remains unknown. It seems, therefore, that chemical identification and physiological characterization of one or more PRFs remain relevant questions, and have proven to be difficult ones, about the neuroendocrine control of prolactin secretion (Fig. 7.1). Perhaps we may come to understand that there are diverse PRF-like molecules more or less specifically involved with the different sorts of stimuli and patterns of prolactin secretion, all in concert with changes in dopamine inhibition. This hypothesis is consistent with the pleiotropic nature of prolactin function and may become another feature distinguishing prolactin from the other hormones of the anterior pituitary gland. Nonetheless, as described in the next section, accumulating evidence indicates a critical role of a PRF in the suckling-induced increase in prolactin secretion during lactation.

7.2.4 Prolactin Secretion During Lactation

Due to its major role in milk production, nipple suckling by the offspring is the most powerful stimulus to increase prolactin secretion and is a common feature of all mammals. Nevertheless, prolactin release also responds to a variety of other environmental and physiological stimuli, which sometimes generate complex patterns of hormonal secretion. This wide range of secretory responses reflects the numerous biological actions of prolactin in different species and, likewise, involves distinct neuroendocrine mechanisms of hormonal control (for references see Freeman et al. 2000; Grattan and Le Tissier 2015).

Before effectively attaining lactation, females display alterations in prolactin secretion in response to estrogens, mating and pregnancy. As a major prolactin secretagogue, estradiol acts at the lactotrophs to increase prolactin production (Yamamoto et al. 1986) and also negatively modulates hypothalamic dopaminergic neurons to increase prolactin secretion (Arbogast and Hyde 2000). Accordingly, prolactin secretion fluctuates during the ovarian cycle in response to changes in estradiol levels, although the profile of secretion varies between species. A pattern of periodic secretion is evident in the rat estrous cycle, in which a major surge of prolactin secretion occurs on the afternoons of proestrus, coinciding with the preovulatory surge of luteinizing hormone. On the following day, a second surge of smaller magnitude is reported in some studies on the afternoon of estrus—the day of the cycle when the ovulation occurs (Smith et al. 1975; Szawka et al. 2005). These prolactin surges depend on estradiol stimulation and are synchronized to the afternoon period by an endogenous circadian rhythm, whose signaling is probably conveyed by daily suppression of NEDA neuronal activity (Mai et al. 1994). On the other hand, a well-defined prolactin surge does not seem to occur during the mouse estrous cycle. Instead, a blunted increase in prolactin secretion has been reported on proestrus, with noticeable differences between mouse strains. In women, it remains controversial whether prolactin displays a mid-cycle surge and, when detected, the rise appears not to be as prominent as that observed in the rat (Djahanbakhch et al. 1984; Phillipps et al. 2022; Vekemans et al. 1977). Thus, there are species differences concerning the estrogenic regulation of prolactin secretion that need to be further clarified. The same is true for the physiological relevance of the estrogenic regulation of prolactin secretion that has not been clearly identified so far. What is clear, however, is that estradiol action contributes to the significant sex difference in circulating prolactin levels seen in many species, with prolactin markedly higher in females.

In mammalian species, prolactin secretion is markedly increased during pregnancy, when it promotes mammary gland development and plays a role in a variety of physiological and behavioral adaptations required for the success of gestation. There are species differences in the pattern of prolactin secretion during pregnancy as well, which are especially remarkable in early pregnancy (see Chap. 1). For example, while in humans the ovulatory corpus luteum and, therefore, progesterone secretion required for gestation, is rescued by human chorionic gonadotropin after implantation, in rodents, prolactin plays a luteotropic role that is critical for progesterone secretion and maintenance of pregnancy. Rats and mice have an ultra-short-lived corpus luteum and, therefore, display luteal phases of only a few hours that would not typically allow sufficient time for embryo implantation. It is the increased prolactin secretion induced by the vaginocervical stimulation during mating that rescues the corpus luteum, supporting a functional luteal phase and subsequent pregnancy, in the case of embryo implantation. In rodents, mating triggers a unique neuroendocrine reflex that generates twice-daily surges of prolactin secretion, one nocturnal and one diurnal, synchronized to the light-dark cycle. Likewise, sterile mating or copulomimetic stimulation of the uterine cervix results in the same pattern of prolactin secretion (Gunnet and Freeman 1983), which maintains functional corpora lutea in a physiological state called pseudopregnancy. This semi-circadian pattern of secretion recurs for 10 days after fertile mating or for 12–13 days in pseudopregnancy (Butcher et al. 1972; Smith et al. 1975). By mid-pregnancy, the placenta of rodents secretes increasing amounts of placental lactogens, which are homologous proteins and can activate the prolactin receptor. Thus, the earlier interruption of the prolactin surges in pregnancy in comparison to pseudopregnancy is attributed to the activation of NEDA neurons by placental lactogens, mimicking the prolactin negative feedback effect. Pituitary prolactin levels then remain low during the second half of pregnancy until they rise again on the day before birth (Tonkowicz and Voogt 1983). This latter surge is induced by the change in ovarian steroid secretion that occurs before parturition, i.e., the increase in estradiol secretion from growing ovarian follicles associated with the decline in progesterone secretion leading up to parturition. Hence, the late-pregnancy surge of prolactin occurs despite the high circulating levels of placental lactogens, which in normal conditions would be expected to stimulate dopamine release from NEDA neurons and, thereby, prevent further prolactin release. However, the occurrence of a secretory surge at this moment indicates that the prolactin negative feedback loop is not operating by the end of pregnancy. Similarly, prolactin secretion progressively increases throughout pregnancy in women, concurrent with the rise in human placental lactogen. These data suggest that there is a progressive loss of the feedback regulation of prolactin in human pregnancy, too. As detailed below, the modification in the homeostatic control of prolactin secretion seems to involve striking plasticity in the secretory activity of NEDA neurons, in which met-enkephalin replaces dopamine, and this alteration persists during lactation to favor high prolactin release.

The increase in prolactin secretion in response to the suckling by the nursing young is defined as a classical neuroendocrine reflex. As such, the mechanical stimulation of the nipple is signaled by the ascending somatosensory neuronal system and, through a polysynaptic pathway, eventually reaches the hypothalamus and pituitary gland to increase prolactin secretion. This same stimulus results in the concomitant release of oxytocin that, by the contraction of the myoepithelial cells surrounding the mammary alveoli, will promote the milk flow through ducts up to the nipple, feeding the offspring ("milk let-down"—see Chap. 6). Despite being driven by the same stimulus, the pattern of secretion of the two hormones is quite different. Oxytocin is released in discrete "milk ejection bursts", whereas prolactin is secreted continuously directly in proportion to the intensity and duration of the suckling stimulation, so milk is produced on demand, reflecting the amount of prolactin released in the previous lactation session. Oxytocin can also become a "conditioned reflex", where sensory cues associated with feeding can induce bursts of oxytocin secretion in preparation for feeding. While in women, the conditioned signals that release oxytocin are not followed by prolactin secretion—this requires the physical suckling stimulus (Noel et al. 1974), an increase in prolactin secretion has been reported in lactating rodents when exposed to exteroceptive stimuli from their offspring (Grosvenor et al. 1981). Essentially, by controlling the total amount of prolactin released, the baby is regulating the amount of milk production (e.g., twins will stimulate twice as much prolactin), and the suckling-induced prolactin will induce the synthesis of the appropriate amount of milk to be available in the next meal. The response of prolactin to suckling has been well characterized in the rat, in which the number of nursing pups and the intensity and duration of nipple suckling influence prolactin release. In this species, prolactin release increases within a few minutes of suckling initiation and reaches maximum levels after approximately 15 min, after which time prolactin levels remain elevated in a plateau for as long as nursing continues. Prolactin will rapidly fall once nursing is finished, reflecting the abrupt cessation of its secretion by the lactotrophs (Grosvenor and Whitworth 1974; Terkel et al. 1972). The pituitary content of prolactin is depleted, while prolactin mRNA expression increases during lactation. Thus, both stored and newly synthesized prolactin are expected to contribute to the remarkable rise in blood levels (Lee et al. 1989; Mena and Grosvenor 1968), and the amount of prolactin secreted in response to suckling is attenuated with the advance of lactation (Shanti et al. 1995). In addition, in different species, such as rats and women, prolactin secretion during lactation continues to be under the influence of endogenous circadian signals and, therefore, the magnitude of the suckling response changes according to the time of the day (Arey et al. 1991; Diaz et al. 1989). It is worth noting that despite the widespread development of genetically modified mouse models over the past decades and their valuable contribution to the advance of basic research, the detailed assessment of prolactin secretion in the mouse has been limited by its small size. However, the pattern of prolactin secretion in lactating mice has been determined recently using an ultrasensitive prolactin enzyme-linked immunosorbent assay (ELISA) (Guillou et al. 2015). This study reported a robust increase in prolactin secretion in response to suckling, as predicted by the current knowledge in rats and other species and, most importantly, it opened the possibility of further investigation on the mechanisms controlling prolactin secretion by the use of genetically modified mouse lines.

Box 7.1 Measuring Prolactin Levels in Tail-Tip Blood

Ultrasensitive ELISA assays can be used to measure prolactin in a very small volume of whole blood (around 5 μL) collected from mice or rats. The assays were developed as sandwich ELISAs using a combination of commercially available anti-prolactin and horseradish peroxidase-conjugated antibodies (Aquino et al. 2017; Guillou et al. 2015). When combined with the method of tail-tip blood sampling, this assay allows for the long-term monitoring of prolactin secretion in freely moving animals, without the stress influence caused by the surgeries of blood vessel cannulation or the withdrawal of large volumes of blood, commonly associated with the traditional methods of blood collection and hormonal measurement by radioimmunoassay. A key element of this approach is the habituation of the animals to the sampling procedures. With many days of training, mice and rats will become used to being handled, and basal levels of prolactin can be detected without apparent stress (which might otherwise elevate prolactin secretion).

The hyperprolactinemic state of lactation is promoted by the suckling stimulation and is associated with the proliferation of the lactotrophs and alterations in the negative feedback control of prolactin secretion—changes initiated earlier during pregnancy. The acute surge of prolactin secretion in response to suckling requires the reduction in the inhibitory dopaminergic tone and, possibly, the additional stimulatory effect of a PRF. Accordingly, during lactation, there is a marked reduction in the biosynthesis and release of dopamine by TIDA neurons, a necessary condition to create the state of chronic hyperprolactinemia. In line with reduced dopaminergic function, prolactin negative feedback action on NEDA neurons is also attenuated during lactation (Arbogast and Voogt 1996; Demarest et al. 1981), which further contributes to the suppression of dopaminergic inhibition. There is evidence that this change in the homeostatic control of prolactin secretion starts in late pregnancy when prolactin loses the property of inhibiting its own secretion (Fliestra and Voogt 1997; Grattan and Averill 1995). Contrarily to what was previously expected, the expression of prolactin receptors on TIDA neurons is not downregulated in lactating rats (Kokay and Grattan 2005). However, STAT5 signaling, required to mediate prolactin's actions, is attenuated (Grattan et al. 2001), while its endogenous inhibitor, suppressor of cytokine signaling (SOCS), is upregulated (Anderson et al. 2006a), indicating alterations in the cell-signaling pathways being recruited by the prolactin receptor in TIDA neurons during lactation. Consistent with the notion of altered intracellular pathways, prolactin can still increase the firing rate of TIDA neurons in lactating mice, but this neuronal activation no longer results in dopamine release at the median eminence (Romano et al. 2013). Remarkably, the TIDA neurons of lactating rats, but not non-lactating females or male rats, co-express enkephalin (Merchenthaler 1993). This change emerges during pregnancy and seems to depend on the chronically elevated placental lactogen levels acting through the prolactin receptor (Merchenthaler 1994; Nahi and Arbogast 2003). Enkephalin, in turn, is likely to act locally to inhibit the activity of TIDA neurons and thereby stimulate prolactin secretion (Arita and Kimura 1988). The same phenotypic alteration is found in the mouse. It has been shown that prolactin induces enkephalin release in the median eminence of lactating mice, and that enkephalin can directly increase Ca^{2+} signaling and prolactin release from the lactotrophs (Yip et al. 2019). Therefore, the action of prolactin to increase enkephalin release by TIDA neurons at this time results in both indirect (TIDA neuron inhibition) and direct (lactotroph stimulation) facilitation of its own secretion, creating a sort of positive feedback loop in contrast to the negative feedback system prevalent in the non-lactating state. This switch in prolactin's auto-feedback enables prolactin levels to be chronically elevated even during the non-suckling periods of lactation, unopposed by negative feedback regulation (Fig. 7.3).

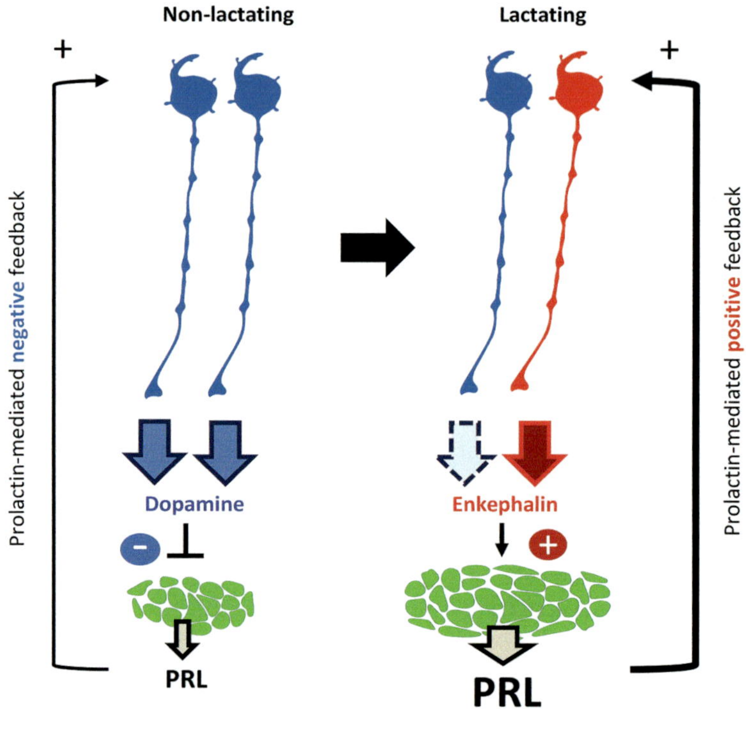

Fig. 7.3 The switch in prolactin (PRL)'s auto-feedback regulation enabling chronically elevated hormonal secretion during lactation. In non-lactating rodents, PRL inhibits its own secretion through a short-loop negative feedback that stimulates TIDA neurons to release dopamine into the portal blood (blue neurons). Under the influence of high PRL levels, a subpopulation of TIDA neurons undergo a phenotypic switch from producing dopamine to producing the endogenous opioid peptide met-enkephalin (red neurons) during lactation. This PRL action to increase enkephalin release by TIDA neurons results in both indirect (via TIDA neuron inhibition) and direct (via lactotroph stimulation) facilitation of its own secretion, creating a positive feedback loop in contrast to the negative feedback loop operating in the non-lactating state. Figure adapted from Phillipps et al. (2020)

In lactating rats, suckling decreases TH gene expression in the arcuate nucleus and phosphorylation of TH in the nerve terminals of the TIDA neurons at the median eminence (Berghorn et al. 2001; Feher et al. 2010), resulting in reduced synthesis and therefore lower release of dopamine into the hypophyseal portal blood (Ben-Jonathan et al. 1980). This response is essential for the hyperprolactinemic state of lactation, which is confirmed by the well-known effect of the D2R agonist, bromocriptine, blocking suckling-induced prolactin release. However, the actual role played by the already reduced levels of dopamine in the massive prolactin surge in response to suckling remains controversial. While there are reports of decreased dopamine turnover in the median eminence in response to suckling in lactating rats (Selmanoff and Wise 1981), measurements in the portal blood during electrical stimulation of the mammary nerve revealed only a brief decrease in dopamine levels that was not coincident with the prolonged increase in prolactin secretion (de Greef et al. 1981). Thus, the subtle additional decrease in dopamine release during nipple suckling does not seem to account for the acute rise in prolactin secretion. This view has led to a long-standing hypothesis that proposes that the reduced dopaminergic tone during lactation would make the lactotrophs responsive to a suckling-induced PRF. This is supported by earlier studies in lactating rats showing that a dose of TRH unable to stimulate prolactin secretion before suckling is effective when injected after it (Grosvenor and Mena 1980) and, likewise, that a transitory suppression of dopamine production would facilitate the increase in prolactin induced by TRH (Plotsky and Neill 1982). In line with these findings, while the pharmacological blockade of pituitary DR2 with domperidone in lactating rats increases basal prolactin secretion, it does not prevent the suckling-induced prolactin rise, and a secretory surge of physiological profile occurs over the elevated baseline of prolactin levels (Silva et al. 2020). Thus, the reduced release of dopamine by NEDA neurons during lactation appears to be required to sensitize the pituitary to a suckling-induced PRF (Fig. 7.4). As mentioned above, it has been demonstrated that posterior pituitary ablation prevents the suckling-induced rise in prolactin, suggesting the existence of a PRF in the posterior pituitary gland distinct from oxytocin or vasopressin (Murai and Ben-Jonathan 1987). Moreover, studies using discrete lesions of the rat paraventricular nucleus have suggested that this region also provides stimulatory factors for suckling-induced prolactin release (Bodnar et al. 2002). Although these pharmacological and lesion studies provide compelling evidence of a suckling-induced PRF, its identity remains to be uncovered.

The neuronal network responsible for transmitting the suckling information from the somatosensory ascendant system to the hypothalamus is still not fully understood. The suckling stimulus travels through somatosensory afferents into the spinal cord and brainstem to reach the hypothalamus. Extensive evidence implicates the hypothalamic endogenous opioid peptides (β-endorphin, dynorphin and enkephalin) in the control of prolactin secretion during lactation. Accordingly, neutralization of the effects of opioid peptides in the brain (e.g., using the broad-spectrum opioid receptor antagonist, naloxone, or a range of more specific approaches) prevents the reduction in TIDA neuronal activity and diminishes the rise in prolactin induced by suckling (Arbogast and Voogt 1998; Callahan et al. 2000). Studies employing

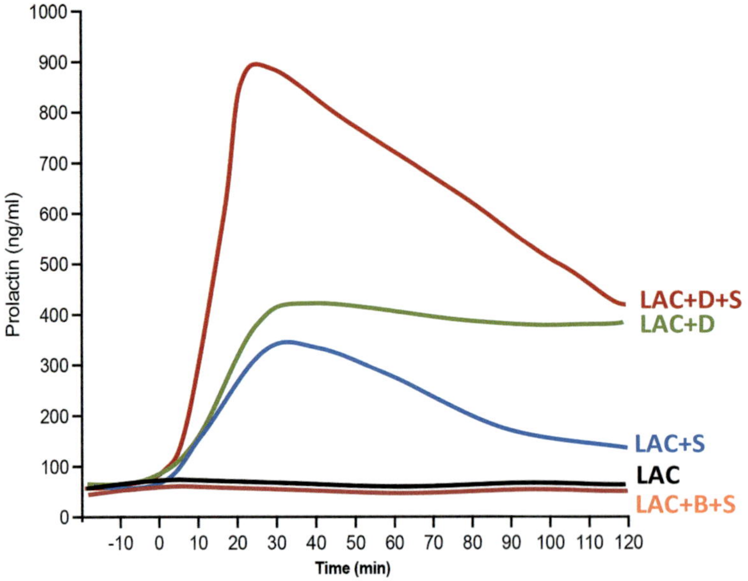

Fig. 7.4 The effects of blockade or activation of dopamine D2 receptors (D2R) on the suckling-induced rise in prolactin. Serial blood samples were obtained from lactating (LAC) rats that received bolus intravenous injection (time zero) of domperidone (D) or bromocriptine (B), selective D2R antagonist and agonist, respectively. LAC rats, separated from their pups for 4 h, were reunited with their litters for suckling 5 min after drug treatments (LAC+S) or remained without pups (LAC). Nipple suckling induced an acute surge of prolactin secretion, which was blocked by the pharmacological activation of DR2 with B (LAC+B+S). While the blockade of pituitary D2R with D increased basal prolactin secretion (LAC+D), it did not prevent the suckling-induced prolactin rise, which occurs over the elevated baseline of prolactin levels (LAC+D+S). This suggests that the reduced release of dopamine during lactation is required to sensitize the lactotrophs to a suckling-induced PRF that stimulates the acute rise in prolactin secretion. Figure adapted from Silva et al. (2020)

selective antagonists indicate that the kappa and mu receptor subtypes (but not the delta subtype) are required for these effects of opioid peptides (Andrews and Grattan 2003) and, thus, suggest the involvement of dynorphin and β-endorphin that predominately bind to these receptors, respectively. In this regard, TIDA neurons are densely innervated by dynorphin neurons (Fitzsimmons et al. 1992), providing an anatomical connection to the relay of the suckling input. Interestingly, the dynorphin-expressing neurons in the arcuate nucleus of different species have been found to co-express two other peptides, kisspeptin and neurokinin-B. On the basis of this, these neurons have been given the acronym KNDy (Lehman et al. 2010). Kisspeptin and neurokinin are essential to the function of the gonadal axis and fertility. KNDy neurons project to gonadotropin-releasing hormone (GnRH)-producing neurons, controlling the pulsatile secretion of luteinizing hormone (Herbison 2016). Moreover, kisspeptin fibers also contact TIDA neurons, and kisspeptin can stimulate prolactin secretion (Szawka et al. 2010). Thus, because KNDy neurons send projections to both GnRH and TIDA neurons, this neuronal population may be an integrating center, coordinating the release of both prolactin and luteinizing hormone. Lactation suppresses kisspeptin expression in the arcuate nucleus, which is associated with the inhibition of luteinizing hormone secretion and the consequent lactational infertility (Araujo-Lopes et al. 2014). However, whether the synaptic contacts of KNDy terminals upon TIDA neurons respond to suckling is an interesting question that is yet to be elucidated.

A neuronal group of the posterior intralaminar complex of the thalamus that expresses the tuberoinfundibular peptide of 39 residues (TIP39), also known as parathyroid hormone 2, have emerged as a possible upstream connection to the arcuate nucleus in the transmission of suckling sensory inputs. Their influence on the TIDA neurons is likely to be indirect, however, because TIDA neurons do not express the TIP39 receptor (the parathyroid hormone 2 receptor) (Dobolyi et al. 2006). On the other hand, TIP39 neurons send projections to dynorphin-containing neurons in the arcuate nucleus and are activated by suckling in lactating rats. The blockade of parathyroid hormone 2 receptor in the arcuate nucleus blunts prolactin secretion induced by suckling (Cservenak et al. 2013; Szabo et al. 2010). Thus, the neuroendocrine circuit of suckling-induced prolactin release seems to encompass the sequential relay of the sensory signal between TIP39, KNDy and TIDA neuronal populations, which eventuates in the stimulus-induced reduction of dopamine release in the portal blood.

7.3 NEDA Neurons

7.3.1 Anatomy and Morphology of the NEDA Neurons

As described above, prolactin secretion from the pituitary gland is primarily controlled by the activation of inhibitory dopamine D2R expressed on the lactotrophs. The following section will briefly consider the NEDA neurons, introduced earlier, which are responsible for releasing dopamine into the median eminence, allowing it to access the anterior pituitary gland via the hypophyseal portal system. Previous reviews have provided a detailed and comprehensive description of these neurons (Freeman et al. 2000; Grattan 2015; Phillipps et al. 2020). The following section will therefore summarize some of their key features, but then primarily focus on recent findings indicating their unexpected heterogeneity, complex morphology and ability to form dynamic signaling networks. The existence of such complexity is likely to have implications for regulation of prolactin secretion under differing physiological conditions (as considered in Sect. 7.2.4). The NEDA neurons comprise three anatomically distinct hypothalamic groupings, the TIDA, tuberohypophyseal dopaminergic (THDA) and periventricular hypophyseal dopaminergic (PHDA) neurons (Fig. 7.5a). The cell bodies of the TIDA neurons are located within the dorsomedial arcuate nucleus, those of the THDA neurons more rostrally within the same nucleus, while as the name suggests, the PHDA population arises from the periventricular nucleus. The TIDA neurons, which are believed to be the primary

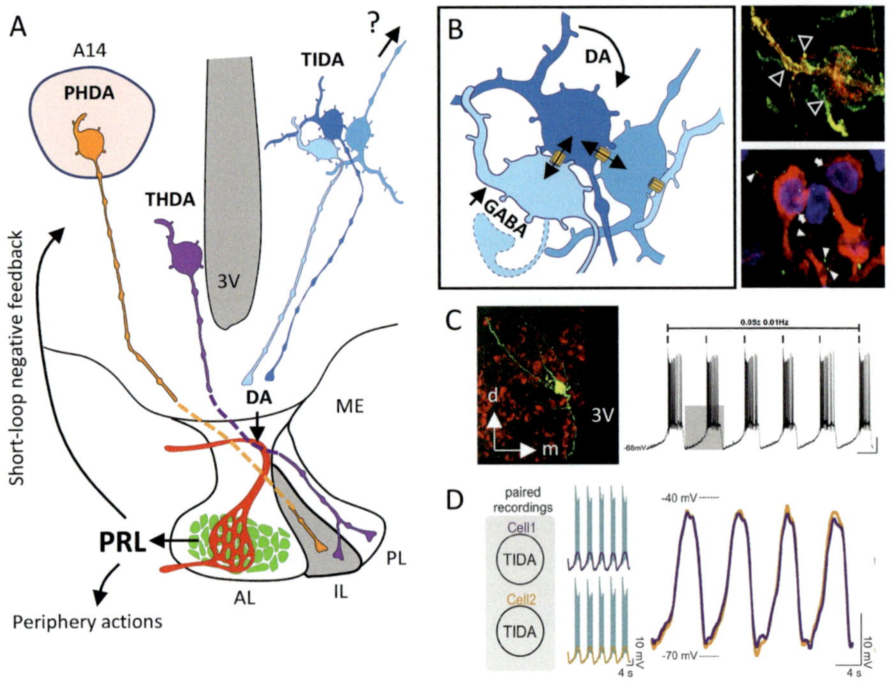

Fig. 7.5 The neuroendocrine dopaminergic (NEDA) neuronal network. (**a**) Summary diagram of the NEDA network. Prolactin (PRL) activation of the TIDA neurons promotes dopamine (DA) release into the median eminence (ME), which then suppresses further PRL secretion from the anterior lobe of the pituitary gland (AL). Similar actions may be mediated by the periventricular hypophyseal dopaminergic (PHDA) and tuberohypophyseal dopaminergic (THDA) neurons, projecting to the intermediate (IL) and posterior (PL) lobes of the pituitary. The question mark indicates possible unknown projection targets. (**b**) Possible interaction between TIDA neurons through dendritic release of DA, synaptic release of GABA or gap-junction interaction. The top right photomicrograph shows a TIDA neuron making possible axo-dendritic, dendro-somatic and dendro-dendritic contacts (open arrows, clockwise from the top, respectively). Image adapted from Yip et al. (2020). Bottom right photomicrograph shows possible gap junctions (arrow heads) stained for Cx35/36 (green) between tyrosine hydroxylase (TH)-positive TIDA neurons (red) in the male rat. Image adapted from Stagkourakis et al. (2018). (**c**) Left panel, a biocytin-filled TIDA neuron (green) extending dendritic arborizations among other TH-positive neurons (red). Right panel, a current-clamp recording of a spontaneously oscillating TIDA neuron with the grey box indicating a transition between **hyperpolarized** and **depolarized** states. d, dorsal; m, medial; 3 V, third ventricle. Images adapted from Lyons et al. (2010). (**d**) Left panel, paired whole-cell recordings performed to test synchronous activity. Middle panel, paired rat TIDA neurons shown superimposed. Right panel, a recording filtered for action potentials, revealing near identical membrane voltage fluctuations. Images adapted from Stagkourakis et al. (2018)

physiological regulators of pituitary prolactin secretion, project to the external zone of the median eminence. Early studies employed retrograde tracing to demonstrate that approximately 70% of these neurons terminated within this region in appositions with the portal capillaries (Kawano and Daikoku 1987). In contrast, the projections of other NEDA populations travel via the hypothalamus-hypophyseal tract passing through the median eminence to terminate, in the case of the THDA neurons in the intermediate and neural lobes of the pituitary, or in the case of PHDA neurons within the intermediate lobe alone. Dopamine released from these latter two NEDA populations may access the lactotrophs via short portal vessels, although the actual contribution their output makes to regulating prolactin secretion is unresolved. While anatomically distinct, evidence suggests that many neurons of each population express the prolactin receptor, suggesting they are all involved in the neuroendocrine feedback pathway and functionally may act as an integrated population.

Studies have found that most TIDA neurons have small round or fusiform soma, with polar, generally long, unbranching dendrites (Chan-Palay et al. 1984; Romero-Fernandez et al. 2014). Typically, these cells contain more than three main dendrites, and there is extensive overlap between the dendritic fields of adjacent neurons. The improved resolution obtained through the selective **neurobiotin** filling of TIDA neurons in hypothalamic slices prepared from TH-green fluorescent protein (GFP) mice revealed that they possess dendritic trees comprised of thick processes decorated with spines, and a separate thin axonal projection, devoid of spines, that branch extensively both before and within the median eminence (Fig. 7.5c) (Herde et al. 2013; Lyons et al. 2010). Furthermore, as considered later in this review, the TIDA neurons receive synaptic inputs from, and send projections to, other hypothalamic regions including the paraventricular, ventromedial and dorsomedial nuclei (Zhang and van den Pol 2016). Recent studies using adeno-associated virus (AAV) **Brainbow**-transfected TIDA neurons in the rat (see Box 7.2) have confirmed the round or fusiform morphology of their soma and the presence of both somatic and dendritic spines with axonal processes arising primarily from either the soma or proximal dendrite (Fig. 7.5b). Interestingly, the anatomical orientation of these axonal projections is highly diverse, with few axons being initially targeted towards the median eminence. Extensive **bouton-like structures** are present on axonal projections and often make close appositions with dendritic spines on adjacent TIDA neurons, thus providing a substrate for interactions within the TIDA neuron population and with other neurons in the arcuate nucleus (Yip et al. 2020).

Box 7.2 Morphological Analysis of TIDA Neurons

A combination of a transgenic rat model expressing Cre-recombinase under the control of the TH promoter (TH-Cre) and Cre-inducible AAV-farnesylated Brainbow markers (Cai et al. 2013) has been used to specifically fill the TIDA neurons following stereotaxic injection into the arcuate nucleus. The Brainbow markers are expressed at random within Cre-positive cells, meaning that individual cells may be a different color, allowing single TIDA neurons to be independently identified and their processes resolved. The farnesylated nature of the markers enhances cell membrane delineation and thus the visualization of fine structures, including neuronal spines. When combined with confocal microscopy, the putative synaptic contacts and thus potential interactions of TIDA neurons can be determined (see Fig. 7.5 for an example of this approach).

The relative complexity of the TIDA neuronal network is likely to arise from underlying molecular heterogeneity within the cell population. Such diversity is evident from single-cell RNA sequence analysis of mouse hypothalamic tissue, with, for example, the identification of 6 distinct TH-expressing cell populations within the arcuate nucleus (Campbell et al. 2017). Evidence for TIDA neuron heterogeneity has also been supported by morphological studies reporting variation in dendritic spine density (Yip et al. 2020), immunohistochemical studies demonstrating differential expression of proteins specifically relevant to TIDA neuron function, including the prolactin receptor (Kokay et al. 2018), DAT (Ciliax et al. 1995; Yip et al. 2018), and vesicular GABA transporter (Brown et al. 2016b), and electrophysiological findings of a range of distinct cell-specific patterns of electrical activity within the TIDA neuron population (Brown et al. 2012; Romano et al. 2013; Stagkourakis et al. 2018; Zhang and van den Pol 2015). Additionally, differential protein and gene expression of, for example, the clock genes CLOCK and PER1/2 (Sellix et al. 2006), and receptors for corticotrophin-releasing hormone (Campbell et al. 2003), leptin and ghrelin and their downstream signaling protein RhoA (Skov et al. 2019), suggests that subsets of TIDA neurons may also play a role in circadian rhythms, stress, body weight, and energy homeostasis. There are, however, two potential caveats regarding this apparent heterogeneity within what could otherwise be considered a "defined" neuronal population. First, the term TIDA should be limited to those neurons that project to, and release dopamine at the median eminence, thus not all TH-expressing neurons within the dorsomedial arcuate should be considered a "true" TIDA neuron. Second, the TIDA neurons are dynamic, changing their protein expression profiles and electrophysiological behavior under differing physiological conditions. Thus, TIDA subpopulations and the boundaries between them are likely to be situation-dependent. Indeed, as detailed in Sect. 7.3.3, further diversity becomes evident during late pregnancy when some TIDA neurons undergo a neurotransmitter switch, replacing dopamine with the neuropeptide enkephalin.

7.3.2 Activity and Regulation of the TIDA Neurons

Electrophysiological investigations have demonstrated that the TIDA neurons of the male rat oscillate between hyperpolarized and depolarized states, with a periodicity of about 20 s. During the depolarized phase they discharge spontaneous action potentials at a frequency of approximately 4 Hz (Fig. 7.5c). Paired electrophysiological recordings, and Ca^{2+}-imaging of multiple TIDA neurons, have revealed that these oscillations exhibit a high degree of synchronization and phase lock (Fig. 7.5d) (Lyons et al. 2010; Stagkourakis et al. 2018). This coordination between TIDA neurons has been attributed to the expression of extensive interconnecting gap junctions. The rat TIDA neurons are therefore proposed to act as a synchronized network in which strong gap junction-mediated interaction harmonizes individual cell firing frequencies to a relatively slow population rhythm. In contrast, TIDA neurons in the male mouse lack extensive gap-junction connectivity and are thus free to oscillate at a faster, and generally unsynchronized, rate (Stagkourakis et al. 2018). This surprising species difference may have profound functional implications in that the slower coordinated TIDA activity in the rat supports sustained dopamine release (Stagkourakis et al. 2019) and thus a more effective suppression of pituitary prolactin secretion than is achieved in the mouse (Stagkourakis et al. 2020). The resultant relatively high levels of circulating prolactin in the male mouse have been proposed to underlie the enhanced level of paternal care observed in this species compared to the rat.

Given the diversity of prolactin's physiological actions, it is perhaps not surprising that the TIDA and presumably the other NEDA neurons are subject to multiple regulatory influences (Fig. 7.6). While the physiological significance of the more recently identified of these supplementary inputs is unclear, the ability of prolactin to activate the TIDA neurons and thus mediate negative feedback on its release is well established. This feedback inhibition ensures the maintenance of low levels of circulating prolactin until it is "overridden" by other influences such as those occurring during lactation or stress. Activation of the prolactin receptor triggers multiple intracellular signaling pathways, which collectively stimulate dopamine output to suppress further prolactin secretion. Application of prolactin to rat hypothalamic slice preparations immediately stimulates the firing of TIDA neurons through a mechanism mediated by activation of a transient receptor potential channel (TRPC) and the inhibition of a Ca^{2+}-dependent BK-type K^+ current (Lyons et al. 2012). Hence, deletion of the TRPC5 channel leads to a failure in prolactin-induced tonic excitation (Blum et al. 2019). While this increase in firing rate will promote immediate neurosecretory release from the TIDA neurons, the capacity to maintain enhanced dopamine secretion is limited by the activity of the catecholamine synthesizing enzyme TH. In this respect, TH activity in the TIDA neurons can be increased acutely by its site-specific phosphorylation and chronically through increased TH gene transcription. Thus, TH phosphorylation has been identified as providing the "rapid" feedback (approximately 2–4 h) (Selmanoff 1985), while increased TH gene expression the "delayed" (approximately 12 h) component of prolactin's short-loop negative feedback mechanism (Demarest et al. 1986). The precise intracellular mechanisms responsible for these actions are less well defined, perhaps in part due to the complexity of the signaling pathways activated by the prolactin receptor. While phosphorylation of the relevant serine residues on TH has been shown to occur (Ma et al. 2005), the kinases responsible, or their time-courses of action, have not been fully resolved (Dunkley and Dickson 2019). Similarly, although prolactin activation of TIDA neurons is known to selectively activate the extracellular signal-regulated kinases (ERK) 1/2 and JAK-STAT5b transcription pathways (Bole-Feysot et al. 1998), exactly how these signals couple to an increase in TH expression is unclear, although it should be noted that the deletion of STAT5b in the mouse results in hyperprolactinemia (Grattan et al. 2001).

In addition to prolactin-mediated short-loop negative feedback, the TIDA neurons are also subject to an "ultra-short" negative feedback by dopamine (see Fig. 7.5b). When released within the somatodendritic region of the TIDA neurons, dopamine may act on D2R to slow their oscillation frequency (Stagkourakis et al. 2016). Similarly, GABA, which is co-released with dopamine from a subpopulation

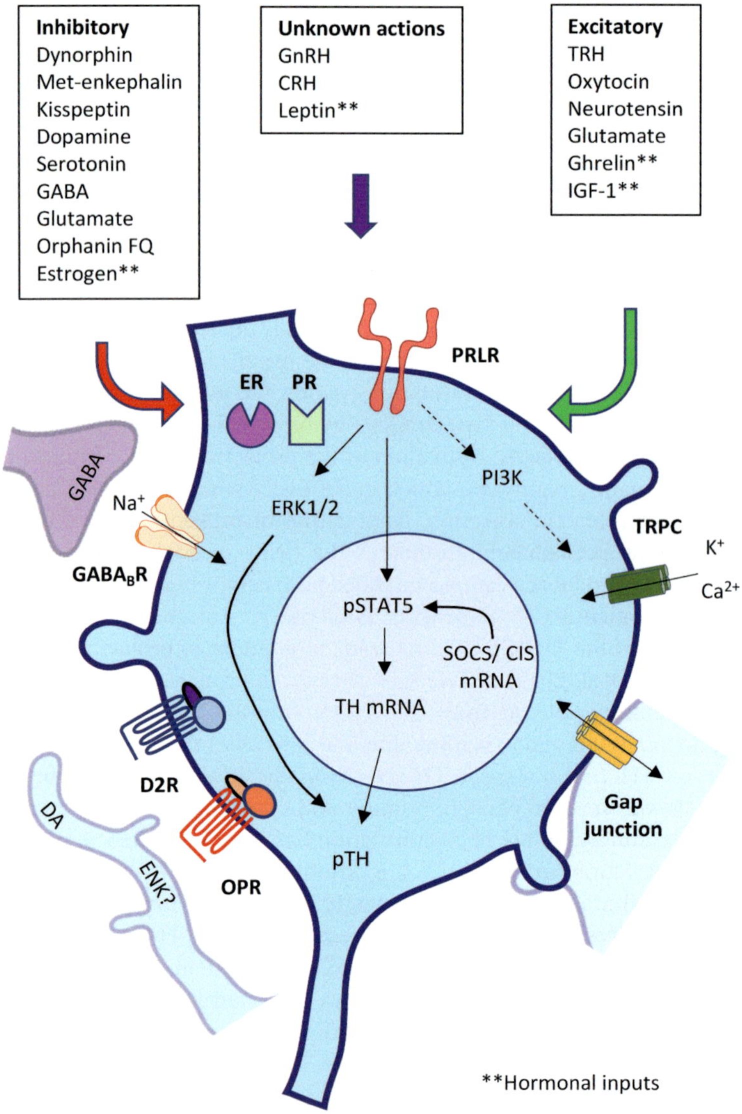

Fig. 7.6 Tuberoinfundibular dopaminergic (TIDA) neuron activity may be regulated by multiple neuronal and hormonal inputs. Prolactin receptor (PRLR) stimulation activates a number of intracellular signaling pathways including those mediated by the phosphorylated signal transducer activator of transcription 5 (pSTAT5), the extracellular signal-regulated kinases 1/2 (ERK1/2), and phosphoinositide 3-kinase (PI3K). While the former pathways are important for expression of tyrosine hydroxylase (TH) mRNA and its phosphorylation (pTH), PI3K may cause excitation of the TIDA neurons via Ca^{2+} influx through transient receptor potential channels (TRPC). pSTAT5 also triggers intracellular negative feedback by increasing the expression of suppressors of cytokine signaling (SOCS) and cytokine-inducible SH2 proteins (CIS). Accumulated evidence from multiple studies also indicates the TIDA neurons are regulated via multiple additional factors, including both excitatory and inhibitory neurotransmitters and a number of hormones, such as estrogen, progesterone, leptin, ghrelin and IGF-1 (indicated by double asterisk). Red arrow, inhibitory inputs. Green arrow, excitatory inputs. *D2R* dopamine D2 receptor, *ER* estrogen receptor, *GABA$_B$R* GABA-B receptor, *OPR* opioid receptor, *PR* progesterone receptor

of NEDA neurons, may also suppress TIDA neuron activity via $GABA_B$ receptors expressed on the TIDA neurons themselves or terminals of excitatory innervation (Fig. 7.5b) (Ammari and Broberger 2020). Evidence for such regulatory interactions within and between TIDA neurons may also extend to adjacent neuronal populations within the arcuate nucleus, strengthening the case for an integrated neuronal network underlying the regulation of prolactin secretion. As discussed above for the control of prolactin secretion during lactation, the regulation of TIDA neuronal behavior is not limited to stimuli arising from within the arcuate nucleus. Such diversity of inputs presumably allows prolactin secretion to be adjusted to a wide variety of physiological parameters, including circadian rhythms, reproductive and metabolic status and level of stress.

7.3.3 Adaptation of TIDA Neuron Morphology and Behavior During Lactation

Given the central role that TIDA neurons play in controlling the circulating level of prolactin, it is not unexpected that their behavior is modified during lactation, although the extent of this adaptation is quite remarkable. In order to achieve the essential high circulating levels of prolactin required during lactation, the TIDA neurons (and presumably the other NEDA neurons) must reduce their dopamine output. This adaptation has been extensively documented as decreased dopaminergic activity, with reduced DOPAC and DOPA levels at the median eminence and decreased levels of dopamine detected in the hypophyseal portal plasma (Ben-Jonathan et al. 1980; Demarest et al. 1981). These changes are mirrored by a fall in TH mRNA within the arcuate nucleus (Kokay and Grattan 2005; Wang et al. 1993), although, interestingly, at the cellular level, lactation was associated with a decrease in TH-phosphorylation status, particularly at its key serine-40 regulatory site, rather than in total enzyme level (Feher et al. 2010; Romano et al. 2013). While the specific mechanisms responsible for the lactation-induced suppression of dopamine output from the TIDA neurons are not fully understood, it is likely that the suckling stimulus plays a major role (Arbogast and Voogt 1996). As summarized in Fig. 7.6, the TIDA neurons are subject to multiple inhibitory inputs in addition to dynorphin. It was therefore surprising to find that a subpopulation of TIDA neurons exhibited a lactation-associated increase in somatic spine density (Fig. 7.7a). Given that it is generally assumed that these structures represent excitatory input sites, this observation raises important, and as yet unanswered, questions regarding the role of excitatory innervation of the TIDA neurons during lactation.

As outlined earlier, the major physiological regulator of TIDA neuron activity involves excitation by circulating prolactin to stimulate their release of dopamine, and thus suppress pituitary prolactin secretion. Multiple studies have shown, however, that elevated serum prolactin concentrations during lactation do not engage this negative feedback loop (Arbogast and Voogt 1996; Demarest et al. 1983). While it was initially assumed that the apparent insensitivity of the TIDA neurons to prolactin at this time was likely to be due to the downregulation of prolactin receptors,

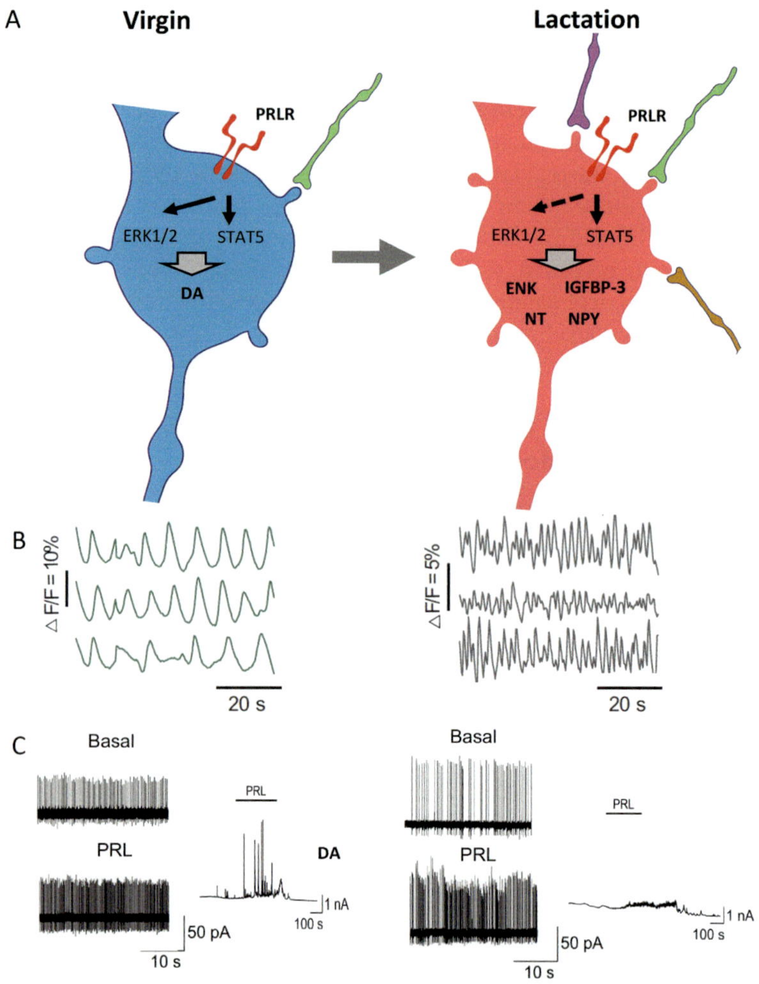

Fig. 7.7 Neurochemical and electrical adaptation of tuberoinfundibular dopaminergic (TIDA) neurons during lactation. (**a**) Diagrammatic comparison between a TIDA neuron from a virgin and a lactating mouse. During lactation the TIDA neurons display a higher somatic spine density suggesting increased synaptic inputs. While the level of prolactin receptor (PRL-R) expression in the TIDA neurons is unchanged during lactation, the signaling pathways activated may be modified, including an insensitivity of the extracellular signal-regulated kinases 1/2 (ERK1/2) pathway (see main text). This, and potentially other alterations, leads to a shift away from dopamine (DA) synthesis towards the expression of other signals including met-enkephalin (ENK), insulin-like growth factor binding protein-3 (IGFBP-3), neurotensin (NT) and neuropeptide Y (NPY). (**b**) Representative Ca^{2+} traces in TIDA neurons of virgin (left) or lactating (right) mice. Note the frequency of the Ca^{2+} signal is increased during lactation. ΔF/F = relative changes in fluorescence. Images adapted from Thorn Perez et al. (2020). (**c**) Representative patch-clamp recordings of a TIDA neuron from a virgin (left) and a lactating (right) mouse. Prolactin (PRL) application after a 15-min basal period resulted in an increased TIDA neuron firing rate in both groups, but while PRL stimulated DA secretion at the median eminence from virgin animals (measured by amperometry), this did not occur during lactation (right-hand panel). Images adapted from Romano et al. (2013)

this does not appear to be the case because prolactin receptor mRNA expression in the arcuate nucleus and on the TIDA neurons is unchanged in lactating rats (Brown et al. 2011; Kokay and Grattan 2005). While there are probably modifications to the intracellular signaling pathways downstream of the prolactin receptor (discussed below), the major players (ERK1/2 and STAT5) are still activated during lactation (Fig. 7.7a). Perhaps most strikingly, *in vitro* electrophysiological studies have shown that prolactin promotes an essentially identical increase in the firing rate of TIDA neurons from both lactating and diestrus mice, but only promotes dopamine secretion in the former (Fig. 7.7c) (Romano et al. 2013). This may be explained by the finding that during lactation the TIDA neurons undergo a neurotransmitter switch and begin releasing the opioid peptide enkephalin, while reducing dopamine secretion. This lactation-associated change in the phenotype of the TIDA neurons was first reported in the early 1990s (Merchenthaler 1993) and has been supported by more recent investigations (Yip et al. 2019). The phenomenon occurs in both lactating rats and mice and is dependent on a suckling-stimulated increase in prolactin and thus there is a decrease in enkephalin expression in the arcuate nucleus following pup removal. Enkephalin induction appears to be solely dependent on increased prolactin levels, since pituitary grafting under the kidney capsule is sufficient to induce such a response in cycling female rats (Merchenthaler et al. 1995). Similarly, inhibition of prolactin secretion by bromocriptine significantly decreases both proenkephalin mRNA levels and met-enkephalin immunostaining in the arcuate nucleus of lactating dams, while the fall in proenkephalin expression seen following pup withdrawal can be prevented by continuous prolactin infusion (Nahi and Arbogast 2003; Yip et al. 2019). Finally, the inability of mice lacking prolactin receptors in forebrain neurons to express enkephalin in the arcuate nucleus strongly suggests that the prolactin receptors located on the TIDA neurons are responsible for driving the neurotransmitter switch (Yip et al. 2019). The mechanism by which activation of the prolactin receptor on the TIDA neurons can stimulate dopamine synthesis and secretion under one set of physiological conditions and enkephalin synthesis and release under another is unknown. As noted earlier, the intracellular signaling pathways activated by the prolactin receptor are very similar under both lactating and non-lactating conditions. However, there is some evidence that an increase in the expression of the endogenous inhibitors of cytokine signaling (SOCS and cytokine-inducible SH2 proteins) within the TIDA neurons may result in a blunting of the prolactin response (Anderson et al. 2006a, b). Similarly, while the level of ERK1/2 phosphorylation in the TIDA neurons is greater in lactating compared to virgin mice, it is relatively insensitive to acute prolactin receptor activation during lactation (Yip et al. 2019). These data suggest that a subtle modification to the prolactin-receptor signaling pathway in the TIDA neurons may be responsible for their lactation-associated increase in proenkephalin expression, although this proposal has not yet been effectively tested (Fig. 7.7a). It is important to note that the chronically elevated levels of placental lactogen during pregnancy and suckling-induced prolactin present during lactation provide a very distinctive temporal pattern of prolactin receptor activation, which may influence its cellular response. Such temporal-dependence has been well characterized for providing a mechanism for sexually dimorphic STAT5-mediated growth hormone signaling in the liver (Lau-Corona et al. 2022).

Gene expression is commonly regulated by the convergence of multiple signaling events, which in many cells will include those reflecting Ca^{2+} dynamics. In this context, it is noteworthy that a recent study employing both electrophysiology and Ca^{2+}-imaging techniques reported that TIDA neurons from lactating mice showed an increased oscillatory frequency under basal conditions, compared to non-lactating animals (Fig. 7.7b) (Thörn Pérez et al. 2020). It would seem likely that such an altered intracellular environment would impact on both gene expression and neuron behavior. Thus, it is tempting to speculate that this altered electrophysiological activity could underlie a prolactin-induced upregulation of enkephalin expression and release. A switch in the TIDA neurotransmitter from dopamine to enkephalin has potentially profound physiological implications, on at least two possible anatomical targets, namely within the arcuate nucleus or at the anterior pituitary gland. Within the arcuate nucleus, enkephalin from the TIDA neurons may play an autocrine or paracrine role. This proposal is supported by the extensive neuronal interaction described above, thus allowing enkephalin to replace dopamine in mediating somatodendritic signaling within, and potentially between, neuronal populations. It is likely that enkephalin released from TIDA neurons would suppress their electrical activity because exogenous met-enkephalin has been shown to act on both mu and delta receptors to hyperpolarize TIDA neurons in mouse hypothalamic slices (Zhang and van den Pol 2015). Immunoneutralization and the use of opioid receptor antagonists also indicate that enkephalin released from TIDA neurons could auto-regulate to suppress the expression and activation of TH (Callahan et al. 2000). Given the extensive interaction and probable heterogeneity of the TIDA neurons, such local actions of enkephalin may help to regulate and coordinate multiple maternal adaptations that originate from within the hypothalamus, such as energy homeostasis and food intake, growth hormone regulation and modulation of the stress response. As described in Sect. 7.2.4, the lactation-induced neurotransmitter switch in the TIDA neurons may effectively convert the homeostatic negative feedback loop into positive feedback, by replacing the inhibitory dopamine signal with enkephalin, which in addition to damping TIDA activity may also directly enhance prolactin secretion from the pituitary gland.

7.4 The Lactotrophs

The major source of prolactin in all mammalian species is the lactotroph cells of the anterior pituitary, a gland that contains a range of other hormone-producing cells with essential roles in physiology, specifically: corticotrophs, secreting adrenocortocotrophic hormone; somatotrophs, secreting growth hormone; gonadotrophs, secreting the gonadotrophins, luteinizing and follicle-stimulating hormone; and thyrotrophs, secreting thyroid-stimulating hormone. The colocalization of cells regulating multiple axes is important, as it allows coordination of the response of these various cell types to physiological and pathological challenges. In addition, a non-hormonal cell type, the folliculostellate cells, are present throughout the gland and may have important roles both in coordinating the function of various cell types and their interactions, as well as potentially acting as stem cells that are capable of replenishing the endocrine cells throughout life (Le Tissier and Mollard 2021). The localization of the anterior pituitary gland at the base of the brain (but outside the blood-brain barrier) is important for both its hypothalamic regulation through the specialized portal blood supply and its exposure to peripheral factors. Below, we describe the development and inter-relationship of these various cell types with lactotroph cells and then focus on the lactotroph function, the role of the anatomical organization of this cell type and how this is altered to support the dramatic increase in prolactin secretion in lactation, as well as the importance of hypothalamic and feedback regulation in preventing pituitary pathology.

7.4.1 Anatomy and Cell Biology

Lactotroph cells of the anterior pituitary develop from oral ectoderm, with a cascade of transcription factors and signaling molecules driving terminal differentiation (Kelberman et al. 2009). Key to lactotroph differentiation is the transcription factor POU1F1 (or PIT1), which is also essential for the generation of somatotrophs and thyrotrophs. Lactotrophs and somatotrophs maintain a close developmental relationship throughout life, with transdifferentiation of the two cell types under certain conditions and the existence of mammosomatotrophs that produce both growth hormone and prolactin. Indeed, the close relationship between the two cell types is apparent in analysis of single-cell RNA sequencing studies of the mouse pituitary gland, with a strong concordance of gene expression in lactotrophs and somatotrophs (Peel et al. 2018). This analysis of gene expression has also identified factors responsible for the final steps of differentiation of somatotrophs and lactotrophs, with the orphan nuclear receptor NR4A2 as a potential factor responsible for lactotroph identity (Peel et al. 2020), whilst other studies have also implicated the zinc finger and BTB-domain containing protein ZBTB20 (Cao et al. 2016) and the microRNA miR7a2 (LaPierre et al. 2021). While POU1F1 drives expression in a pituitary-specific fashion, the extra-pituitary expression of prolactin driven by the supra-distal promoter can occur in the absence of the transcription factor (Gellersen et al. 1994). Whether there is a role for POU1F1 in extra-pituitary expression of prolactin remains unclear and would be worthy of further investigation.

Prolactin mRNA expression is detected in the pituitary gland of rats at embryonic day 18, although the protein can only be detected in a few cells at birth and expression remains low until 2 weeks of age, with a gradual increase up until 5 weeks of age (Bjelobaba et al. 2015). Interestingly, there is significant expression of prolactin receptors in the embryo and neonate, which may respond to maternal placental lactogen and maternal prolactin, respectively. Thereafter, in females, but not in male animals, there is a rapid increase in lactotroph cell numbers and hormone production, consistent with the higher secretion of prolactin in females from 5 to 8 weeks of age, after which it remains relatively static in the absence of pathology or pregnancy (see below). The population of lactotrophs may be functionally diverse, however, with sub-types described based on morphology and function (Grattan and Le Tissier 2015), as well, more recently as on clustering of cells based on gene expression (Cheung et al. 2018). From current single-cell RNA sequencing studies, it is not possible to say whether these are distinct sub-populations or simply cells at different phases of activity. The power of single-cell RNA sequencing also identifies the functional differences between lactotrophs of males and females, with 288 genes found to be differentially expressed between the sexes in rats (Fletcher et al. 2019), emphasizing the distinct roles for prolactin in males and females.

Two-dimensional histological analysis of thin pituitary sections shows a heterogeneous distribution of lactotrophs, although their morphological association with gonadotrophs can be noted, as well as large-scale preferential localization within specific regions of the gland. The use of transgenic animals with fluorophores identifying lactotrophs has transformed this view and has identified a network organization, similar to that described for other pituitary cell types (for references see Le Tissier et al. 2012). However, lactotrophs are organized as a honeycomb structure (Fig. 7.8a, b), which differs from the topological organization of other cell types. Network organization has been shown to have a role in the regulation of hormone output in somatotroph cells, which suggests this is the case for lactotrophs (see below), but as the network motifs of the two cell types are distinct, this also suggests that the organization may have a different functional consequence. A further important feature of the anatomical organization of lactotrophs within the gland is their relationship with the vasculature, as this will affect both the exposure of the cells to hypothalamic and peripheral factors, as well as clearance of secreted prolactin from the gland. In this respect, lactotrophs form a close apposition with the capillary vasculature that pervades the gland, which again is distinct from the relationship of other pituitary cell types (Le Tissier et al. 2017) and may also have a role in sustaining the metabolically demanding process of secretion, as well as the requirement for persistent tonic exposure to dopamine to maintain the inhibition required over extended periods of time for appropriate physiological regulation.

7.4.2 Regulation and Behavior Under Non-lactating Conditions

The concentration and pattern of circulating prolactin is dependent on the accumulated rate of secretion of the lactotroph population within the anterior pituitary gland. While short-term modification may be mediated by changes in blood flow and capillary permeability, in the long-term, the amount of hormone secreted is dependent on the number of lactotrophs, the proportion of these cells that respond to a stimulus and the strength of this response, as well the amount of stored hormone. Multiple studies have shown that dopamine and estradiol have important regulatory roles in each of these processes, although the temporal effects of these differ with age and sex, suggesting that a range of other factors must influence each process (Grattan and Le Tissier 2015). Included in these potential regulatory factors

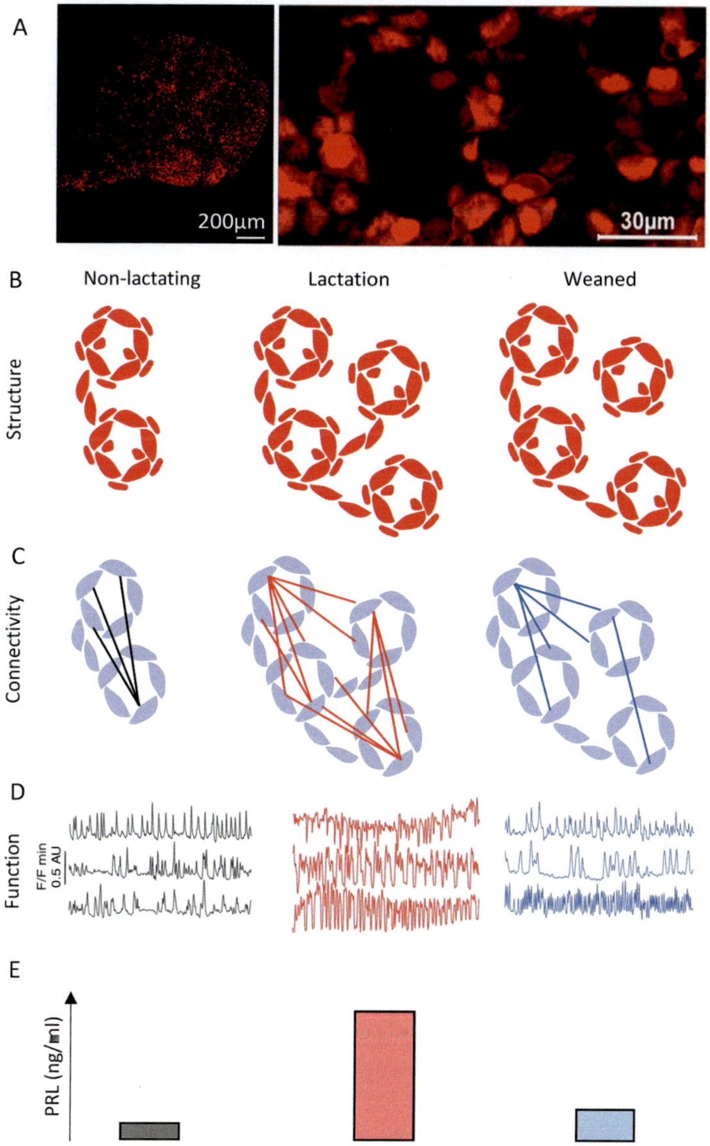

Fig. 7.8 Lactotroph cells are organized as a structural and functional network within the anterior pituitary gland. (**a**) Imaging of lactotroph cells identified by the transgenic expression of dsRed fluorescent protein in a hemi-pituitary (left), showing large-scale organization with an increased density of cells in the ventral part of the gland. At higher resolution (right) the organization into a honeycomb network structure becomes apparent. Images adapted from Le Tissier et al. (2012). (**b**) Depiction of the structural changes in network organization that occur in lactation and following weaning. Lactotroph cell (red) organization is enhanced in lactation and remains after cessation of demand at weaning. (**c**) The increased structural organization leads to an increased functional network activity with some lactotrophs acting as hub cells with coordinating activity, despite a return to virgin secretory activity (depicted in (**d**), with each trace representing the calcium activity of individual lactotrophs) and prolactin output from the gland (**e**). Figure adapted from Le Tissier et al. (2015)

are those produced within the anterior pituitary gland itself, with paracrine actions allowing cross-talk between the prolactin system and other pituitary axes (Denef 2008). This could include local production of dopamine (Orrillo et al. 2020). An autocrine action of prolactin itself has also been suggested, with specific deletion of the prolactin receptor in mouse lactotrophs resulting in an alteration of hypothalamic dopamine tone, suggesting that this compensates for a loss of prolactin feedback at the level of the pituitary in these mice (Orrillo et al. 2020).

The impact of a range of factors is demonstrated by the regulation of lactotroph proliferation by dopamine. As mentioned above (see Sect. 7.2.1), a dramatic lactotroph hyperplasia occurs in older D2R knockout mice. However, in younger D2R knockout animals, there is no such increase. Although dopamine suppresses lactotroph proliferation, it is other factors, such as insulin-like growth factor I, that may be the principal regulator of lactotroph numbers in younger animals (Hikake et al. 2009). Prolactin itself may also have a role in the regulation of lactotroph cell number, with loss of both prolactin receptor and D2R in knockout mouse models resulting in markedly increased proliferation of lactotrophs, compared with the loss of D2R alone (Schuff et al. 2002). Estradiol also has a role in lactotroph proliferation, with hypoplasia resulting from the genetic loss of estrogen receptor alpha, but this may be indirect and mediated by hypothalamic factors, as anesthesia blocks the normal lactotroph proliferation that occurs on estrus in female rats (Hashi et al. 1995). Furthermore, estradiol may also have a role in apoptosis of this cell type (Seilicovich 2010).

The region of the prolactin gene that encodes the pituitary promoter includes binding sites for a range of proteins that modify prolactin gene expression, again revealing that lactotroph function is regulated by complex interactions of signaling pathways responding to a range of hypothalamic, intra-pituitary and peripheral factors (for references see Featherstone et al. 2012). The POU1F1 transcription factor is absolutely required for prolactin expression in lactotrophs, and the interaction with a range of binding partners modifies its transcriptional activity. This modification and the integration of a range of other intracellular signaling pathways (including cAMP, phosphatidylinositol and the mitogen-activated protein) alter prolactin gene expression. Thus, dopamine activation of D2R results in lowering of cAMP and inhibition of prolactin gene expression, while any of the putative PRFs may modulate one or more of these pathways and result in increased prolactin expression. An important role for estradiol has been described in increasing gene expression through synergistic interaction with POU1F1, and there are estrogen receptor binding sites within the prolactin promoter. Recent studies have also suggested a role for microRNAs in the estrogen promotion of prolactin gene expression, with loss of miR375 resulting in a reduction of estrogen receptor alpha and consequent suspension of prolactin synthesis (Zhang et al. 2021).

In accordance with the secretion of prolactin being primarily under inhibitory regulation by dopamine, lactotrophs have high spontaneous electrical activity and calcium influx driving prolactin secretion in the absence of dopamine. In addition, hypothalamic somatostatin and calcitonin may also have inhibitory roles, whilst, as described above, there is a wide range of potential hypothalamic PRFs. Estradiol has a stimulatory role, primarily ascribed to its modification of lactotroph responsiveness to other factors (Grattan and Le Tissier 2015), although rapid actions of this steroid, possibly mediated through the G-protein-coupled estrogen receptor, have been described (Camilletti et al. 2019b). Similar rapid actions mediated by membrane steroid receptors may also mediate the negative effects of progesterone on prolactin secretion (Camilletti et al. 2019a). It is also possible that paracrine factors have an important role in regulating prolactin secretion and may mediate some of the effects of peripheral and hypothalamic regulators. For example, pituitary transforming growth factor beta, which is produced in lactotroph cells and inhibits prolactin output, is regulated by dopamine and estradiol and may account for some of the sex differences in prolactin secretion (Abeledo-Machado et al. 2020).

In addition to the mechanisms regulating lactotroph function at the single-cell level, the organization of cells into a structural network also enables another level of regulatory interactions. Indeed, in studies of pituitary slices with lactotrophs, identified by the transgenic expression of dsRed fluorescent protein, it has been shown that the structural organization mediates a functional connectivity that coordinates intracellular calcium activity (Hodson et al. 2012). A key feature of this functional network is the presence of "hub cells", defined as cells that appear to be acting as nodes driving coordinated network activity (Fig. 7.8c). The identity of these hub cells changes over time, suggesting that there is some plasticity in which cells can fulfill this role, but it is possible that these are a sub-type of lactotroph with transient coordinating activity. A role for networks in regulating prolactin gene expression has also been described (Featherstone et al. 2016), suggesting that these networks may have a role in regulating all of the processes related to the amount of hormone that is secreted.

7.4.3 Regulation and Behavior Under Lactating Conditions

As described in Sect. 7.4.2, there is a limited range of factors that can affect sustained changes in prolactin secretion. Current evidence would suggest that an alteration in lactotroph cell number may have an important role in the sustained secretion required during lactation, although this may be species-dependent (Grattan and Le Tissier 2015). In both humans and rats, an increase in lactotroph cell number has been reported in lactation, which in humans begins early in pregnancy. This increase in cell number rapidly regresses at weaning, further supporting the important role for this hyperplasia in supporting high prolactin secretion. However, in mice there are somewhat conflicting reports of altered cell number, with either no increase or a modest increase of around 25% (Ho et al. 2020; Hodson et al. 2012). The increase in lactotroph cell number may occur by a number of mechanisms, including the proliferation of lactotrophs, transdifferentiation from other hormone positive cell types, or differentiation from pituitary stem cells, and it is likely that a combination of these may underlie population expansion (Grattan and Le Tissier 2015). Analysis of the number of somatotrophs, lactotrophs and mammosomatotrophs in rats as pregnancy and lactation progress (Frawley and Boockfor 1991) shows a dynamic that is consistent with transdifferentiation of somatotrophs to lactotrophs. However, the drivers of this transdifferentiaition and the role of the transcription factors, other proteins and micro-RNAs identified as potentially defining lactotrophs is currently unclear, requires definition and may provide insight into the mechanisms underlying plasticity and pathology of all pituitary axes.

It is unclear what changes in lactotrophs occur to produce the increased gene expression that must occur to support sustained, high prolactin secretion, or an altered response to stimulation or reduced inhibition of secretion. Comparison of lactotroph gene expression in the pituitaries of virgin and lactating mice using single-cell RNA sequencing identified a limited range of gene expression changes, with alterations noted for secretory proteins but not regulatory factors (Ho et al. 2020). This does not, of course, preclude post-translational modifications fulfilling a role in increased prolactin production, but the drivers of this are not identified in studies to date. However, changes to the pituitary structural and functional network have been described in lactation and are likely to have an important role in supporting the high level of secretion over a protracted period. In lactation, the organization of lactotrophs as a honeycomb structure becomes more apparent, with an increase in the contacts between lactotrophs. This structural remodeling is accompanied by a dramatic increase in the coordination of cell activity and the proportion of lactotrophs acting as hub cells, functional changes that are mediated by an increase in homotypic lactotroph gap-junction coupling (Hodson et al. 2012). A surprising feature of this altered structural and functional network is its persistence after weaning, which supports further augmented function with a higher level of prolactin secreted in a second lactation through enhanced network coordination. This suggests that the network organization both supports the high levels of prolactin required for lactation and allows plasticity of the prolactin axis to ensure higher levels of secretion in subsequent lactations (Fig. 7.8c, d).

7.5 Perspectives

The neuroendocrine control of prolactin secretion is unique among the anterior pituitary gland hormones. The lactotrophs display a high spontaneous capacity for proliferation and secretion of prolactin that is tonically restrained by dopamine, released into the pituitary portal blood by the NEDA neurons. Prolactin is a pleiotropic hormone with multiple functions, which seem to have evolved to provide physiological adaptations required for the success of reproduction, perhaps with a major function in parental behavior that is conserved across many species. In mammals, this role includes pregnancy and lactation, aided by the additional support of placental lactogens. Prolactin's primary role is to support lactation, and mammals show complex adaptations in the control of prolactin secretion to enable sustained elevations in prolactin required for milk production. It has become evident that phenotypic and functional modifications in the NEDA neurons, alongside changes in the number and organization of lactotrophs, are essential to support the hyperprolactinemic state of lactation. However, the mechanisms responsible for these changes are only beginning to be understood. Moreover, the extensive change in the nature of NEDA neurons during lactation and whether this sort of adaptation can occur in other conditions of high prolactin are questions still not answered. The identity and role of PRFs in increasing prolactin secretion are long-standing questions that remain. In conclusion, the control of prolactin secretion provides an interesting model of how a neuroendocrine system can adapt to support different functional modalities, resulting in important physiological adaptations to cope with the challenges of maternal care.

Key References (see main list for reference details)

Cservenak et al. (2013) Identification of a thalamic-hypothalamic pathway mediating suckling action on TIDA neurons and prolactin secretion.

Grattan et al. (2001) Demonstration that classical JAK/STAT signal transduction was required for negative feedback actions of prolactin on TIDA neurons.

Hodson et al. (2012) Demonstration that the increased prolactin secretion by the lactotroph population in lactation results from a functional change in the output of individual cells through homotypic cell network coordination; furthermore, the homotypic networks allow a functional memory of demand and hormone secretion in subsequent lactations.

Murai and Ben-Jonathan (1987). First demonstration that a PRF originating in the neurointermediate lobe of the pituitary is required for the suckling-induced release of prolactin during lactation.

Romano et al. (2013) First demonstration that the TIDA neurons are still electrically sensitive to prolactin during lactation. However, prolactin is unable to stimulate the release of dopamine as seen in the non-lactation state. This essentially reduces the inhibitory tone to prolactin secretion during lactation.

Stagkourakis et al. (2020) Elucidation of how differential patterns of activity of TIDA neurons have different consequences for the amount of pituitary prolactin secretion and its behavioral effects.

Wang et al. (1993) Description of molecular changes in TIDA neurons resulting in the suppression of TH gene expression in the arcuate nucleus of the hypothalamus.

Yip et al. (2019) A full description that the TIDA neuron phenotypic switch from dopamine to enkephalin, which is important in shifting the prolactin regulatory system from a negative- to a positive-feedback loop during lactation.

References

Abeledo-Machado A, Perez PA, Camilletti MA, Faraoni EY, Picech F, Petiti JP, Gutierrez S, Diaz-Torga G (2020) TGFbeta1 regulates prolactin secretion during postnatal development: gender differences. J Endocrinol 246:29–39

Ammari R, Broberger C (2020) Pre- and post-synaptic modulation by GABAB receptors of rat neuroendocrine dopamine neurones. J Neuroendocrinol 32:e12881

Anderson GM, Beijer P, Bang AS, Fenwick MA, Bunn SJ, Grattan DR (2006a) Suppression of prolactin-induced signal transducer and activator of transcription 5b signaling and induction of suppressors of cytokine signaling messenger ribonucleic acid in the hypothalamic arcuate nucleus of the rat during late pregnancy and lactation. Endocrinology 147:4996–5005

Anderson ST, Barclay JL, Fanning KJ, Kusters DH, Waters MJ, Curlewis JD (2006b) Mechanisms underlying the diminished sensitivity to prolactin negative feedback during lactation: reduced STAT5 signaling and up-regulation of cytokine-inducible SH2 domain-containing protein (CIS) expression in tuberoinfundibular dopaminergic neurons. Endocrinology 147:1195–1202

Andrews ZB, Grattan DR (2003) Opioid receptor subtypes involved in the regulation of prolactin secretion during pregnancy and lactation. J Neuroendocrinol 15:227–236

Aoki M, Wartenberg P, Grunewald R, Phillipps HR, Wyatt A, Grattan DR, Boehm U (2019) Widespread cell-specific prolactin receptor expression in multiple murine organs. Endocrinology 160:2587–2599

Aquino NSS, Araujo-Lopes R, Henriques PC, Lopes FEF, Gusmao DO, Coimbra CC, Franci CR, Reis AM, Szawka RE (2017) alpha-Estrogen and progesterone receptors modulate kisspeptin effects on prolactin: role in estradiol-induced prolactin surge in female rats. Endocrinology 158:1812–1826

Aquino NSS, Kokay IC, Perez CT, Ladyman SR, Henriques PC, Silva JF, Broberger C, Grattan DR, Szawka RE (2019) Kisspeptin stimulation of prolactin secretion requires kiss1 receptor but not in tuberoinfundibular dopaminergic neurons. Endocrinology 160:522–533

Araujo-Lopes R, Crampton JR, Aquino NS, Miranda RM, Kokay IC, Reis AM, Franci CR, Grattan DR, Szawka RE (2014) Prolactin regulates kisspeptin neurons in the arcuate nucleus to suppress LH secretion in female rats. Endocrinology 155:1010–1020

Arbogast LA, Hyde JF (2000) Estradiol attenuates the forskolin-induced increase in hypothalamic tyrosine hydroxylase activity. Neuroendocrinology 71:219–227

Arbogast LA, Voogt JL (1996) The responsiveness of tuberoinfundibular dopaminergic neurons to prolactin feedback is diminished between early lactation and midlactation in the rat. Endocrinology 137:47–54

Arbogast LA, Voogt JL (1998) Endogenous opioid peptides contribute to suckling-induced prolactin release by suppressing tyrosine hydroxylase activity and messenger ribonucleic acid levels in tuberoinfundibular dopaminergic neurons. Endocrinology 139:2857–2862

Arey BJ, Kanyicska B, Freeman ME (1991) The endogenous stimulatory rhythm regulating prolactin secretion is present in the lactating rat. Neuroendocrinology 53:35–40

Arita JA, Kimura K (1988) Enkephalin inhibits dopamine synthesis in vitro in the median eminence portion of rat hypothalamic slices. Endocrinology 123:694–699

Ben-Jonathan N, Hnasko R (2001) Dopamine as a prolactin (PRL) inhibitor. Endocr Rev 22:724–763

Ben-Jonathan N, Neill MA, Arbogast LA, Peters LL, Hoefer MT (1980) Dopamine in hypophysial portal blood: relationship to circulating prolactin in pregnant and lactating rats. Endocrinology 106:690–696

Ben-Jonathan N, Lapensee CR, Lapensee EW (2008) What can we learn from rodents about prolactin in humans? Endocr Rev 29:1–41

Berghorn KA, Le WW, Sherman TG, Hoffman GE (2001) Suckling stimulus suppresses messenger RNA for tyrosine hydroxylase in arcuate neurons during lactation. J Comp Neurol 438:423–432

Bjelobaba I, Janjic MM, Kucka M, Stojilkovic SS (2015) Cell type-specific sexual dimorphism in rat pituitary gene expression during maturation. Biol Reprod 93:21

Blum T, Moreno-Perez A, Pyrski M, Bufe B, Arifovic A, Weissgerber P, Freichel M, Zufall F, Leinders-Zufall T (2019) Trpc5 deficiency causes hypoprolactinemia and altered function of oscillatory dopamine neurons in the arcuate nucleus. Proc Natl Acad Sci U S A 116:15236–15243

Bodnar I, Banky ZS, Toth BE, Nagy GM, Halasz B (2002) Brain structures mediating the suckling stimulus-induced release of prolactin. J Neuroendocrinol 14:384–396

Bole-Feysot C, Goffin V, Edery M, Binart N, Kelly PA (1998) Prolactin (PRL) and its receptor: actions, signal transduction pathways and phenotypes observed in PRL receptor knockout mice. Endocr Rev 19:225–268

Boyd AE, Spencer E, Jackson IM, Reichlin S (1976) Prolactin-releasing factor (PRF) in porcine hypothalamic extract distinct from TRH. Endocrinology 99:861–871

Brown RS, Herbison AE, Grattan DR (2011) Differential changes in responses of hypothalamic and brainstem neuronal populations to prolactin during lactation in the mouse. Biol Reprod 84:826–836

Brown RS, Piet R, Herbison AE, Grattan DR (2012) Differential actions of prolactin on electrical activity and intracellular signal transduction in hypothalamic neurons. Endocrinology 153:2375–2384

Brown RS, Wyatt AK, Herbison RE, Knowles PJ, Ladyman SR, Binart N, Banks WA, Grattan DR (2016a) Prolactin transport into mouse brain is independent of prolactin receptor. FASEB J 30:1002–1010

Brown RS, Kokay IC, Phillipps HR, Yip SH, Gustafson P, Wyatt A, Larsen CM, Knowles P, Ladyman SR, LeTissier P et al (2016b) Conditional deletion of the prolactin receptor reveals functional subpopulations of dopamine neurons in the arcuate nucleus of the hypothalamus. J Neurosci 36:9173–9185

Butcher RL, Fugo NW, Collins WE (1972) Semicircadian rhythm in plasma levels of prolactin during early gestation in the rat. Endocrinology 90:1125–1127

Cai D, Cohen KB, Luo T, Lichtman JW, Sanes JR (2013) Improved tools for the Brainbow toolbox. Nat Methods 10:540–547

Callahan P, Klosterman S, Prunty D, Tompkins J, Janik J (2000) Immunoneutralization of endogenous opioid peptides prevents the suckling-induced prolactin increase and the inhibition of tuberoinfundibular dopaminergic neurons. Neuroendocrinology 71:268–276

Camilletti MA, Abeledo-Machado A, Faraoni EY, Thomas P, Diaz-Torga G (2019a) New insights into progesterone actions on prolactin secretion and prolactinoma development. Steroids 152:108496

Camilletti MA, Abeledo-Machado A, Ferraris J, Perez PA, Faraoni EY, Pisera D, Gutierrez S, Diaz-Torga G (2019b) Role of GPER in the anterior pituitary gland focusing on lactotroph function. J Endocrinol 240:99–110

Campbell RE, Grove KL, Smith MS (2003) Distribution of corticotropin releasing hormone receptor immunoreactivity in the rat hypothalamus: coexpression in neuropeptide Y and dopamine neurons in the arcuate nucleus. Brain Res 973:223–232

Campbell JN, Macosko EZ, Fenselau H, Pers TH, Lyubetskaya A, Tenen D, Goldman M, Verstegen AM, Resch JM, McCarroll SA et al (2017) A molecular census of arcuate hypothalamus and median eminence cell types. Nat Neurosci 20:484–496

Cao D, Ma X, Cai J, Luan J, Liu AJ, Yang R, Cao Y, Zhu X, Zhang H, Chen YX et al (2016) ZBTB20 is required for anterior pituitary development and lactotrope specification. Nat Commun 7:11121

Chan-Palay V, Zaborszky L, Kohler C, Goldstein M, Palay SL (1984) Distribution of tyrosine-hydroxylase-immunoreactive neurons in the hypothalamus of rats. J Comp Neurol 227:467–496

Cheung LYM, George AS, McGee SR, Daly AZ, Brinkmeier ML, Ellsworth BS, Camper SA (2018) Single-cell RNA sequencing reveals novel markers of male pituitary stem cells and hormone-producing cell types. Endocrinology 159:3910–3924

Ciliax BJ, Heilman C, Demchyshyn LL, Pristupa ZB, Ince E, Hersch SM, Niznik HB, Levey AI (1995) The dopamine transporter: immunochemical characterization and localization in brain. J Neurosci 15:1714–1723

Cooke NE, Coit D, Shine J, Baxter JD, Martial JA (1981) Human prolactin. cDNA structural analysis and evolutionary comparisons. J Biol Chem 256:4007–4016

Cowie AT, Tindal JS, Benson GK (1960) Pituitary grafts and milk secretion in hypophysectomized rats. J Endocrinol 21:115–123

Cservenak M, Szabo ER, Bodnar I, Leko A, Palkovits M, Nagy GM, Usdin TB, Dobolyi A (2013) Thalamic neuropeptide mediating the effects of nursing on lactation and maternal motivation. Psychoneuroendocrinology 38:3070–3084 de Greef WJ, Plotsky PM, Neill JD (1981) Dopamine levels in hypophysial stalk plasma and prolactin levels in peripheral plasma of the lactating rat: effects of a simulated suckling stimulus. Neuroendocrinology 32:229–233

Demarest KT, Johnston CA, Moore KE (1981) Biochemical indices of catecholaminergic neuronal activity in the median eminence during the estrous cycle of the rat. Neuroendocrinology 32:24–27

Demarest KT, McKay DW, Riegle GD, Moore KE (1983) Biochemical indices of tuberoinfundibular dopaminergic neuronal activity during lactation: a lack of response to prolactin. Neuroendocrinology 36:130–137

Demarest KT, Riegle GD, Moore KE (1986) The rapid 'tonic' and the delayed 'induction' components of the prolactin-induced activation of tuberoinfundibular dopaminergic neurons following the systemic administration of prolactin. Neuroendocrinology 43:291–299

Denef C (2008) Paracrinicity: the story of 30 years of cellular pituitary crosstalk. J Neuroendocrinol 20:1–70

Denef C, Baes M, Schramme C (1984) Stimulation of prolactin secretion after short term or pulsatile exposure to dopamine in superfused anterior pituitary cell aggregates. Endocrinology 114:1371–1378

Diaz S, Seron-Ferre M, Cardenas H, Schiappacasse V, Brandeis A, Croxatto HB (1989) Circadian variation of basal plasma prolactin, prolactin response to suckling, and length of amenorrhea in nursing women. J Clin Endocrinol Metab 68:946–955

Djahanbakhch O, McNeilly AS, Warner PM, Swanston IA, Baird DT (1984) Changes in plasma levels of prolactin, in relation to those of FSH, oestradiol, androstenedione and progesterone around the preovulatory surge of LH in women. Clin Endocrinol 20:463–472

Dobolyi A, Irwin S, Wang J, Usdin TB (2006) The distribution and neurochemistry of the parathyroid hormone 2 receptor in the rat hypothalamus. Neurochem Res 31:227–236

Dobolyi A, Olah S, Keller D, Kumari R, Fazekas EA, Csikos V, Renner E, Cservenak M (2020) Secretion and function of pituitary prolactin in evolutionary perspective. Front Neurosci 14:621

Dunkley PR, Dickson PW (2019) Tyrosine hydroxylase phosphorylation in vivo. J Neurochem 149:706–728

Everett JW (1956) Functional corpora lutea maintained for months by autografts of rat hypophyses. Endocrinology 58:786–796

Featherstone K, White MR, Davis JR (2012) The prolactin gene: a paradigm of tissue-specific gene regulation with complex temporal transcription dynamics. J Neuroendocrinol 24:977–990

Featherstone K, Hey K, Momiji H, McNamara AV, Patist AL, Woodburn J, Spiller DG, Christian HC, McNeilly AS, Mullins JJ et al (2016) Spatially coordinated dynamic gene transcription in living pituitary tissue. elife 5:e08494

Feher P, Olah M, Bodnar I, Hechtl D, Bacskay I, Juhasz B, Nagy GM, Vecsernyes M (2010) Dephosphorylation/inactivation of tyrosine hydroxylase at the median eminence of the hypothalamus is required for suckling-induced prolactin and adrenocorticotrop hormone responses. Brain Res Bull 82:141–145

Fitzsimmons MD, Olschowka JA, Wiegand SJ, Hoffman GE (1992) Interaction of opioid peptide-containing terminals with dopaminergic perikarya in the rat hypothalamus. Brain Res 581:10–18

Fletcher PA, Smiljanic K, Maso Previde R, Iben JR, Li T, Rokic MB, Sherman A, Coon SL, Stojilkovic SS (2019) Cell type- and sex-dependent transcriptome profiles of rat anterior pituitary cells. Front Endocrinol (Lausanne) 10:623

Fliestra RJ, Voogt JL (1997) Lactogenic hormones of the placenta and pituitary inhibit suckling-induced prolactin (PRL) release but not the ante-partum PRL surge. Proc Soc Exp Biol Med 214:258–264

Frawley LS, Boockfor FR (1991) Mammosomatotropes: presence and functions in normal and neoplastic pituitary tissue. Endocr Rev 12:337–355

Freeman ME, Kanyicska B, Lerant A, Nagy G (2000) Prolactin: structure, function, and regulation of secretion. Physiol Rev 80:1523–1631

Gellersen B, Kempf R, Telgmann R, DiMattia GE (1994) Nonpituitary human prolactin gene transcription is independent of Pit-1 and differentially controlled in lymphocytes and in endometrial stroma. Mol Endocrinol 8:356–373
Grattan DR (2015) 60 years of neuroendocrinology: the hypothalamo-prolactin axis. J Endocrinol 226:T101–T122
Grattan DR, Averill RL (1995) Absence of short-loop autoregulation of prolactin during late pregnancy in the rat. Brain Res Bull 36:413–416
Grattan DR, Le Tissier P (2015) Hypothalamic control of prolactin secretion, and the multiple reproductive functions of prolactin. In: Plant TM, Zelezni AJ (eds) *Knobil and Neil's physiology of reproduction*, 4th edn. Elsevier, Amsterdam
Grattan DR, Xu J, McLachlan MJ, Kokay IC, Bunn SJ, Hovey RC, Davey HW (2001) Feedback regulation of PRL secretion is mediated by the transcription factor, signal transducer, and activator of transcription 5b. Endocrinology 142:3935–3940
Grosvenor CE, Mena F (1980) Evidence that thyrotropin-releasing hormone and a hypothalamic prolactin-releasing factor may function in the release of prolactin in the lactating rat. Endocrinology 107:863–868
Grosvenor CE, Whitworth N (1974) Evidence for a steady rate of secretion of prolactin following suckling in the rat. J Dairy Sci 57:900–904
Grosvenor CE, Whitworth NS, Mena F (1981) Evidence that the depletion and release phases of prolactin secretion in the lactating rat have different activation thresholds in response to exteroceptive stimulation from rat pups. Endocrinology 108:820–824
Guillou A, Romano N, Steyn F, Abitbol K, Le Tissier P, Bonnefont X, Chen C, Mollard P, Martin AO (2015) Assessment of lactotroph axis functionality in mice: longitudinal monitoring of PRL secretion by ultrasensitive-ELISA. Endocrinology 156:1924–1930
Gunnet JW, Freeman ME (1983) The mating-induced release of prolactin: a unique neuroendocrine response. Endocr Rev 4:44–61
Hashi A, Mazawa S, Kato J, Arita J (1995) Pentobarbital anesthesia during the proestrous afternoon blocks lactotroph proliferation occurring on estrus in female rats. Endocrinology 136:4665–4671
Hashizume T, Watanabe R, Inaba Y, Sawai K, Fulop F, Nagy GM (2017) Hypothalamic dopamine is required for salsolinol-induced prolactin secretion in goats. Anim Sci J 88:1588–1594
He Z, Fernandez-Fuente M, Strom M, Cheung L, Robinson IC, Le Tissier P (2011) Continuous on-line monitoring of secretion from rodent pituitary endocrine cells using fluorescent protein surrogate markers. J Neuroendocrinol 23:197–207
Herbison AE (2016) Control of puberty onset and fertility by gonadotropin-releasing hormone neurons. Nat Rev Endocrinol 12:452–466
Herde MK, Iremonger KJ, Constantin S, Herbison AE (2013) GnRH neurons elaborate a long-range projection with shared axonal and dendritic functions. J Neurosci 33:12689–12697
Hikake T, Hayashi S, Iguchi T, Sato T (2009) The role of IGF1 on the differentiation of prolactin secreting cells in the mouse anterior pituitary. J Endocrinol 203:231–240
Ho Y, Hu P, Peel MT, Chen S, Camara PG, Epstein DJ, Wu H, Liebhaber SA (2020) Single-cell transcriptomic analysis of adult mouse pituitary reveals sexual dimorphism and physiologic demand-induced cellular plasticity. Protein Cell 11:565–583
Hodson DJ, Schaeffer M, Romano N, Fontanaud P, Lafont C, Birkenstock J, Molino F, Christian H, Lockey J, Carmignac D et al (2012) Existence of long-lasting experience-dependent plasticity in endocrine cell networks. Nat Commun 3:605
Horseman ND, Yu-Lee LY (1994) Transcriptional regulation by the helix bundle peptide hormones: growth hormone, prolactin, and hematopoietic cytokines. Endocr Rev 15:627–649
Kanematsu S, Sawyer CH (1973) Elevation of plasma prolactin after hypophysial stalk section in the rat. Endocrinology 93:238–241
Kawano H, Daikoku S (1987) Functional topography of the rat hypothalamic dopamine neuron systems: retrograde tracing and immunohistochemical study. J Comp Neurol 265:242–253
Kelberman D, Rizzoti K, Lovell-Badge R, Robinson IC, Dattani MT (2009) Genetic regulation of pituitary gland development in human and mouse. Endocr Rev 30:790–829

Kobayashi T, Usui H, Tanaka H, Shozu M (2018) Variant prolactin receptor in agalactia and hyperprolactinemia. N Engl J Med 379:2230–2236
Kokay IC, Grattan DR (2005) Expression of mRNA for prolactin receptor (long form) in dopamine and pro-opiomelanocortin neurones in the arcuate nucleus of non-pregnant and lactating rats. J Neuroendocrinol 17:827–835
Kokay IC, Wyatt A, Phillipps HR, Aoki M, Ectors F, Boehm U, Grattan DR (2018) Analysis of prolactin receptor expression in the murine brain using a novel prolactin receptor reporter mouse. J Neuroendocrinol 30:e12634
LaPierre MP, Godbersen S, Torres Esteban M, Schad AN, Treier M, Ghoshdastider U, Stoffel M (2021) MicroRNA-7a2 Regulates Prolactin in Developing Lactotrophs and Prolactinoma Cells. Endocrinology 162:bqaa220
Lau-Corona D, Ma H, Vergato C, Sarmento-Cabral A, Del Rio-Moreno M, Kineman RD, Waxman DJ (2022) Constitutively active STAT5b feminizes mouse liver gene expression. Endocrinology 163:bqac046
Le Tissier PR, Mollard P (2021) Renewing an old interest: pituitary folliculostellate cells. J Neuroendocrinol 33:e13053
Le Tissier PR, Hodson DJ, Lafont C, Fontanaud P, Schaeffer M, Mollard P (2012) Anterior pituitary cell networks. Front Neuroendocrinol 33:252–266
Le Tissier PR, Hodson DJ, Martin AO, Romano N, Mollard P (2015) Plasticity of the prolactin (PRL) axis: mechanisms underlying regulation of output in female mice. Adv Exp Med Biol 846:139–162
Le Tissier P, Campos P, Lafont C, Romano N, Hodson DJ, Mollard P (2017) An updated view of hypothalamic-vascular-pituitary unit function and plasticity. Nat Rev Endocrinol 13:257–267
Lee LR, Haisenleder DJ, Marshall JC, Smith MS (1989) The role of the suckling stimulus in regulating pituitary prolactin mRNA in the rat. Mol Cell Endocrinol 64:243–249
Lehman MN, Coolen LM, Goodman RL (2010) Minireview: kisspeptin/neurokinin B/dynorphin (KNDy) cells of the arcuate nucleus: a central node in the control of gonadotropin-releasing hormone secretion. Endocrinology 151:3479–3489
Leong DA, Frawley LS, Neill JD (1983) Neuroendocrine control of prolactin secretion. Annu Rev Physiol 49:109–127
Liu B, Arbogast LA (2008) Phosphorylation state of tyrosine hydroxylase in the stalk-median eminence is decreased by progesterone in cycling female rats. Endocrinology 149:1462–1469
Lookingland KJ, Jarry HD, Moore KE (1987) The metabolism of dopamine in the median eminence reflects the activity of tuberoinfundibular neurons. Brain Res 419:303–310
Lyons DJ, Horjales-Araujo E, Broberger C (2010) Synchronized network oscillations in rat tuberoinfundibular dopamine neurons: switch to tonic discharge by thyrotropin-releasing hormone. Neuron 65:217–229
Lyons DJ, Hellysaz A, Broberger C (2012) Prolactin regulates tuberoinfundibular dopamine neuron discharge pattern: novel feedback control mechanisms in the lactotrophic axis. J Neurosci 32:8074–8083
Ma FY, Grattan DR, Goffin V, Bunn SJ (2005) Prolactin-regulated tyrosine hydroxylase activity and messenger ribonucleic acid expression in mediobasal hypothalamic cultures: the differential role of specific protein kinases. Endocrinology 146:93–102
Mai LM, Shieh KR, Pan JT (1994) Circadian changes of serum prolactin levels and tuberoinfundibular dopaminergic neuron activities in ovariectomized rats treated with or without estrogen: the role of the suprachiasmatic nuclei. Neuroendocrinology 60:520–526
Mena F, Grosvenor CE (1968) Effect of number of pups upon suckling-induced fall in pituitary prolactin concentration and milk ejection in the rat. Endocrinology 82:623–626
Merchenthaler I (1993) Induction of enkephalin in tuberoinfundibular dopaminergic neurons during lactation. Endocrinology 133:2645–2651
Merchenthaler I (1994) Induction of enkephalin in tuberoinfundibular dopaminergic neurons of pregnant, pseudopregnant, lactating and aged female rats. Neuroendocrinology 60:185–193

Merchenthaler I, Lennard DE, Cianchetta P, Merchenthaler A, Bronstein D (1995) Induction of proenkephalin in tuberoinfundibular dopaminergic neurons by hyperprolactinemia: the role of sex steroids. Endocrinology 136:2442–2450

Murai I, Ben-Jonathan N (1987) Posterior pituitary lobectomy abolishes the suckling-induced rise in prolactin (PRL): evidence for a PRL-releasing factor in the posterior pituitary. Endocrinology 121:205–211

Nahi F, Arbogast LA (2003) Prolactin modulates hypothalamic preproenkephalin, but not proopiomelanocortin, gene expression during lactation. Endocrine 20:115–122

Nicoll CS (1980) Ontogeny and evolution of prolactin's functions. Fed Proc 39:2563–2566

Noel GL, Suh HK, Frantz AG (1974) Prolactin release during nursing and breast stimulation in postpartum and nonpostpartum subjects. J Clin Endocrinol Metab 38:413–423

Oliver C, Mical RS, Porter JC (1977) Hypothalamic-pituitary vasculature: evidence for retrograde blood flow in the pituitary stalk. Endocrinology 101:598–604

Orrillo SJ, de Dios N, Asad AS, De Fino F, Imsen M, Romero AC, Zarate S, Ferraris J, Pisera D (2020) Anterior pituitary gland synthesises dopamine from l-3,4-dihydroxyphenylalanine (l-dopa). J Neuroendocrinol 32:e12885

Peel MT, Ho Y, Liebhaber SA (2018) Transcriptome analyses of female somatotropes and lactotropes reveal novel regulators of cell identity in the pituitary. Endocrinology 159:3965–3980

Peel MT, Ho Y, Liebhaber SA (2020) The transcription factor NR4A2 plays an essential role in driving prolactin expression in female pituitary lactotropes. Endocrinology 161:bqaa046

Phillipps HR, Yip SH, Grattan DR (2020) Patterns of prolactin secretion. Mol Cell Endocrinol 502:110679

Phillipps HR, Khant Aung Z, Grattan DR (2022) Elevated prolactin secretion during proestrus in mice: absence of a defined surge. J Neuroendocrinol 34:e13129

Plotsky PM, Neill JD (1982) Interactions of dopamine and thyrotropin-releasing hormone in the regulation of prolactin release in lactating rats. Endocrinology 111:168–173

Riddle O, Bates RW, Dykshorn SW (1933) The preparation, identification and assay of prolactin - a hormone of the anterior pituitary. Am J Phys 105:191–216

Romano N, Yip SH, Hodson DJ, Guillou A, Parnaudeau S, Kirk S, Tronche F, Bonnefont X, Le Tissier P, Bunn SJ et al (2013) Plasticity of hypothalamic dopamine neurons during lactation results in dissociation of electrical activity and release. J Neurosci 33:4424–4433

Romanov RA, Zeisel A, Bakker J, Girach F, Hellysaz A, Tomer R, Alpar A, Mulder J, Clotman F, Keimpema E et al (2017) Molecular interrogation of hypothalamic organization reveals distinct dopamine neuronal subtypes. Nat Neurosci 20:176–188

Romero-Fernandez W, Borroto-Escuela DO, Vargas-Barroso V, Narvaez M, Di Palma M, Agnati LF, Larriva Sahd J, Fuxe K (2014) Dopamine D1 and D2 receptor immunoreactivities in the arcuate-median eminence complex and their link to the tubero-infundibular dopamine neurons. Eur J Histochem 58:2400

Schuff KG, Hentges ST, Kelly MA, Binart N, Kelly PA, Iuvone PM, Asa SL, Low MJ (2002) Lack of prolactin receptor signaling in mice results in lactotroph proliferation and prolactinomas by dopamine-dependent and -independent mechanisms. J Clin Invest 110:973–981

Seilicovich A (2010) Cell life and death in the anterior pituitary gland: role of oestrogens. J Neuroendocrinol 22:758–764

Sellix MT, Egli M, Poletini MO, McKee DT, Bosworth MD, Fitch CA, Freeman ME (2006) Anatomical and functional characterization of clock gene expression in neuroendocrine dopaminergic neurons. American Journal of Physiology. Regulatory, Integrative and Comparative Physiology 290:R1309–R1323

Selmanoff M (1985) Rapid effects of hyperprolactinemia on basal prolactin secretion and dopamine turnover in the medial and lateral median eminence. Endocrinology 116:1943–1952

Selmanoff M, Wise PM (1981) Decreased dopamine turnover in the median eminence in response to suckling in the lactating rat. Brain Res 212:101–115

Shanti AS, Subramanian MG, Savoy-Moore RT, Kruger ML, Moghissi KS (1995) Attenuation of the magnitude of suckling-induced prolactin release with advancing lactation: mechanisms. Life Sci 56:259–266

Shin SH (1980) Physiological evidence for the existence of prolactin releasing factor: stress-induced prolactin secretion is not linked to dopaminergic receptors. Neuroendocrinology 31:375–379

Silva KSC, Aquino NSS, Gusmao DO, Henriques PC, Reis AM, Szawka RE (2020) Reduced dopaminergic tone during lactation is permissive to the hypothalamic stimulus for suckling-induced prolactin release. J Neuroendocrinol 32:e12880

Skov LJ, Ratner C, Hansen NW, Thompson JJ, Egerod KL, Burm H, Dalboge LS, Hedegaard MA, Brakebusch C, Pers TH et al (2019) RhoA in tyrosine hydroxylase neurones regulates food intake and body weight via altered sensitivity to peripheral hormones. J Neuroendocrinol 31:e12761

Smith MS, Freeman ME, Neill JD (1975) The control of progesterone secretion during the estrous cycle and early pseudopregnancy in the rat: prolactin, gonadotropin and steroid levels associated with rescue of the corpus luteum of pseudopregnancy. Endocrinology 96:219–226

Stagkourakis S, Kim H, Lyons DJ, Broberger C (2016) Dopamine autoreceptor regulation of a hypothalamic dopaminergic network. Cell Rep 15:735–747

Stagkourakis S, Perez CT, Hellysaz A, Ammari R, Broberger C (2018) Network oscillation rules imposed by species-specific electrical coupling. elife 7:e33144

Stagkourakis S, Dunevall J, Taleat Z, Ewing AG, Broberger C (2019) Dopamine release dynamics in the tuberoinfundibular dopamine system. J Neurosci 39:4009–4022

Stagkourakis S, Smiley KO, Williams P, Kakadellis S, Ziegler K, Bakker J, Brown RSE, Harkany T, Grattan DR, Broberger C (2020) A neuro-hormonal circuit for paternal behavior controlled by a hypothalamic network oscillation. Cell 182:960–975

Szabo FK, Snyder N, Usdin TB, Hoffman GE (2010) A direct neuronal connection between the subparafascicular and ventrolateral arcuate nuclei in non-lactating female rats. Could this pathway play a role in the suckling-induced prolactin release? Endocrine 37:62–70

Szawka RE, Helena CV, Rodovalho GV, Monteiro PM, Franci CR, Anselmo-Franci JA (2005) Locus coeruleus norepinephrine regulates the surge of prolactin during oestrus. J Neuroendocrinol 17:639–648

Szawka RE, Ribeiro AB, Leite CM, Helena CV, Franci CR, Anderson GM, Hoffman GE, Anselmo-Franci JA (2010) Kisspeptin regulates prolactin release through hypothalamic dopaminergic neurons. Endocrinology 151:3247–3257

Terkel J, Blake CA, Sawyer CH (1972) Serum prolactin levels in lactating rats after suckling or exposure to ether. Endocrinology 91:49–53

Thörn Pérez C, Ferraris J, van Lunteren JA, Hellysaz A, Iglesias MJ, Broberger C (2020) Adaptive Resetting of Tuberoinfundibular Dopamine (TIDA) Network Activity During Lactation in Mice. J Neurosci 40:3203–3216

Tierney T, Robinson IC (2002) Increased lactotrophs despite decreased somatotrophs in the dwarf (dw/dw) rat: a defect in the regulation of lactotroph/somatotroph cell fate? J Endocrinol 175:435–446

Tonkowicz PA, Voogt JL (1983) Termination of prolactin surges with development of placental lactogen secretion in the pregnant rat. Endocrinology 113:1314–1318

Toth BE, Homicsko K, Radnai B, Maruyama W, DeMaria JE, Vecsernyes M, Fekete MI, Fulop F, Naoi M, Freeman ME et al (2001) Salsolinol is a putative endogenous neuro-intermediate lobe prolactin-releasing factor. J Neuroendocrinol 13:1042–1050

Valverde RC, Chieffo V, Reichlin S (1972) Prolactin releasing factor in porcine and rat hypothalamic tissue. Endocrinology 91:982–993

Vekemans M, Delvoye P, L'Hermite M, Robyn C (1977) Serum prolactin levels during the menstrual cycle. J Clin Endocrinol Metab 44:989–993

Wang HJ, Hoffman GE, Smith MS (1993) Suppressed tyrosine hydroxylase gene expression in the tuberoinfundibular dopaminergic system during lactation. Endocrinology 133:1657–1663

Yamamoto N, Seo H, Suganuma N, Matsui N, Nakane T, Kuwayama A, Kageyama N (1986) Effect of estrogen on prolactin mRNA in the rat pituitary. Analysis by in situ hybridization and immunohistochemistry. Neuroendocrinology 42:494–497

Yip SH, York J, Hyland B, Bunn SJ, Grattan DR (2018) Incomplete concordance of dopamine transporter Cre (DAT(IREScre))-mediated recombination and tyrosine hydroxylase immunoreactivity in the mouse forebrain. J Chem Neuroanat 90:40–48

Yip SH, Romano N, Gustafson P, Hodson DJ, Williams EJ, Kokay IC, Martin AO, Mollard P, Grattan DR, Bunn SJ (2019) Elevated prolactin during pregnancy drives a phenotypic switch in mouse hypothalamic dopaminergic neurons. Cell Rep 26(1787–1799):e1785

Yip SH, Araujo-Lopes R, Szawka RE, York J, Hyland B, Grattan DR, Bunn SJ (2020) Morphological plasticity of the tuberoinfundibular dopaminergic neurones in the rat during the oestrous cycle and lactation. J Neuroendocrinol 32:e12884

Zhang X, van den Pol AN (2015) Dopamine/tyrosine hydroxylase neurons of the hypothalamic arcuate nucleus release GABA, communicate with dopaminergic and other arcuate neurons, and respond to dynorphin, met-enkephalin, and oxytocin. J Neurosci 35:14966–14982

Zhang X, van den Pol AN (2016) Hypothalamic arcuate nucleus tyrosine hydroxylase neurons play orexigenic role in energy homeostasis. Nat Neurosci 19:1341–1347

Zhang J, Gao J, Zhang D, Liu H, Gou K, Cui S (2021) miR-375 acts as a novel factor modulating pituitary prolactin synthesis through Rasd1 and Esr1. J Endocrinol 250:25–35

Neuroendocrine Control of Maternal Behavior

8

Joseph S. Lonstein, Mayra L. Almanza-Sepulveda, Frédéric Lévy, and Alison S. Fleming

Abstract

Maternal caregiving is an essential component of successful mammalian reproduction. Indeed, almost all mothers exhibit a species-typical repertoire and pattern of care that is sensitive to the needs of their young and essential for healthy offspring development. Although different mammalian mothers show different types of care, the onset of their nurturant behaviors at the time of giving birth involves the same collection of steroid (i.e., estradiol, progesterone, glucocorticoids) and peptide (i.e., prolactin, oxytocin) hormones. These hormones underlie the physiology of pregnancy and lactation, and simultaneously act on the brain to instill motherhood. This chapter provides an overview of the species-specific behaviors and hormonal regulation of maternal caregiving in three representative groups of mammals: (1) laboratory rats and mice, which give birth to many underdeveloped young that need tremendous care, (2) sheep, which give birth to one or two highly developed lamb(s), and (3) humans and some other primates, which care for one or two intermediately developed young. We also briefly discuss the influence of hormones on affective state in human mothers, given that mood and emotion are critical modulators of the relationship between hormones and infant caregiving in our own species.

J. S. Lonstein (✉)
Department of Psychology and Neuroscience Program, Michigan State University, East Lansing, MI, USA
e-mail: lonstein@msu.edu

M. L. Almanza-Sepulveda · A. S. Fleming
Department of Psychology, University of Toronto at Mississauga, Mississauga, ON, Canada

F. Lévy
CNRS, IFCE, INRAE, Université de Tours, PRC, Nouzilly, France

P. J. Brunton, D. R. Grattan (eds.), *Neuroendocrine Regulation of Mammalian Pregnancy and Lactation*, Masterclass in Neuroendocrinology 15,
https://doi.org/10.1007/978-3-031-51138-7_8

Keywords

Maternal behavior · Estrogen · Human · Glucocorticoids · Progesterone · Prolactin · Oxytocin · Sheep · Rats

8.1 Introduction

Behaviors associated with delivering and caring for young are essential for survival in mammals, with the mother providing necessary food, warmth, shelter and protection. Mothers also provide their young with their first social experiences and help them to integrate into their social world. Maternal caregiving behaviors in mammals assume a wide variety of patterns, some of which are common across many species, while other patterns rely more on the maturity of the young at birth. These patterns of caregiving also depend on the ecology and social structure of the species, and among humans, on our cultural norms and desired child developmental outcomes. Importantly, these caregiving behaviors also almost always depend on a complex profile of endocrine fluctuations that occur across pregnancy, **parturition** (i.e., giving birth) and the early postpartum period.

The onset of maternal behavior must be synchronized with the end of pregnancy and parturition. At the appearance of the young, almost all mothers express a seemingly sudden interest in them. In laboratory rodents and many other mammals (Box 8.1), mothers avidly lick pups while consuming amniotic fluids and placenta, which facilitates continued expulsion of the fetuses, has an analgesic effect, maintains the cleanliness of the birth site, and minimizes the attraction of predators. Mothers also sometimes display retrieval, the gathering or carrying behaviors that keep the young safe and in close proximity. Nursing is the most common maternal behavior in mammals, as only mothers can provide milk as the earliest form of food. Nursing generally occurs within the first hours after parturition, with mothers adopting a characteristic nursing posture—many female rodents and other small mammals arch their back over the litter in what is termed **kyphosis** (Fig. 8.1a), female ungulates instead usually adopt a parallel-inverse nursing position with their young (Fig. 8.1b), and female primates hold their infants in their arms and bring them to the breast (Fig. 8.1c, d).

In addition, parental care is influenced by whether the neonates are physiologically and behaviorally immature (i.e., **altricial**), highly mature (i.e., **precocial**), or of intermediate maturity at birth. Altricial mammalian young are born blind and deaf, and have limited locomotor ability. Because they are born without fur, they are also unable to thermoregulate. Their care usually involves the mother preparing a soft, heat-retaining nest before giving birth to the numerous young. This nesting can take place underground, as is found in many small rodents and in rabbits; at ground level, as found among carnivores; or in the forest canopy, as in some small primates. The nest is then the central element around which maternal care and mother-infant interaction is organized. Mothers almost always nurse their young in the nest, retrieve them back to the nest if they are displaced, and in many species clean them there by licking and grooming. In most of these altricial species, the young stay huddled together within the nest for the first days or weeks of life while mothers

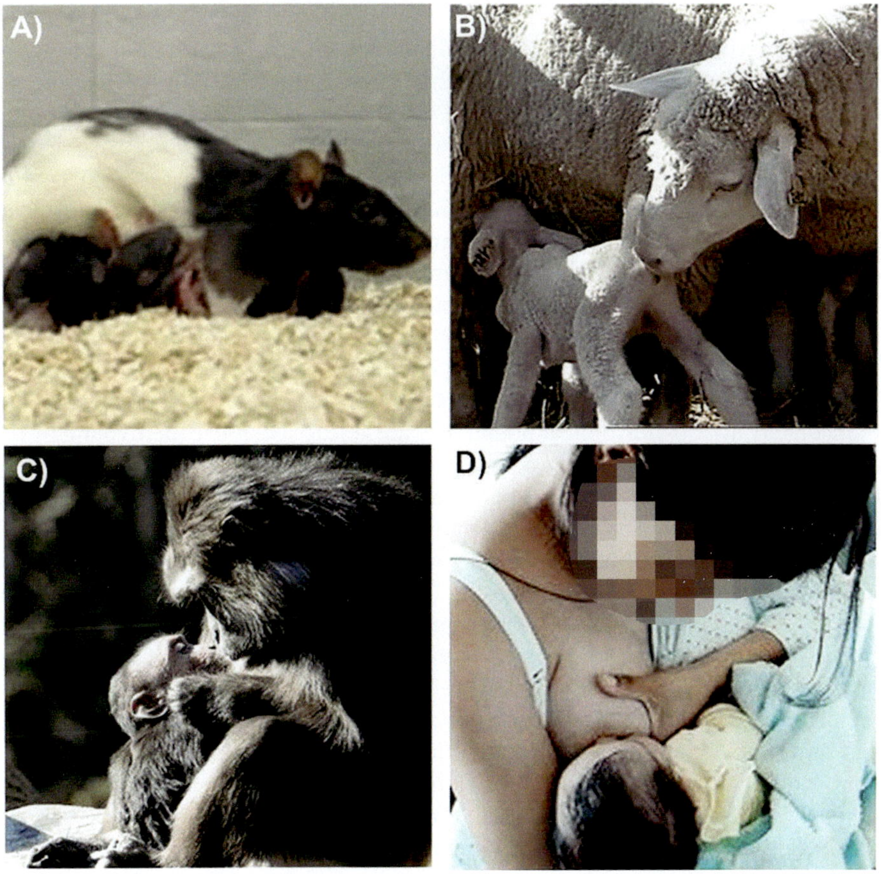

Fig. 8.1 Nursing behavior in representative mammals. (**a**) Laboratory rats and other small rodents arch their back over the litter in a posture called kyphosis; (**b**) female sheep and other **ungulates** usually adopt a parallel-inverse position with their young; (**c**, **d**) female primates hold their infants in their arms and bring them to their chest. All photos are the property of the authors

vigilantly defend the nest from intruders. Thus, the presence of completely unrelated young becomes very unlikely and there is little need for the mother to recognize individual members of her litter.

In contrast, species of mothers that give birth to precocial young (most ungulates and **pinnipeds**, as well as a few species of rodents) tend to have only one offspring or a small litter of fully developed young that are capable of following the mother shortly after birth. These newborns have functional hearing and vision, rapid locomotor development, and the ability to thermoregulate on their own. These species are usually herd animals, with the offspring following their mother as she moves along with conspecifics of the group (sheep, horses). In some species (goats, deer), however, the neonates are hidden, to protect them from predators and the mother returns periodically to nurse. In most of the precocial species living in large groups, mothers' own young interact with unrelated neonates, so there is a risk for mothers of having their milk usurped by unrelated offspring. These mothers, therefore, develop discriminative maternal care, allowing their own young to suckle while rejecting unrelated (i.e., "alien") young that approach the udder. In this respect, the establishment of a selective bond within the first few hours after parturition represents one of the essential characteristics of maternal behavior in most precocial species that cannot be appropriately studied in the indiscriminate laboratory rat or mouse. Individual recognition of young has been demonstrated in sheep, goat, cattle, horses and pinnipeds. It relies on the mother recognizing their young's olfactory cues at a short distance and their visual and auditory cues at a long distance.

Between the extremes of altriciality and precociousness lie a continuum of newborn developmental states and thus mothering styles. In many species, neonates have functional sensory systems but limited thermoregulation (pigs) or locomotion (primates). Sows build a nest and do not develop selectivity towards their piglets immediately after parturition, although they are able to recognize their young by the next day. Finally, most primates do not build nests but instead the mother carries the offspring in her arms or on her back. Like other primates, humans can discriminate between their own infants and others, and may show a preference for their own and develop a strong bond, but they do not necessarily reject infants belonging to other mothers. Human mothers also engage in behaviors that soothe and/or stimulate the infant through nursing, rocking, massaging, singing and talking to their infants in the sing-song pattern of "motherese" almost exclusively found when mothers and other sensitive adults talk to infants (Barrett and Fleming 2011; Fig. 8.1d).

It is clear that mammalian mothers provide a broad range of caregiving patterns and activities. As will be detailed below, however, the endocrine fluctuations across reproduction are almost the same for all of these mammalian mothers. The present chapter is an overview of these endocrine changes that mothers undergo across pregnancy and the early postpartum period in representative mammals that give birth to altricial (laboratory rats and mice), precocial (sheep), and intermediate (humans and some other primates) young, and then details the role of those hormones in regulating the onset of their maternal caregiving. We also include a discussion of some influences of peripartum hormones on affective state in human mothers, given that mood and emotion are such critical modulators of the relationship between hormones and infant caregiving in our own species.

Box 8.1 Methods for Studying Maternal Caregiving Behavior in Laboratory Rodents

Laboratory rodents are by far the best-studied model for understanding the biological mechanisms underlying mammalian maternal care. Over the years, many researchers studying rodent maternal caregiving in the laboratory have converged upon a few behavioral observation methods to conduct their experiments. One method is to use periodic "spot-check" observations of a mother's undisturbed behavior with her litter. This involves the researcher very quietly standing and watching (or a digital camera recording) the subject and litter in their familiar home cage, and the simple presence or absence of each

(continued)

Box 8.1 (continued)

behavior of interest noted in successive 5- to 20-s bins or intervals. These spot-check observations can last from as little as 5 min (brief snapshots) to an hour or more. This type of test can be repeated many times throughout the day, depending on feasibility; conducting 30- to 90-min observations more than once a day requires a devoted staff with a great deal of time, while a single 5- to 15-min observation each day is much easier but probably does not provide a sufficient amount of data to expect reliability, so are best when done multiple times across a day. This spot-check method is the simplest to learn, and the least expensive, because it can be performed with just a stopwatch, pen and paper (see Lonstein and Fleming 2001 for more details).

More detail-rich data on maternal caregiving can be obtained from continuous observation of subject-litter interactions, either without disturbing the dams before testing or after a period of separation between the dam and litter. In the latter case, if young litters are used, they are usually kept in an incubator set at nest temperature (34 °C). Both continuous observation methods require a computerized data acquisition system, such as Noldus Observer™, or a number of freely available programs available online that can be used with one's own laptop. Short 5- to 10-min separations and a reunion are often used because mothers will usually show all caregiving behaviors within a 15-min observation period, with the likely exception of quiescent kyphotic nursing, given that it has a much longer latency before beginning. However, if the subject and litter are separated for 2–4 h before testing, the entire repertoire of parental behaviors is likely to be observed within a single 30-min observation. Scoring can be done contemporaneously by a quiet observer of the home cage or can be digitally-recorded for later scoring.

Because maternal behavior is a complex category of different activities with different determinants, almost all studies measure numerous behaviors. At the very minimum, almost all studies use one of the methods above to record the latency and duration for mothers to retrieve one or all the pups after the pups have been experimentally displaced. Many other studies also include some measure of whether the mother is or is not on the nest with the pups. More sophisticated studies interested in the larger suite of postpartum caregiving behaviors, and the details of how they are displayed and patterned, will usually score additional behaviors. These include the duration and/or frequency of mothers' actively hovering over the litter in the nest (which is distinct from quiescent nursing), various quiescent nursing postures, licking the pups, digging or oral movement of nesting material, self-grooming, cage exploration, ingestive behaviors and resting alone away from the nest. Other factors to consider that increase data validity and replicability within and between laboratories include training observers to a very high level of inter-observer reliability (>90%) for every behavior scored, understanding that the mother's postpartum stage and the age of litter strongly influence the behaviors occurring, that the size and sex ratio of the litter affect how mothers interact with them, and that different strains of rats and mice can show quite different maternal caregiving (see Lonstein and Fleming 2001).

8.2 Major Endocrine Changes Associated with Pregnancy and Parturition

8.2.1 Laboratory Rats

In laboratory rats, plasma concentrations of estradiol and other estrogens are relatively low until about day 16 of the 22-day pregnancy, after which they slowly rise. Associated with the postpartum ovulation and behavioral estrus in these females, which allows them to reproduce again most efficiently, is a notable rise in estrogens during the final 24–48 h of pregnancy that reaches its peak between 3 h and 9 h after parturition. Concentrations then fall to premating titers again by 12 h postpartum (Fig. 8.2a). Circulating estrogens continue to be low through mid-lactation, after which they begin to rise again and continue to do so through weaning of the pups and the mother's eventual resumption of ovulation (see Lonstein et al. 2015b; Numan 2021).

A different pattern is seen for circulating progesterone, which rises within days after insemination to reach concentrations as high, or many times higher, than those seen in the estrous cycle. Progesterone then sharply falls beginning on day 19 of pregnancy, to reach a nadir during or soon after parturition. Thus, the peripartum period in laboratory rats (and many other mammals, as will be evident below) involves a dramatic shift in the ratio of circulating estrogen to progesterone. Progesterone temporarily rises again between 9 h and 12 h after giving birth, as a result of the postpartum estrus, and is then low through the first few days postpartum. A mostly understudied rise in progesterone subsequently occurs, reaching its peak around postpartum day 10, and then falling through the remainder of lactation. While it has been traditionally thought that hormones have little role in maternal caregiving once they establish its onset at parturition, this postpartum rise in progesterone does regulate the duration of time rat mothers spend with their pups as lactation progresses (Grieb et al. 2017).

The vaginocervical stimulation during mating causes prolactin to be released from the anterior pituitary gland in twice-daily pulses for the first 10 days of pregnancy. These prolactin pulses are required for the ovaries' corpora lutea to secrete high blood concentrations of progesterone and maintain the first half of pregnancy, but thereafter most circulating prolactin in the mother instead originates from the placenta (i.e., *placental lactogens*). Through a negative feedback loop stimulated by these placental lactogens, pituitary prolactin secretion is then low throughout most of the second half of pregnancy, but the inhibition ceases in the last few days of pregnancy and results in a large prepartum surge of prolactin of pituitary origin. After parturition, plasma prolactin concentrations are generally high until postpartum day 10 while mothers are in suckling or even non-suckling contact with their pups, but then decline through the remainder of lactation (Bridges 2018).

Consistent with the uterine muscles needing to refrain from contracting and prematurely expelling the fetuses, plasma oxytocin concentrations are low for the first 2 weeks of pregnancy in rats but quadruple by day 17. Circulating oxytocin continues to rise through parturition as each pup is born, and then oxytocin drops back to

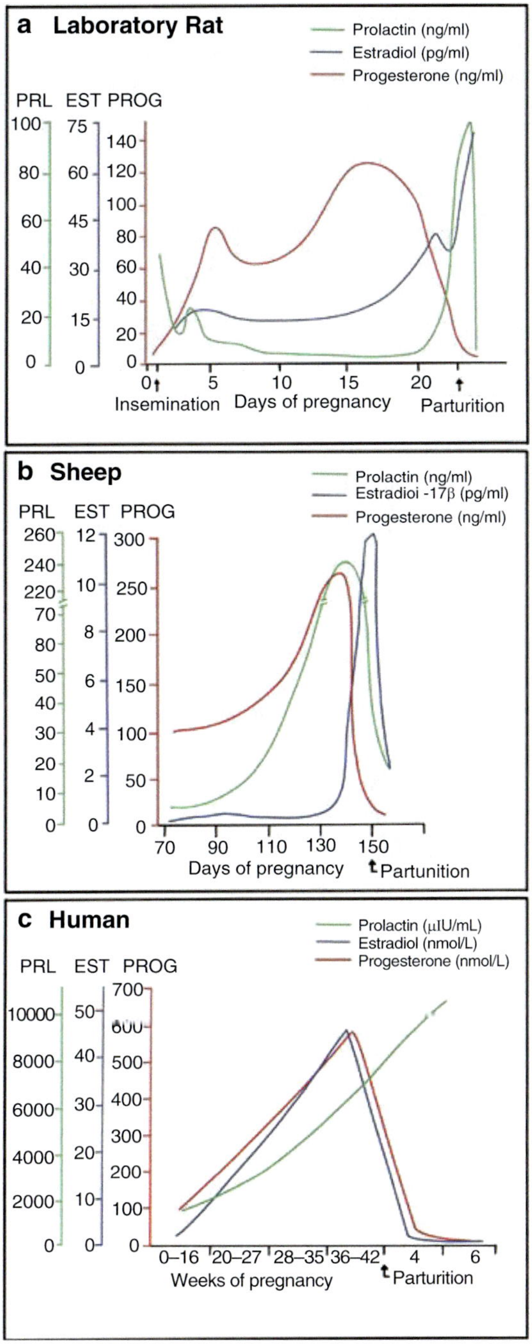

Fig. 8.2 Circulating plasma concentrations of prolactin (green line), estradiol (blue line), and progesterone (red line) across pregnancy and parturition in (**a**) laboratory rats, (**b**) sheep, (**c**) humans. Note the generally similar patterns across all species. Modified with permission from Lonstein et al. (2015a)

early-pregnancy concentrations across the subsequent 24 h. These concentrations continue to remain low during the postpartum period until suckling by the young elicits brief boluses of oxytocin released into the blood, resulting in milk ejection (Anderson et al. 2015).

Corticosterone, the primary circulating glucocorticoid in many non-primates, rises from mid-pregnancy and peaks at parturition (Fig. 8.3a). This pattern is necessary to meet the increased metabolic demands of the fetus and for the secretory function of the mammary gland and thus the onset and maintenance of lactation

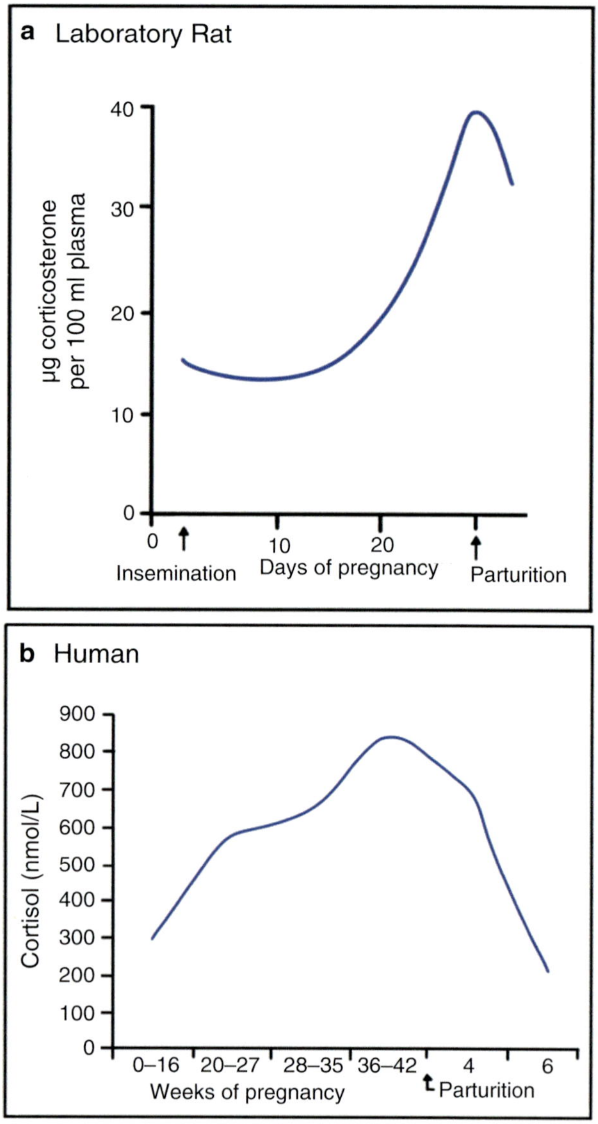

Fig. 8.3 Circulating plasma concentrations of corticosterone in laboratory rats (**a**) and cortisol in women (**b**) both rise across pregnancy, peak at parturition, and fall within the first few weeks postpartum. Modified with permission from Voogt et al. (1969) and Fleming et al. (1997)

(Anderson et al. 2015). After parturition in rats, corticosterone titers decline, with the circadian peaks lower compared to premating patterns for at least the first few days postpartum. Mother rats also show reduced hypothalamic-pituitary-adrenal (HPA) axis responses toward many stressors through mid-lactation (see Chap. 4), which helps prevent them from over-reacting to environmental insults that are not overly threatening. If the stressor is likely to directly affect the pups (e.g., a predator or an infanticidal male conspecific near the nest site), though, the mothers do display increased corticosterone secretion.

8.2.2 Sheep

Similar to laboratory rats, sheep also exhibit a shift across their 152-day pregnancy in the ratio of circulating estradiol to progesterone. One day before parturition, plasma concentrations of estradiol rise to reach a temporary periparturitional peak, which can be contrasted to the almost undetectable amounts found throughout the rest of pregnancy. Circulating progesterone progressively increases during pregnancy to reach concentrations five times higher than those measured during the estrous cycle. Two to six days before parturition, plasma progesterone concentrations then dramatically decline to reach low titers at parturition and soon after (Fig. 8.2b). This change in steroid hormone balance is accompanied by a rise in circulating prolactin, which occurs 24–48 h before parturition, and these hormone concentrations are maintained by the lamb's suckling stimulation. Plasma cortisol concentrations gradually increase from around 5 days before term and peak at parturition. There is also a rise in plasma oxytocin secreted from the posterior pituitary, contemporaneous with expulsion of the neonate, and induced by the vaginocervical stimulation ewes receive as the lambs travel through the birth canal.

8.2.3 Humans and Non-human Primates

The profiles of circulating estrogens and progestins in pregnant and parturient humans and other Homininae are generally similar to each other, but do differ in some ways from those of other mammals such as the rodents and sheep discussed above. First, women undergo rising estrogens in addition to progesterone, starting during early pregnancy, and both hormones remain high throughout almost all of the remainder of pregnancy (Fig. 8.2c) (Lonstein et al. 2015b). It is also unclear in women whether estradiol or the "weaker" estrogen, estriol, is the primary circulating estrogen in pregnancy. While a number of studies have disputed that humans show a periparturitional withdrawal of progesterone that triggers delivery such as that seen in rats and sheep, recent analyses with very large sample sizes found that plasma progesterone does decrease at the end of pregnancy for most women and that concentrations of estriol surge (Smith et al. 2009). Thus, this results in an elevated peripartum estrogen:progestin ratio that is quite reminiscent of that found in other mammals. Similar to other mammals, suckling by infants postpartum strongly suppresses women's hypothalamic secretion of gonadotropin-releasing hormone (GnRH) and hence pituitary gonadotropin secretion, meaning regular breastfeeding can suppress ovulation for a few years, while also stimulating prolactin and oxytocin secretion.

Pregnant gorillas also experience rising plasma estrogens and progestins that peak a month or two before parturition, then notably decline a day or two before parturition. Estriol is probably the primary circulating estrogen during much of pregnancy in gorillas, but this is not true of all non-human primates, because estriol is not even detectable in pregnant rhesus monkeys and baboons. Blood concentrations of all estrogens and progesterone are very low after parturition in gorillas and remain that way for a few weeks postpartum. These hormone patterns are also very similar to those observed in pregnant and parturient chimpanzees.

With regards to the HPA axis hormones, circulating adrenocorticotropic hormone (ACTH) and cortisol increase tremendously from the first to third trimesters and fall during the first few days postpartum in women (Almanza-Sepulveda et al. 2020; Fig. 8.3b). This is also true in non-human primates including cotton-top tamarins, white-faced capuchins and baboons. The two peptide hormones often considered "hallmarks" of motherhood, prolactin and oxytocin, also fluctuate across pregnancy and postpartum in primates. Basal circulating prolactin concentrations rise from mid-pregnancy in women to the final few weeks of pregnancy and later spike in response to suckling and other infant cues. Circulating oxytocin is also elevated near the end of pregnancy, but it is unclear whether it is due to a gradual rise or a more abrupt one. Oxytocin concentrations measured in blood in the early postpartum period are generally higher than during pregnancy in some, but not all studies, and vary widely in women due to many factors including the presence or absence of breastfeeding. In terms of other primates, there is a rise in the secretion of prolactin and other lactogens during pregnancy and high concentrations are maintained during early lactation. In non-human primates, circulating oxytocin rises rapidly during the peripartum period and is low postpartum, other than during milk letdown (Saltzman and Maestripieri 2011).

In summary, the peripheral endocrine profiles observed across pregnancy and into the early postpartum period are remarkably similar in mammals, ranging from laboratory rodents to ungulates to non-human and human primates. This is despite the fact that these animals show different patterns of maternal caregiving, and indicates that species differences in maternal care must not be regulated by different patterns of hormones across female reproduction, but rather that the mostly same pattern uniquely affects the brain and behavior in different species (see below).

8.3 Hormonal Control of Maternal Caregiving and Associated Behaviors

There are at least two ways to answer the question as to how hormones exert their effects on mothering. The first is to study it at the cellular, systemic and molecular levels and focuses on how hormones influence the brain network for caregiving in the new mother. Here the issue is where and how these hormones act on the brain to stimulate nurturant behaviors. The second way to study it is not the main purpose of this chapter, but should be kept in mind, and asks what are the broader classes of behaviors affected by hormones that could indirectly affect mothering. That is, in what ways do the hormones that activate mothering also affect a mother's attraction to infant cues and promote approach to young; or a mother's ability to attend to the environment and their ability to easily shift attention in ways necessary for sensitive mothering; or a mother's ability to learn about the offspring and as a result become more efficient at mothering? Indeed, the broader behavioral dimensions of hedonics, affective state, attention and memory are all influenced by hormones and in turn affect mothering (see Barrett and Fleming 2011; Lonstein et al. 2015a). Of course, the brain systems underlying these more generalized behaviors must be able to respond to the hormones and interact with the brain systems involved in maternal care.

8.3.1 Estrogens and Progesterone

8.3.1.1 Laboratory Rats and Mice

Early studies on the role of hormones in maternal caregiving demonstrated that transfusing blood between virgin and periparturient female rats elicited pup retrieval in the virgins (Terkel and Rosenblatt 1972). The fact that the blood-borne factors responsible for this behavioral change in the virgins was the receipt of high circulating progesterone concentrations in the face of low-to-moderate estradiol, followed by their reversal at parturition, was established by many subsequent studies administering similar regimens of exogenous hormones to groups of virgin, ovariectomized females. For example, implanting subcutaneous capsules containing estradiol and progesterone, followed by later removal of the progesterone capsule, can produce a rapid onset of high maternal caregiving in virgin female rats (Bridges 1984). Although estradiol treatment alone also instills pup retrieval in virgin rats, it does so only if given in supraphysiological doses, indicating that progesterone synergizes with estradiol to maximize its effects on motherhood. Notably, progesterone's effects on caregiving are biphasic, because after it synergizes with estradiol to promote motherhood it must be withdrawn, or else the onset of caretaking behavior is strongly inhibited.

More recent evidence of the critical role for estrogens in rapidly stimulating maternal caregiving in laboratory rodents includes studies in mice with a null mutation of both major isoforms of the estrogen receptor—alpha and beta—or just a null mutation of estrogen receptor alpha. Since these mutations render female mice infertile, the effects of estrogen receptor loss on responsiveness to pups have only been tested in **nulliparous** mice from a strain that shows high spontaneous retrieval. These estrogen receptor mutant females show lower responsiveness to pups compared to wild-type controls (Ogawa et al. 1998). It is surprising, then, that nulliparous female mice with whole-body knockout of the aromatase enzyme (which is responsible for synthesizing the estradiol that binds to estrogen receptors) show maternal responsiveness similar to their wild-type sisters (Stolzenberg and Champagne 2016). It was also recently found that gonadally intact, virgin estrogen receptor alpha knockout rats can also be induced to show retrieval after continuous exposure to pups as easily as wild-type females. Systemic injection of an aromatase inhibitor to adult ovariectomized mice of the same C57/BL6 strain, however, does depress retrieval of pups and nest building. These apparent inconsistent results suggest that either lifelong loss of estradiol synthesis permits some compensation that is not possible after acute loss of estradiol synthesis, that some functional estrogen receptors remain in the estrogen receptor mutants that can support maternal responding, and/or that estrogen receptors promote caregiving through some type of ligand-independent activation of the receptors that do not require estradiol.

Many brain sites express estrogen receptors, and their expression in hypothalamic and limbic regions such as the amygdala, is strongly regulated by female reproduction. The only singular site where estradiol infusion has been found to promote maternal behaviors is the medial preoptic area (mPOA), a site lying just rostral to the hypothalamus that is involved in many hormone-sensitive social behaviors. Implanting estradiol-filled cannulae into the mPOA elicits a rapid onset of pup retrieval in virgin or pregnancy-terminated female rats (Numan et al. 1977; Numan 2021). Conversely, disrupting estrogen receptor signaling in the mPOA just before parturition, by implanting cannulae filled with the mixed estrogen receptor antagonist, tamoxifen, slightly delays their onset of maternal behavior. It has also been found that short hairpin RNA interference of estrogen receptor alpha mRNA in the mPOA beginning before mating greatly interferes with later postpartum caregiving in mice (Ribeiro et al. 2012). The ability of estrogenic activity in the mPOA to stimulate maternal retrieval in nulliparous rats suggests that estrogens (along with other hormones) "release" the mPOA from a tonic inhibition of caregiving imposed by other brain sites, which include numerous areas of the hypothalamus (i.e., dorsal, anterior, and ventromedial nucleus) and olfactory-processing regions of the amygdala (i.e., medial nucleus) (Fig. 8.4).

Estrogen-receptor expressing cells in the mPOA project to many brain sites involved in maternal caregiving, including the ventral tegmental area (VTA), which is the source of the mesolimbocortical dopamine system that is involved in reinforcement learning and reward. As with other goal-directed behaviors, dopamine release from VTA projections to the nucleus accumbens is required for maternal motivation and disinhibition of motoric activity required for mothers to interact with pups (Numan and Stolzenberg 2009; Olazábal et al. 2013) (Fig. 8.4). Optogenetic inhibition of the mPOA's estrogen receptor α-expressing cells that project to the VTA severely impairs pup retrieval in female mice, while stimulating these cells causes females to stop nursing and leave the nest to scan the cage, which could be interpreted as the mothers searching for more pups (Fang et al. 2018). Thus, the mPOA-to-VTA pathway is a critical target for estrogen's influence on maternal motivation and pup-seeking and carrying behaviors.

Progestin receptor expression in numerous hypothalamic sites is also affected by female reproduction, but how progesterone first promotes and then inhibits maternal care is mostly unknown. Mice with life-long null mutations of the A and/or B

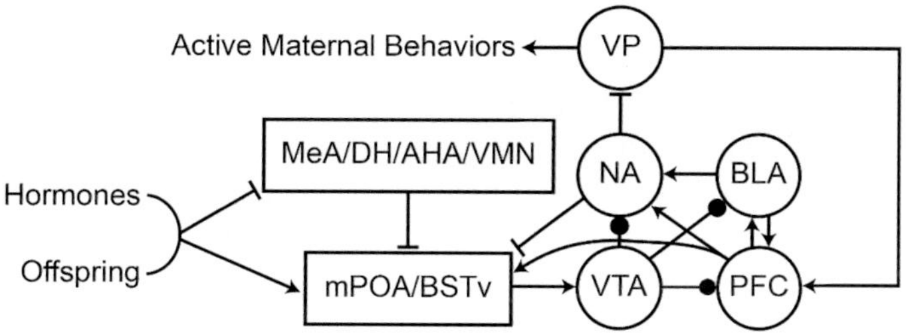

Fig. 8.4 Hypothetical neural model for the stimulation of active maternal behaviors. The hormones of pregnancy and sensory cues of the offspring suppress inhibitory input from the MeA/DH/AHA/VMN to the mPOA/BSTv, while simultaneously stimulating mPOA/BSTv output to the VTA. This VTA activity elicits dopamine release in the NA as well as the PFC and BLA. DA release in the NA inhibits VP output, thus promoting active components of maternal behavior including retrieval. The NA, VP, PFC and BLA can modulate this pathway by their connections to the MPOA/vBST or NA. *AHA* anterior hypothalamic area, *BLA* basolateral amygdala, *BSTv* ventral bed nucleus of the stria terminalis, *DH* dorsal hypothalamus, *DA* dopamine, *MeA* medial amygdala, *mPOA* medial preoptic area, *NA* nucleus accumbens, *PFC* prefrontal cortex, *VP* ventral pallidum, *VMN* ventromedial nucleus, *VTA* ventral tegmental area. Lines ending in arrows = excitatory input; lines ending in vertical bars = inhibitory input; lines ending in circles = DAergic signaling. Modified with permission from Olazabal et al. (2013) and Lonstein et al. (2015b)

isoforms of the progestin receptor have been generated, but not yet tested for maternal caregiving. Additionally, only one study has examined where in the brain progesterone affects caregiving—it found that implanting progesterone-filled cannulae into the mPOA, hypothalamus or midbrain of pregnant female rats does not delay the onset of caregiving (Numan 2021). Either the most relevant sites of progesterone action have not been identified, or progesterone's effects are spread across multiple brain sites that need to be simultaneously targeted to reveal effects on caregiving behavior.

8.3.1.2 Sheep

In precocial mammals, such as sheep, some studies have reported a facilitatory effect of estradiol on maternal responsiveness. Acceptance of a lamb only occurs when plasma estrogen concentrations are high, such as at oestrus or the very end of pregnancy. Furthermore, plasma estradiol concentrations alone and the estradiol:progesterone ratio in late pregnancy are positively correlated with maternal grooming behavior (Dwyer et al. 1999). Additionally, the period during which parturient ewes are positively responsive to newborn lambs can be extended for one day by administering exogenous estradiol and a single high dose of estradiol injected into non-pregnant ewes induces maternal responding in the majority of animals (Poindron et al. 1988), although this is only effective in experienced mothers. Steroid treatments using more physiological doses fail to induce maternal behavior in non-pregnant ewes. Thus, contrary to rodents, steroid hormones alone are not sufficient to induce maternal responsiveness in sheep. They instead have a priming role that allows expulsion of the neonate at birth to trigger the expression of maternal caregiving behavior. Precisely where in the ewe brain estradiol and progesterone act for this priming role is unknown. However, changes in the density of estradiol receptor alpha-labeled neurons in late pregnancy and parturition in the mPOA and the hypothalamic oxytocin system are consistent with the hypothesis that these sites are involved in estradiol's priming ability (Meurisse et al. 2005).

After steroid priming, the relevant process occurring at parturition that is necessary for the expression of caregiving is the vaginocervical stimulation that ewes receive as their lamb(s) moves through the birth canal. Full maternal responsiveness can be induced in non-pregnant **multiparous** ewes primed with progesterone and estradiol if they receive 5 min of vaginocervical stimulation (Keverne et al. 1983). These dramatic effects of vaginocervical stimulation have been further confirmed in numerous breeds of sheep, with low doses of estradiol, and even in cycling ewes at estrus (Poindron et al. 1988). On the other hand, providing vaginocervical stimulation without the steroid pretreatment is ineffective, which confirms the importance of the steroidal priming (Poindron et al. 1988). Experimental vaginocervical stimulation is able to re-induce maternal responses in mothers even if given after the first day postpartum (Kendrick et al. 1991), but if stimulation of the genital tract at parturition is prevented by peridural anesthesia, maternal behavior is disrupted.

8.3.1.3 Humans and Non-human Primates

How circulating estrogens and progestins during pregnancy and parturition are related to later maternal caregiving in primates is not very well understood. The few studies in non-human primates measuring estrogens and progesterone have found that increased interest in infants is positively correlated with circulating estradiol, as well as the estradiol:progesterone ratio, in late-pregnant pigtail macaques. This high interest in infants could also be instilled in non-pregnant females by chronic exogenous estradiol (Saltzman and Maestripieri 2011). However, not all studies find significant relationships between estrogen and/or progesterone and infant-directed behavior in monkeys (Bardi et al. 2003).

In women, positive attitudes about infants and attraction to their sensory cues increase across mid-to-late pregnancy (e.g., Fleming et al. 1997; Saltzman and Maestripieri 2011). Estrogen and progesterone concentrations during pregnancy, however, are unrelated to their contemporaneous attitudes about the fetus and future infant (Fleming et al. 1997). Hormone concentrations during the first few days postpartum are similarly unrelated to mothers' current caregiving behaviors. Pregnancy hormone concentrations do predict other aspects of women's postpartum mothering, such that women with the lowest estradiol:progesterone ratio or lowest circulating estrogen early in pregnancy, but then showing a greater increase in that ratio from early to late pregnancy, have greater attachment with their infants early postpartum (Fleming et al. 1997; Fig. 8.5a). Interestingly, women with a lower estradiol:progesterone ratio early in pregnancy and a slower rise in progesterone through the end of pregnancy were the most maternally sensitive even one year postpartum, but only if they were gestating a daughter (Glynn et al. 2016).

As mentioned above, a new mother's affective state can mediate the relationship between their hormonal profile and their ability to sensitively care for their infants. Some studies have suggested that the etiology of postpartum depression (PPD) and

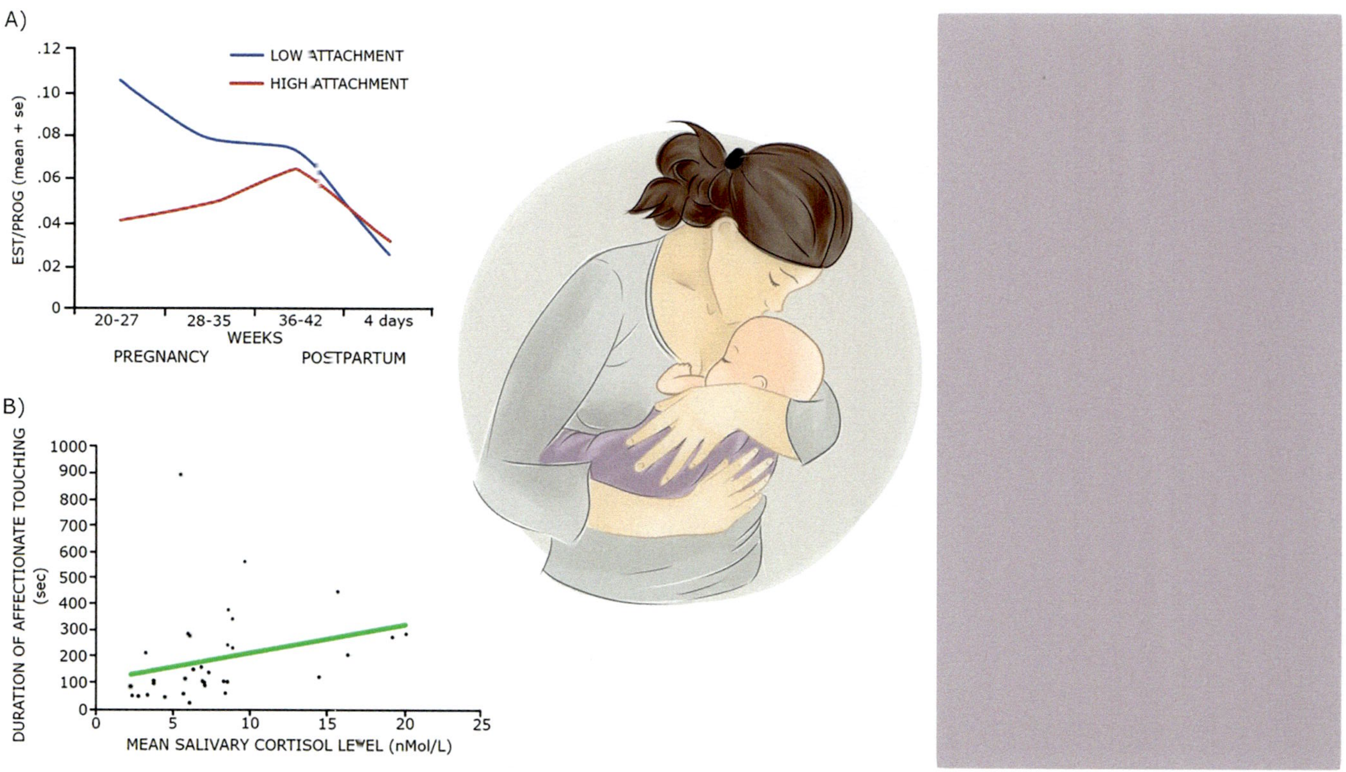

Fig. 8.5 (**a**) Women with the smallest change across pregnancy in their ratio of circulating estrogen and progesterone have the highest maternal attachment to their infants early postpartum. (**b**) Relationship between circulating cortisol and affective touch behaviors in early-postpartum women. Women with higher salivary cortisol concentrations displayed longer durations of affective touch with their infants. Modified with permission from Fleming et al. (1997)

possibly anxiety can be found in the precipitous reduction in estrogen and progesterone secretion during the peripartum period (see Brummelte and Galea 2016 and Chap. 10), with the mental health dysregulation negatively affecting maternal care. For instance, Kikuchi et al. (2021) showed that postpartum plasma progesterone concentrations of depressed mothers measured at 1 month postpartum were significantly lower than those of non-depressed mothers. However, despite the fact that most studies find that depression and/or anxiety derail motherhood, a number of other studies conducted over many years have found no association between the concentrations or ratios of reproductive hormones and PPD or anxiety (Lonstein 2007; Brummelte and Galea 2016), suggesting that non-hormonal factors must also be responsible.

8.3.2 Prolactin

8.3.2.1 Laboratory Rats and Mice

Prolactin is probably best known for its critical role at the level of the mammary gland for producing milk, but prolactin released into blood plasma by the anterior pituitary gland can also enter the brain in substantial amounts through a specific transporter mechanism. While its actions on the brain are probably not absolutely essential for the onset of caregiving behaviors in laboratory rats and other rodents, prolactin can often help the transition to motherhood. For instance, peripheral administration of prolactin reduces the latency for nulliparous, ovariectomized rats to become maternally sensitized following chronic estradiol and progesterone treatment. Conversely, impairing endogenous pituitary prolactin secretion with a dopamine receptor agonist delays their onset of caregiving (Bridges 2018). This pharmacological inhibition is possible because it mimics the actions of prolactin-monitoring neurons in the arcuate and periventricular nuclei of the hypothalamus that release dopamine into the median eminence of the pituitary stalk, which suppresses pituitary lactotroph activity. However, neither administration of a dopamine receptor agonist that suppresses pituitary prolactin secretion, nor **hypophysectomy**, affects maternal caregiving in naturally pregnant and parturient rats. This is because even in the absence of pituitary prolactin, sufficient prolactin is secreted by the placentae in the form of placental lactogens that act on the maternal brain. Placental lactogens are probably not sufficient to establish maternal behaviors in all species though, as inhibition of prolactin secretion from the pituitary does disrupt caregiving in parturient hamsters and rabbits (Lonstein et al. 2015b).

There are two forms of the prolactin receptor found in the brain, the long form and the short form, with the long form best studied for a role in maternal caregiving. Its expression is upregulated in female laboratory rodents by exogenous estradiol, during late pregnancy and at parturition in brain sites involved in maternal caregiving, including the mPOA. Relatively low doses of prolactin, which cannot stimulate maternal behavior in ovarian steroid-primed virgin rats when given peripherally, become effective when given directly into the cerebral ventricles of the brain. Moreover, even lower doses of prolactin, as well as placental lactogens, can stimulate caregiving if infused directly into the mPOA of steroid-primed virgins. Chronic prolactin receptor antagonism specifically in the mPOA delays the onset of maternal behaviors in steroid-primed virgin rats (Bridges et al. 2001), but the effects of disrupting prolactin receptor signaling in the mPOA are much more severe in laboratory mice. Viral vector-mediated knockdown of prolactin receptors in the mPOA caused female mice to never show maternal care after displaying normal parturition and pup cleaning, and most dams eventually abandoned their nests (Brown et al. 2017).

8.3.2.2 Sheep

In contrast to altricial species such as rats and mice, a causal link between prolactin and the onset of maternal behavior in sheep has so far failed to be demonstrated. In parturient ewes, injection of a dopamine receptor agonist to suppress pituitary prolactin secretion does not prevent the onset of maternal responsiveness or affect the length of time when the mother is responsive to the neonate. Similar to naturally pregnant rats, perhaps placental lactogens are compensating in sheep. However, there are some significant associations between likely functional polymorphisms in the prolactin receptor gene and ewes' licking and nursing behaviors shown during the first 3 days postpartum (Wang et al. 2015).

8.3.2.3 Humans and Non-human Primates

Given how much literature there is on a positive role for prolactin in rodent maternal behavior, it is surprising that the relationship between peripartum or postpartum prolactin and maternal caregiving in female primates has mostly been ignored. In fact, more attention has been reported in male primates' prolactin and their responses to newborns. There is some evidence from women that plasma prolactin concentrations during the first few days postpartum are not correlated with instrumental maternal activities with the infant (Fleming et al. 1987); other measures of maternal caregiving or responsiveness have not been adequately studied. In the two studies of non-human primates, there are also no significant associations between plasma prolactin concentrations and maternal behavior in lactating rhesus macaques or common marmosets, although systemic dopamine receptor agonism eliminated or at least reduced infant carrying by parentally inexperienced female marmosets (Roberts et al. 2001). Clearly, too few studies have been performed to date to determine any role(s) prolactin plays in regulating primate maternal care.

8.3.3 Oxytocin

8.3.3.1 Laboratory Rats and Mice

As with prolactin, the neuropeptide oxytocin has also been long associated with maternity in mammals, particularly peripheral functions such as uterine contractions at parturition and ejection of milk from the mammary glands. Critically, circulating oxytocin in blood plasma secreted by the posterior pituitary gland at physiological concentrations has no transporter mechanism to allow much of it to cross the blood-brain barrier (Jurek and Neumann 2018). Instead, the oxytocin relevant for brain functions mainly comes from neurons of the paraventricular nucleus (PVN) of the hypothalamus that release the peptide intracerebrally through axon terminals or by non-synaptic diffusion through brain tissue (Jurek and Neumann 2018). Thus, the centrally acting oxytocin relevant for maternal caregiving is a less traditional "hormone" compared to estradiol, progesterone and prolactin, because its secretion into the blood is not the source of the peptide necessary for its effects on the brain and behavior.

The first work suggesting a central role for oxytocin related to females' behavior showed that many gonadally intact or ovariectomized/estrogen-primed female rats showed maternal behavior within 2 h of oxytocin infusion into the cerebral ventricles (see Yoshihara et al. 2018 for review). The presence of natural or exogenous estradiol is almost always necessary for oxytocin's pro-maternal effects because estradiol strongly upregulates oxytocin receptors in many brain sites involved in motherhood. Consistent with such pro-maternal effects, central infusion of an oxytocin receptor antagonist can greatly disrupt the onset of maternal behaviors in either estradiol-primed, pregnancy-terminated, or recently parturient rats. On the other hand, central oxytocin receptor blockade during the postpartum period produces no gross disruption in maternal behaviors, suggesting that oxytocin has much greater relevance for the onset rather than the postpartum maintenance of motherhood. The mPOA, VTA and olfactory bulbs were early known targets of oxytocin's pro-maternal effects in female laboratory rats, as seen by disrupted caregiving after peri-parturitional infusion of an oxytocin receptor antagonist into any of those brain regions (Pedersen et al. 1994; Yu et al. 1996). A more recent study demonstrated that viral vector-mediated knockdown of oxytocin receptors in the serotonergic dorsal raphe nucleus starting in pregnancy led to infanticide at or soon after parturition and reduced nursing through early lactation (Grieb et al. 2021).

Despite the findings discussed above, the role of oxytocin in the onset of maternal behavior is still equivocal, because not all laboratories report a rapid onset of maternal behavior in female rats after central oxytocin infusion (Yoshihara et al. 2018). Furthermore, studies using transgenic mice demonstrate that the oxytocin system is not absolutely required for maternal caregiving behaviors at all, as constitutive loss of the ability to produce oxytocin has surprisingly little effect on dams, and they show normal parturition and only fairly minor impairments in their behavioral interactions with pups (although they cannot eject milk to feed them). Whole-body or forebrain knockout of oxytocin receptors (OTRs) in mice also does not affect parturition and has mixed effects on maternal behavior towards pups, depending on the background strain of mouse and their level of maternal experience.

8.3.3.2 Sheep

Parturition in sheep is characterized by the synchronized synthesis and release of oxytocin in numerous brain sites, with its effects amplified by greater numbers of oxytocin receptors (for review see Kendrick 2000). Oxytocin concentrations are greatly increased in the cerebrospinal fluid of ewes at parturition, but not in animals given epidural anesthesia that blocks uterine contractions. Oxytocin synthesis increases at parturition in the hypothalamic PVN and, consequently, higher oxytocin release occurs from oxytocin terminals in brain areas, including the main olfactory bulb and mPOA. Elevated expression of OTRs in various brain regions, in particular the mPOA, increases their sensitivity to the peptide.

Astonishingly, injecting oxytocin into the cerebral ventricles of non-pregnant, steroid-primed ewes induces a full maternal response to a test lamb within 1 min. Similar to the need for vaginocervical stimulation discussed previously, oxytocin is ineffective when given without estradiol priming. In parturient sheep, inhibiting genital feedback with peridural anesthetic prevents both central oxytocin release and maternal behavior, but infusions of oxytocin into the cerebral ventricles reverse the behavioral deficit. The specific sites in the brain where oxytocin accomplishes this include the PVN, where site-specific oxytocin infusions in steroid primed multiparous ewes induce maternal responses towards newborn lambs comparable to that observed following cerebroventricular oxytocin infusion. However, infusing oxytocin into brain sites containing oxytocin nerve terminals stimulates only some elements of the maternal response, indicating that that oxytocin receptor stimulation is needed in multiple brain sites simultaneously to elicit the full complement of maternal activities. For instance, infusing oxytocin into the mPOA or the main olfactory bulb of steroid-primed, non-pregnant ewes reduces rejection of lambs, but fails to induce maternal acceptance. In addition, central administration of oxytocin alters olfactory neurogenesis (Lévy et al. 2019), which is involved in establishing maternal behavior and recognition of the familiar lamb (Corona et al. 2018). Thus, oxytocin is a physiological signal of primary importance in sheep to synchronize maternal responsiveness with the birth of the young.

8.3.3.3 Humans and Non-human Primates

Despite regular media attention to the "love" (or trust) hormone, the evidence for a role of oxytocin in primate maternal caregiving is quite mixed. It is also essential to note up front that studies most often measure plasma oxytocin concentrations, which do not necessarily reflect oxytocin concentrations in the brain, so interpretations of the extant data beyond the peripheral effects of oxytocin should be viewed with caution.

Plasma or urinary oxytocin is elevated during the peripartum and postpartum periods in many, but not all, primates studied (Hirst and Thorburn 1996; Finkenwirth et al. 2016) but only a few studies have tested the relationship between oxytocin and maternal caregiving in non-human primates. One study in marmosets found positive correlations between urinary oxytocin and licking the infant during early lactation and with food sharing during late lactation (Finkenwirth et al. 2016). However, intranasal oxytocin did not increase marmoset mothers' (or other caregivers') interest in infant-related visual and auditory cues, or increase food sharing by mothers to juvenile young (Taylor et al. 2020).

In women, higher pre- or postnatal oxytocin concentrations in blood or saliva in response to infant cues are generally associated with more positive maternal caregiving behaviors (Kim et al. 2014), but not in all studies. Reports have begun to suggest that theory of mind, which is the ability to think about the mental state of one's self and of others, may explain the link between oxytocin and maternal behavior. Mothers who performed well on a theory of mind task showed less intrusive maternal behaviors than mothers who performed poorly, and the effect was mediated by oxytocin (MacKinnon et al. 2018). If or how these results may translate to behavioral interventions for at-risk mothers are unclear.

Only a few studies have addressed the behavioral effects of oxytocin administration on maternal behavior and sensitivity in women, but one such study found that oxytocin treatment altered maternal responses to infants in mothers with postpartum depression. Following intranasal oxytocin administration, the women were asked to engage in a play session with their infant and then listened to pre-recorded infant cries. There were no differences observed in **maternal sensitivity**, but mothers given oxytocin rated the recorded cries as more urgent than those in the placebo group, although the oxytocin-treated mothers were also more likely to say they would choose a harsh caregiving response after listening to the cry (Mah et al. 2017). Given the small sample size used in this study and ongoing debates regarding the amount of intranasally administered oxytocin that reaches the brain (Quintana et al. 2021), the clinical implications are difficult to determine, but these findings agree with the sentiment that oxytocin influences the emotional salience of social cues, including in mothers (Bakermans-Kranenburg and van IJzendoorn 2018). An alternate tripartite function of oxytocin in maternal sensitivity and maternal protectiveness overlaps with the notion that oxytocin affects stress, prosocial behavior and attachment (Bakermans-Kranenburg and van IJzendoorn 2018), affecting the latter two by increasing social salience and reward sensitivity. Again, one should be cautioned against a magic bullet concept of oxytocin's function, stressing that positive oxytocin effects may not occur if mothers have experienced early trauma and poor parenting themselves or high levels of proximal stress (Feldman and Bakermans-Kranenburg 2017).

There is also an association between altered oxytocin function and postpartum mental health (see Lonstein et al. 2015a; Mah 2016, and Chap. 10). For instance, compared with non-depressed mothers, mothers with postpartum mental health disorders in the early postpartum period show lower plasma oxytocin concentrations during pregnancy. Contrary to these findings, other studies have found associations between higher third-trimester oxytocin concentrations and the severity of postpartum mental health symptoms in the early postpartum period (see Mah 2016). An important methodological difference that could explain inconsistencies in the results is the pre-existence of affective disorders, since women with a history of major depressive disorder show significantly higher plasma oxytocin concentrations (Cochran et al. 2013). Also, an association has been found among high levels of life stress, high concentrations of plasma oxytocin, and fewer postpartum mental health symptoms at 8 weeks postpartum. Thus, while more research on the relationships between oxytocin and maternal caregiving is clearly needed, the limited available literature mostly suggests positive associations between primate oxytocin system function and infant caregiving (or other neurobehavioral systems influencing caregiving) in the early postpartum period.

8.3.4 Glucocorticoids

8.3.4.1 Laboratory Rats and Mice

A role for adrenal corticosterone in early mothering has been demonstrated in laboratory rats. In new mother rats, **adrenalectomy** and corticosterone replacement decreased and then enhanced licking of newborn pups, although other behaviors reflecting maternal motivation such as retrieval and nursing responses are unaffected (Rees et al. 2004). Furthermore, female rats exhibited a better memory or retention of their postpartum experience interacting with pups when exposed to additional corticosterone, indicating that corticosterone also enhances maternal memory of how to behave with young (Graham et al. 2006). However, Brummelte and Galea (2010) found that rat mothers who were administered very high doses of exogenous corticosterone (over and above their endogenous corticosterone) nursed their pups less and spent more time away from the nest compared with controls. These results could suggest a biphasic response to corticosterone in rats, with low-to-moderate concentrations facilitating components of maternal responding, whereas high concentrations that might model chronic stress may simply derail maternal capabilities.

8.3.4.2 Sheep

Despite the fact that parturition is accompanied with a peak in circulating cortisol in sheep, no information is available about any role for cortisol or a stress effect on maternal behavior in this species.

8.3.4.3 Humans and Non-human Primates

While in non-pregnant populations higher cortisol has been associated with stress and sometimes depressive states, the 'meaning' or function of this increase in glucocorticoids may be different in pregnancy and the postpartum period. Both mothering and maternal affective state during the early postpartum period may be influenced by the maternal cortisol profile.

Evidence for a role of the HPA axis in human mothering is in fact more plentiful than it is for other hormones, with the clear exception of oxytocin, although the relationships at first seem counter-intuitive (Almanza-Sepulveda et al. 2020). Immediately after parturition, mothers with higher plasma concentrations of baseline cortisol tended to display more contact and affectionate approach responses towards their nursing infants (Almanza-Sepulveda et al. 2020; Fig. 8.5b). When hormone and affective state were included in additional analyses, much of the variance in postpartum maternal behaviors exhibited could be accounted for by day 2 postpartum cortisol concentrations and mothers' positive attitudes during their pregnancies (Fleming et al. 1997; Almanza-Sepulveda et al. 2020). More recently, a positive relationship between cortisol concentrations (prepartum and postpartum) across a dyadic interaction and intensity of affectionate behaviors was also found in young primiparous mothers but not in multiparous mothers. Cortisol is also positively associated with new mothers' attraction to infant cues (Fleming et al. 1997). In terms of discriminating between one's own and other infants' odors on a T-shirt, as opposed to attraction to infant cues, the cortisol-parity relationships are reversed. Multiparous mothers with higher cortisol show significantly higher accuracy scores, whereas accuracy scores and cortisol concentrations were unrelated in primiparous mothers (Fleming et al. 1997). It may be that for individual recognition, prior experience with infants permits mothers to learn about the specific features of neonates. In that case, cortisol enhances maternal attention to the most meaningful features of infant stimuli.

Finally, cortisol effects vary as a function not only of parity, but also how mothers respond emotionally to infant stimuli (Almanza-Sepulveda et al. 2020). When 3- to 6-month postpartum mothers were presented with various infant stimuli and salivary cortisol was measured before and after each stimulus period, primiparous mothers had higher concentrations of cortisol than multiparous mothers, but neither group showed a differential cortisol response to any of the stimuli. When mothers' emotional/affective reactions to these stimuli were considered, though, mothers of both parity conditions who were more sympathetic to the cry stimuli had significantly higher baseline cortisol as well as higher heart rate responses during the infant cry exposure. Thus, by 3- to 6-months postpartum, when pregnancy concentrations of cortisol and other hormones have already declined, the relationship between cortisol and measures of mothering is no longer necessarily a positive one; it could be negative, absent or variable, depending on the mother's prepartum or postpartum stress experience and mental health (Almanza-Sepulveda et al. 2020).

There are only a few studies exploring cortisol and maternal caregiving in other primates, and these report mixed results. For instance, among female marmosets exogenous corticotropin-releasing hormone (CRH) or cortisol given during pregnancy reduces later maternal carrying (Saltzman and Maestripieri 2011), whereas postpartum endogenous glucocorticoid levels are positively associated with maternal caregiving in chimpanzees (Stanton et al. 2015).

A role for altered HPA axis and cortisol function in postpartum anxiety and postpartum disorders has also been proposed, with consequences for mothering (Brummelte and Galea 2016). Although depressive-like symptoms can be produced in a rat model through exogenous corticosterone, in human mothers the relationship between circulating cortisol and postpartum mental health remains ambiguous. Some studies measuring cortisol across pregnancy through the early postpartum period report an association between hypocortisolemia and postpartum anxiety and depression, whereas others report an association between hypercortisolemia and postpartum affective disorders, and yet others report no significant associations between the two (Brummelte and Galea 2016).

8.4 Conclusions/Perspectives

Motherhood in all mammalian species involves an essential, but complex, suite of behaviors that target the specific needs of the young and prepares them for independent living within their unique environment. We hope we have conveyed that, despite the particulars of the caregiving behaviors displayed and the natural ecology of the animals being studied, the major steroid and peptide endocrine regulators of the peripheral events of reproduction (e.g., pregnancy, parturition, lactation) are the same as those involved with the onset of maternal caregiving (Table 8.1). That is, mammals have evolved to use the very same hormone-generating organs (i.e., ovaries, adrenal glands, placenta, pituitary gland, hypothalamus) and their endocrine products to coordinate reproductive physiology with behavior, and have done so in species-specific ways that generate styles of maternal care appropriate for a particular species' young.

While our goal was to provide an overview of the major endocrine regulators of motherhood, there are certainly other relevant centrally acting factors. Indeed, a number of other neuropeptides released within the brain (e.g., CRH, orexin, galanin, melanin-concentrating hormone, endogenous opioids) have been reported in studies mostly performed on laboratory rodents to affect various aspects of motherhood. While release of these factors within the brain from neuronal axons may not make them traditional "endocrine" factors, similar to oxytocin, they are also likely released in large amounts that take them by diffusion and the cerebro spinal fluid to act at distant targets within the brain. Furthermore, whereas most classic neurotransmitters rapidly and directly affect neuronal activity by acting on their cognate receptors, hormones often instead act as longer-term neuromodulators that increase or decrease the propensity of neurons to respond to other fast-acting neurochemicals such as the classic neurotransmitters (e.g., dopamine, norepinephrine, serotonin, GABA, glutamate) that themselves have a rich scientific literature demonstrating their particular roles in maternal caregiving (see Lonstein et al. 2015a; Numan 2021).

We also mention that hormones do not always deterministically drive motherhood but that maternal affective state is sensitive to hormones and regulates how she mothers her young. But in addition to affective state, the expression of maternal care is also influenced by other factors beyond the scope of this chapter including one's early-life experience with their own mother, how often a female has given birth, and how motivated one is to provide care (Fig. 8.6). This motivational effect is probably best exemplified by adoptive mothers, who do not undergo the dramatic endocrine changes of parenting but are often extremely motivated to parent and become strongly attached to their infant. Potential smaller changes in endogenous hormones that may help promote mothering in some way has not been adequately studied in adoptive mothers, though.

There are also strong, presumably non-endocrine, effects of previous experience with caregiving on future caregiving in females. In many animals living in their natural environment, young juvenile and peripubescent females receive maternal experience through taking care of their own younger sisters; this experience is known to produce long-term changes in the female brain and promote future care of their own neonates. Not only do young females not require the endocrine fluctuations of pregnancy and parturition in order to show caregiving behaviors, but even many adult inexperienced nulliparous females that would typically avoid neonates can be induced to become maternal by repeated exposure to them. This process, termed maternal sensitization, has been best studied in laboratory rats and involves 5–7 days of exposure to cohorts of young pups before the full effect is observed.

Table 8.1 Effects of endocrine factors on the expression of maternal behavior around parturition in selected mammalian species *P* progesterone, *E2* estradiol, *C* cortisol, *PRL* prolactin, *KO* genetic knockout, *OT* oxytocin, *mPOA* medial preoptic area, *PVN* paraventricular nucleus of the hypothalamus, *VTA* ventral tegmental area

	Mice	Rat	Sheep	Non-human primates	Humans
Progesterone	Not determined	Not effective by itself; synergizes with E2 to have effect; inhibits behavior if high concentrations at parturition	Not effective by itself; synergizes with E2 to have effect	Positive correlation between E2 and P concentrations or ratio and interest in infants in pigtail macaques	Lowest E2/P ratio associated with greater maternal feelings and attachment with infants. Lower prepartum E2/P ratio predicts maternal behavior at 1-year postpartum
Estradiol	Induction via estrogen receptor alpha	Effective but requires P withdrawal; Induction by itself only at supraphysiological doses; mPOA-to-VTA a critical target of E2 action	Priming effect for OT actions; induces behavior at supra-physiological doses	Positive correlation between E2 concentrations and interest in infants in pigtail macaques	Negative correlation between E2 concentrations and feelings of attachment during the first days postpartum
Prolactin	KO of PRL receptors abolishes retrieval and crouching	Effective if given centrally in steroid-treated females; same for placental lactogens; mPOA is a major brain site of action	No effect	PRL inhibition reduces infant carrying by inexperienced marmosets	PRL concentration not correlated with maternal activities
Oxytocin	Not necessary	Has positive effect if given centrally in steroid-treated females; stress may be necessary for OT induction; mPOA and VTA are some central targets	Effective if given centrally in steroid-treated females; PVN is a central target	Not determined	OT both pre- and postnatal associated with more positive maternal caregiving; negative correlations between OT concentrations and maternal avoidant attachment and anxiety
Glucocorticoids	Not determined	At lower levels, enhances pup licking and maternal memory	Not determined	C concentrations positively associated with maternal caregiving in chimpanzees; C reduces maternal carrying in marmosets	Positive correlation between C concentrations and maternal responsiveness very early postpartum; inversely related to responsiveness after 6 months postpartum

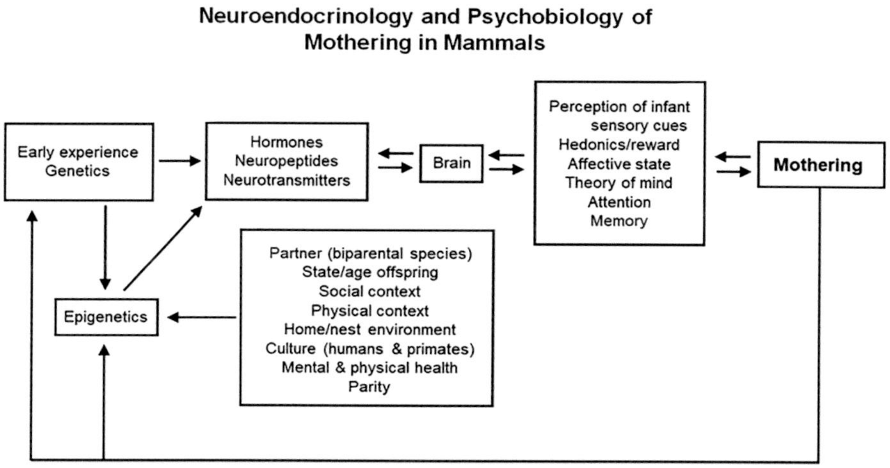

Fig. 8.6 Schematic indicating the many, often reciprocally interacting, psychobiological regulators of mammalian parental care

Sensitization occurs even after the pituitary, ovaries, and adrenal glands have been removed (Rosenblatt 1967). The fact that sensitization can occur indicates that neonates themselves must be able to tap into the same neural substrates necessary for maternal care even without hormones being present. It has been speculated that the hormonal and other neurochemical influences on neuronal function that compel parous females to become maternal may very well be the same mechanisms accessed by repeated exposure to neonates alone, possibly in both cases involving epigenetic changes to the brain (Stolzenberg and Champagne 2016). Thus, through the dramatic hormonal changes involved in female reproduction or through a variety of experiences during early or later life, there are neurobiological mechanisms that provide caregivers of all types with the ability to exhibit sensitive caregiving behaviors and contribute to offspring development and propagation of their species.

Key References

Bridges RS (1984) Foundational series of studies establishing the doses and timing required for exogenous steroid and peptide hormones to rapidly induce the onset of maternal caregiving behaviors in laboratory rats.

Fleming AS et al. (1997) The first, and still one of only two, longitudinal studies to establish a relationship between the trajectories of steroid hormones across pregnancy and aspects of later postpartum caregiving in women.

Keverne EB, et al. (1983) First study to demonstrate that vagino-cervical stimulation was essential for the onset of maternal acceptance of lambs and the formation of the selective ewe-lamb bond at parturition.

Numan M, et al. (1977) Critical localization of function study establishing the medial preoptic area as an essential site where estradiol promotes the rapid onset on maternal caregiving.

Yoshihara C, et al. (2017) Comprehensive, critical review of the influence of oxytocin on the onset and maintenance of maternal caregiving, with important suggestions for experimental design and appropriate data interpretation.

References

Almanza-Sepulveda ML, Fleming AS, Jonas W (2020) Mothering revisited: a role for cortisol? Horm Behav 121:104679

Anderson SM, MacLean PS, McManaman JL, Neville MC (2015) Lactation and its hormonal control. In: Knobil and Neill's physiology of reproduction, 4th edn. Elsevier, New York, pp 2055–2105

Bakermans-Kranenburg MJ, van Ijzendoorn MH (2018) Oxytocin and human sensitive and protective parenting. In: Behavioral pharmacology of neuropeptides: oxytocin, pp 421–448

Bardi M, Shimizu K, Barrett GM, Borgognini-Tarli SM, Huffman MA (2003) Peripartum sex steroid changes and maternal style in rhesus and Japanese macaques. Gen Comp Endocrinol 133(3):323–331

Barrett J, Fleming AS (2011) Annual research review: all mothers are not created equal: neural and psychobiological perspectives on mothering and the importance of individual differences. J Child Psychol Psychiatry 52(4):368–397

Bridges RS (1984) A quantitative analysis of the roles of dosage, sequence, and duration of estradiol and progesterone exposure in the regulation of maternal behavior in the rat. Endocrinology 114(3):930–940

Bridges RS (2018). Prolactin: regulation and actions. In: Oxford Research Encyclopedia of Neuroscience. www.oxford.re.com

Bridges RS, Rigero BA, Byrnes EM, Yang L, Walker AM (2001) Central infusions of the recombinant human prolactin receptor antagonist, S179D-PRL, delay the onset of maternal behavior in steroid-primed, nulliparous female rats. Endocrinology 142(2):730–739

Brown RS, Aoki M, Ladyman SR, Phillipps HR, Wyatt A, Boehm U, Grattan DR (2017) Prolactin action in the medial preoptic area is necessary for postpartum maternal nursing behavior. Proc Natl Acad Sci U S A 114(40):10779–10784

Brummelte S, Galea LA (2010) Chronic corticosterone during pregnancy and postpartum affects maternal care, cell proliferation and depressive-like behavior in the dam. Horm Behav 58(5):769–779

Brummelte S, Galea LA (2016) Postpartum depression: etiology, treatment and consequences for maternal care. Horm Behav 77:153–166

Cochran D, Fallon D, Hill M, Frazier JA (2013) The role of oxytocin in psychiatric disorders: a review of biological and therapeutic research findings. Harv Rev Psychiatr 21(5):219–247

Corona R, Meurisse M, Cornilleau F, Moussu C, Keller M, Lévy F (2018) Disruption of adult olfactory neurogenesis induces deficits in maternal behavior in sheep. Behav Brain Res 347:124–131

Dwyer CM, Dingwall WS, Lawrence AB (1999) Physiological correlates of maternal and offspring behaviour in sheep. Physiol Behav 67(3):443–454

Fang YY, Yamaguchi T, Song SC, Tritsch NX, Lin D (2018) A hypothalamic midbrain pathway essential for driving maternal behaviors. Neuron 98(1):192–207

Feldman R, Bakermans-Kranenburg MJ (2017) Oxytocin: a parenting hormone. Curr Opin Psychol 15:13–18

Finkenwirth C, Martins E, Deschner T, Burkart JM (2016) Oxytocin is associated with infant-care behavior and motivation in cooperatively breeding marmoset monkeys. Horm Behav 80:10–18

Fleming AS, Steiner M, Anderson V (1987) Hormonal and attitudinal correlates of maternal behaviour during the early postpartum period in first-time mothers. J Reprod Infant Psychol 5:193–205

Fleming AS, Ruble D, Krieger H, Wong PY (1997) Hormonal and experiential correlates of maternal responsiveness during pregnancy and the puerperium in human mothers. Horm Behav 31(2):145–158

Glynn LM, Davis EP, Sandman CA, Goldberg WA (2016) Gestational hormone profiles predict human maternal behavior at 1-year postpartum. Horm Behav 85:19–25

Graham MD, Rees SL, Steiner M, Fleming AS (2006) The effects of adrenalectomy and corticosterone replacement on maternal memory in postpartum rats. Horm Behav 49:353–361

Grieb ZA, Tierney SM, Lonstein JS (2017) Postpartum inhibition of ovarian steroid action increases aspects of maternal caregiving and reduces medial preoptic area progesterone receptor expression in female rats. Horm Behav 96:31–41. https://doi.org/10.1016/j.yhbeh.2017.08.007

Grieb ZA, Ford EG, Yagan M, Lau BY, Manfredsson FP, Krishnan K, Lonstein JS (2021) Oxytocin receptors in the midbrain dorsal raphe are essential for postpartum maternal social and affective behaviors. Psychoneuroendocrinology 131:105332

Hirst JJ, Thorburn GD (1996) Initiation of parturition in non-human primates. In: Advances in organ biology, vol 1. Elsevier, pp 61–87

Jurek B, Neumann ID (2018) The oxytocin receptor: from intracellular signaling to behavior. Physiol Rev 98:1805–1908

Kendrick KM (2000) Oxytocin, motherhood and bonding. Exp Physiol 85:111–124

Kendrick KM, Lévy F, Keverne EB (1991) Importance of vaginocervical stimulation for the formation of maternal bonding in primiparous and multiparous parturient ewes. Physiol Behav 50(3):595–600

Keverne EB, Lévy F, Poindron P, Lindsay DT (1983) Vaginal stimulation: an important determinant of maternal bonding in sheep. Science 219:81–83

Kikuchi S, Kobayashi N, Watanabe Z, Ono C, Takeda T, Nishigori H, Yaegashi N, Arima T, Nakai K, Tomita H (2021) The delivery of a placenta/fetus with high gonadal steroid production contributes to postpartum depressive symptoms. Depress Anxiety 38(4):422–430

Kim S, Soeken TA, Cromer SJ, Martinez SR, Hardy LR, Strathearn L (2014) Oxytocin and postpartum depression: delivering on what's known and what's not. Brain Res 1580:219–232

Lévy F, Batailler M, Meurisse M, Keller M, Cornilleau F, Moussu C, Poissenot K, Migaud M (2019) Differential effects of oxytocin on olfactory, hippocampal and hypothalamic neurogenesis in adult sheep. Neurosci Lett 713:134520

Lonstein JS (2007) Regulation of anxiety during the postpartum period. Front Neuroendocrinol 28(2–3):115–141

Lonstein JS, Fleming AS (2001) Parental behaviors in rats and mice. Curr Protoc Neurosci 17(1):8–15

Lonstein JS, Lévy F, Fleming AS (2015a) Common and divergent psychobiological mechanisms underlying maternal behaviors in non-human and human mammals. Horm Behav 73:156–185

Lonstein JS, Pereira M, Morrell JI, Marler CA (2015b). Parenting behavior. In Physiology of reproduction. New York, pp 2371–2437

MacKinnon AL, Carter CS, Feeley N, Gold I, Hayton B, Santhakumaran S, Zelkowitz P (2018) Theory of mind as a link between oxytocin and maternal behavior. Psychoneuroendocrinology 92:87–94

Mah BL (2016) Oxytocin, postnatal depression, and parenting: a systematic review. Harv Rev Psychiatr 24(1):1–13

Mah BL, Van Ijzendoorn MH, Out D, Smith R, Bakermans-Kranenburg MJ (2017) The effects of intranasal oxytocin administration on sensitive caregiving in mothers with postnatal depression. Child Psychiatry Hum Dev 48(2):308–315

Meurisse M, Gonzalez A, Delsol G, Caba M, Lévy F, Poindron P (2005) Estradiol receptor-alpha expression in hypothalamic and limbic regions of ewes is influenced by physiological state and maternal experience. Horm Behav 48(1):34–43

Numan M (2021) The parental brain: mechanisms, development, and evolution. Oxford University Press, Oxford

Numan M, Stolzenberg DS (2009) Medial preoptic area interactions with dopamine neural systems in the control of the onset and maintenance of maternal behavior in rats. Front Neuroendocrinol 30(1):46–64

Numan M, Rosenblatt JS, Komisaruk BR (1977) Medial preoptic area and onset of maternal behavior in the rat. J Comp Physiol Psychol 91(1):146–164

Ogawa S, Eng V, Taylor J, Lubahn DB, Korach KS, Pfaff DW (1998) Roles of estrogen receptor-α gene expression in reproduction-related behaviors in female mice. Endocrinology 139(12):5070–5081

Olazábal DE, Pereira M, Agrati D, Ferreira A, Fleming AS, González-Mariscal G, Lévy F, Lucion AB, Morrell JI, Numan M, Uriarte N (2013) New theoretical and experimental approaches on maternal motivation in mammals. Neurosci Biobehav Rev 37(8):1860–1874

Pedersen CA, Caldwell JD, Walker C, Ayers G, Mason GA (1994) Oxytocin activates the postpartum onset of rat maternal behavior in the ventral tegmental and medial preoptic areas. Behav Neurosci 108(6):1163

Poindron P, Lévy F, Krehbiel D (1988) Genital, olfactory, and endocrine interactions in the development of maternal behaviour in the parturient ewe. Psychoneuroendocrinology 13(1–2):99–125

Quintana DS, Lischke A, Grace S, Scheele D, Ma Y, Becker B (2021) Advances in the field of intranasal oxytocin research: lessons learned and future directions for clinical research. Mol Psychiatry 26:80–91

Rees SL, Panesar S, Steiner M, Fleming AS (2004) The effects of adrenalectomy and corticosterone replacement on maternal behavior in the postpartum rat. Horm Behav 46(4):411–419

Ribeiro AC, Musatov S, Shteyler A, Simanduyev S, Arrieta-Cruz I, Ogawa S, Pfaff DW (2012) siRNA silencing of estrogen receptor-α expression specifically in medial preoptic area neurons abolishes maternal care in female mice. Proc Natl Acad Sci U S A 109(40):16324–16329

Roberts RL, Jenkins KT, Lawler T, Wegner FH, Newman JD (2001) Bromocriptine administration lowers serum prolactin and disrupts parental responsiveness in common marmosets (*Callithrix j. jacchus*). Horm Behav 39(2):106–112

Rosenblatt JS (1967) Nonhormonal basis of maternal behavior in the rat. Science 156(3781):1512–1514

Saltzman W, Maestripieri D (2011) The neuroendocrinology of primate maternal behavior. Prog Neuro-Psychopharmacol Biol Psychiatry 35(5):1192–1204

Smith R, Smith JI, Shen X, Engel PJ, Bowman ME, McGrath SA et al (2009) Patterns of plasma corticotropin-releasing hormone, progesterone, estradiol, and estriol change and the onset of human labor. J Clin Endocrinol Metab 94(6):2066–2074

Stanton MA, Heintz MR, Lonsdorf EV, Santymire RM, Lipende I, Murray CM (2015) Maternal behavior and physiological stress levels in wild chimpanzees (*Pan troglodytes schweinfurthii*). Int J Primatol 36(3):473–488

Stolzenberg DS, Champagne FA (2016) Hormonal and non-hormonal bases of maternal behavior: the role of experience and epigenetic mechanisms. Horm Behav 77:204–210

Taylor JH, Carp SB, French JA (2020) Vasopressin, but not oxytocin, modulates responses to infant stimuli in marmosets providing care to dependent infants. Dev Psychobiol 62:932–940

Terkel J, Rosenblatt JS (1972) Humoral factors underlying maternal behavior at parturition: Cross transfusion between freely moving rats. J Comp Physiol Psychol 80(3):365

Voogt JL, Sar M, Meites J (1969) Influence of cycling, pregnancy, labor, and suckling on corticosterone-ACTH levels. Am J Physiol 216:655–658

Wang LP, Geng RQ, Zhang XN, Sun W (2015) Identification of SNPs within the PRLR gene and effects on maternal behavior in sheep. Gene Mol Res 14(4):17536–17543

Yoshihara C, Numan M, Kuroda KO (2018) Oxytocin and parental behaviors. In: Behavioral pharmacology of neuropeptides: oxytocin, pp 119–153

Yu GZ, Kaba H, Okutani F, Takahashi S, Higuchi T (1996) The olfactory bulb: a critical site of action for oxytocin in the induction of maternal behaviour in the rat. Neuroscience 72:1083–1088

Further Recommended Reading

González-Mariscal G, Caba M, Hoffman KL, Melo AI (2016) Parental behavior. In: Pfaff DW, Joëls M (eds) Hormones, brain and behavior, 3rd edn, 2017. v. 1. Academic, Amsterdam, pp 83–116

Lévy F (2016) Neuroendocrine control of maternal behavior in non-human and human animals. Ann Endocrinol 7(2):114–125

Lonstein JS, Pereira M, Morrell JI, Marler CA (2015) Parenting behavior. In: Plant TM, Zeleznik AJ (eds) Physiology of reproduction, 4th edn. Elsevier, New York, pp 2371–2438

Cognition and Neuroplasticity During Pregnancy and Postpartum

9

Samantha A. Blankers, Kimberly A. Go, Dana C. Surtees, Tallinn F. L. Splinter, and Liisa A. M. Galea

Abstract

Many physiological changes occur during pregnancy and in the postpartum, including dramatic fluctuations in circulating hormones. These changes affect the maternal brain on behavioral, molecular and cellular levels. Pregnancy in the short- and long-term influences cognition in both humans and rodents, with both improvements and declines seen in different types of memory dependent on the stage of pregnancy or postpartum. In addition to changes in cognition, brain changes during pregnancy and the postpartum are discussed. In particular, we focus on the hippocampus, a brain area important for certain forms of memory, that is plastic in adulthood, and is affected by reproductive experience. Pregnancy

S. A. Blankers
Graduate Program in Neuroscience, Djavad Mowifaghian Centre for Brain Health, University of British Columbia, Vancouver, BC, Canada
e-mail: sblanker@student.ubc.ca

K. A. Go
Djavad Mowifaghian Centre for Brain Health, University of British Columbia, Vancouver, BC, Canada
e-mail: kimberly.go@camh.ca

D. C. Surtees · T. F. L. Splinter
Centre for Addiction and Mental Health, Campbell Family Mental Health Research Institute, Toronto, ON, Canada
e-mail: dana.surtees@queensu.ca; tallinn.splinter@camh.ca

L. A. M. Galea (✉)
Graduate Program in Neuroscience, Department of Psychology, Djavad Mowifaghian Centre for Brain Health, Vancouver, BC, Canada

Treliving Family Chair in Women's Mental Health, Centre for Addiction and Mental Health, Toronto, ON, Canada
e-mail: liisa.galea@camh.ca

P. J. Brunton, D. R. Grattan (eds.), *Neuroendocrine Regulation of Mammalian Pregnancy and Lactation*, Masterclass in Neuroendocrinology 15,
https://doi.org/10.1007/978-3-031-51138-7_9

experience influences neuroplasticity, including the production and survival of new neurons in the hippocampus, known as neurogenesis. Lastly, the long-term impacts of reproductive experience on the hippocampus and cognition are discussed.

Keywords

Cognition · Hippocampus · Neurogenesis · Plasticity · Spatial memory

9.1 Introduction

During pregnancy and the postpartum, there are dramatic changes in hormone levels, which contribute to changes in multiple systems across the body, including cardiovascular, pulmonary, immunological and metabolic systems. Each of these physiological changes are crucial for the maintenance of the pregnancy and survival of the offspring, but also affect the maternal brain. Many pregnancy-related steroid and peptide hormones are instrumental in brain changes observed during pregnancy and in the postpartum period (Calcia et al. 2016; Galea et al. 2017; Klein and Flanagan 2016; Leuner and Sabihi 2016; Lucassen et al. 2015; Duarte-Guterman et al. 2019; Puri et al. 2023). In this chapter, the short- and long-term effects of pregnancy on cognition and **brain plasticity** will be covered, with a focus on changes noted in the hippocampus, a region integral for certain forms of memory and stress resilience. Both of these functions are altered during pregnancy and in the postpartum.

9.2 Cognition During Pregnancy and Postpartum

The terms "mommy brain", "pregnancy brain", "baby brain" and "momnesia" have appeared in the popular media. These terms reflect the fact that up to 80% of mothers experience forgetfulness, disorientation and short-term memory loss during the **peripartum** period. However, as McCormack and colleagues have suggested, the terminology around "mommy brain" perpetuates a misconception in the media that such changes reflect impaired function, rather than the reality that these changes result from adaptive, natural and important changes to the maternal brain that are necessary for adoption of maternal behaviors (McCormack et al. 2023). Studies identify both deficits and improvements to different forms of memory across pregnancy and the postpartum (Fig. 9.1). Pregnancy and the postpartum period are times of abundant brain plasticity and this plasticity aids in learning new behaviors as parents. **Subjective memory** complaints during pregnancy and the postpartum do not always align with **objective memory** scores (Cuttler et al. 2011; Orchard et al. 2022). Studies reflecting changes in objective cognition measures during pregnancy and the early postpartum suggest that trimester, fetal sex, and type of memory assessed influence the findings.

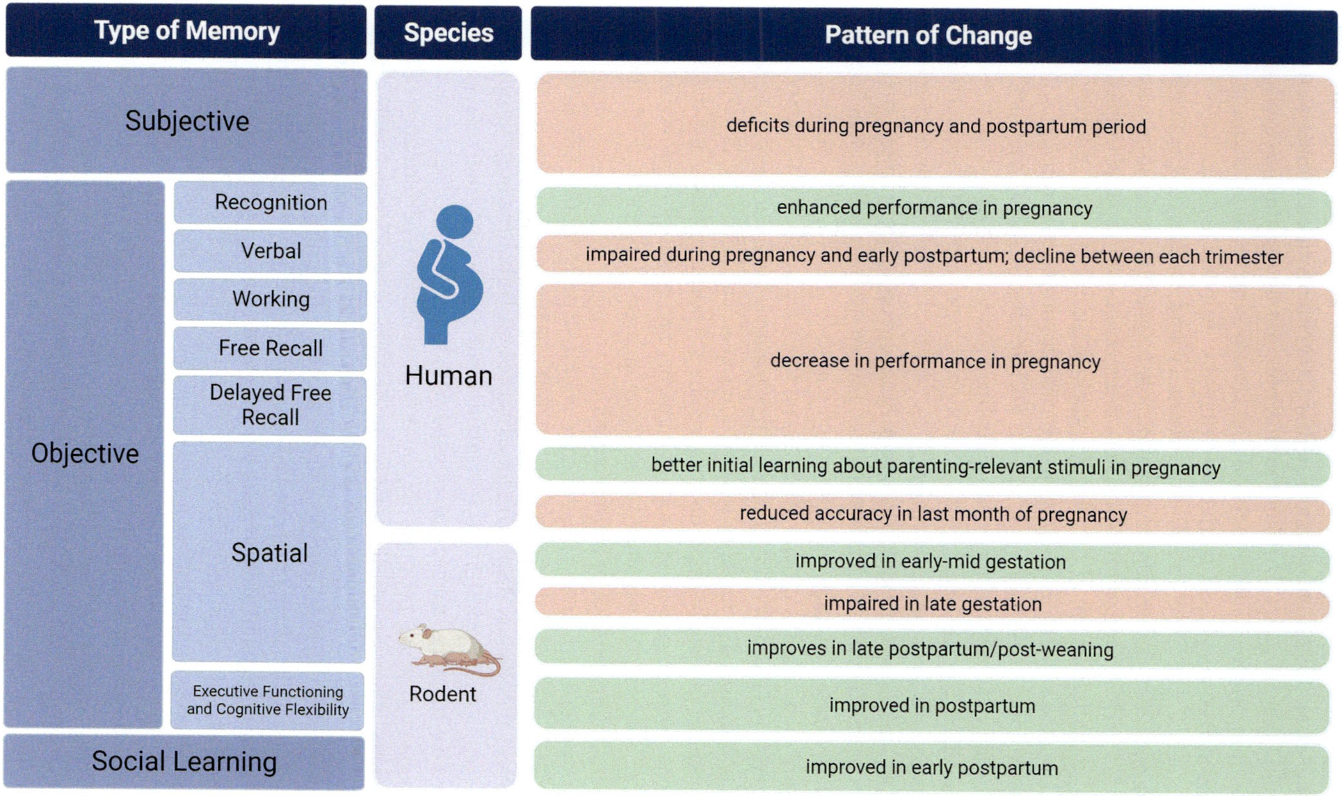

Fig. 9.1 A summary of the types of memory and how they are affected in both humans and rodents throughout pregnancy and the postpartum. Red indicates a deficit relative to controls and green indicates an enhancement relative to controls. Memory is assessed in humans through various questionnaires and cognitive tasks. Rodent memory is assessed through behavioral tasks such as the Morris Water Maze (see Box 9.1). Pregnancy does not uniformly interfere with memory or cognition and is associated with a number of neural plastic changes that indicate brain adaption to the offspring

There are reported deficits in **verbal memory** during pregnancy and the early postpartum compared to non-pregnancy (Barda et al. 2021; Henry and Sherwin 2012; Keenan et al. 1998). Henry and Rendell (2007) performed a meta-analysis, which revealed that pregnant or postpartum people did not perform as well when compared to non-pregnant individuals on tasks evaluating working memory, **free recall**, **delayed free recall** and **subjective memory**. Another meta-analysis found deficits in general cognitive functioning, memory and executive functioning in pregnant people compared to non-pregnant people, especially in the third trimester (Davies et al. 2018). Although no differences were observed in spatial or executive functioning in the last month of pregnancy, pregnant people showed reduced accuracy using a spatial strategy in solving a task, which correlated positively with the left striatum volume (Lisofsky et al. 2016). Differences in verbal memory were observed throughout the trimesters, with a decline observed between each trimester (Davies et al. 2018; Keenan et al. 1998). As alluded to above, not all forms of memory are equally affected during pregnancy, as pregnant people perform better in other memory tasks when compared to non-pregnant people. For example, pregnant people demonstrate better performance in recognition memory tasks, including the ability to recognize novel faces (Anderson and Rutherford 2012; McCormack et al. 2023; Orchard et al. 2023). Another study found that pregnancy was not associated with any deficits in laboratory-based prospective memory tests, but there were some deficits in field prospective memory tests (Cuttler et al. 2011) and the deficits were correlated with the physical symptoms of pregnancy. Thus, pregnancy and the postpartum can be associated with memory decrements and memory improvements in performance; however, as we shall show, these changes depend on a number of parameters including the type of cognitive task.

Parity, hormones and fetal sex are linked to cognitive performance. Glynn (2010) showed that bigravid (pregnant for the second time) and multigravid (pregnant more than two times) individuals display reduced scores on verbal memory throughout pregnancy compared to primigravid (pregnant for the first time) individuals. The decline in verbal memory has been associated with lower cortisol levels and higher estradiol levels in pregnant people with poorer verbal memory performance (Glynn 2010). However, other studies have failed to find an association between steroid hormones and cognitive decline during pregnancy, with some evidence of a relationship emerging postpartum (Buckwalter et al. 1999; Keenan et al. 1998). People who are pregnant with a male fetus showed better performance on working memory and mental rotation (spatial ability) tests when compared to people pregnant with a female fetus (Vanston and Watson 2005). This may be due to the bidirectional flow of cell material between the mother and fetus, as cells from the fetus may accumulate in brain tissue of the mother throughout pregnancy in the phenomenon known as **fetal microchimerism** (Adams Waldorf et al. 2010; Chan et al. 2012; Gammill and Harrington 2017; Zeng et al. 2010). Fetal sex also influences maternal inflammation, as male fetuses are associated with higher levels of **proinflammatory cytokines** (Enninga et al. 2015). Inflammatory signaling influences cognition as well as neuroplasticity (Tronson and Collette 2017; Wang et al. 2021), highlighting the potential role of the immune system in cognition during pregnancy and the postpartum.

It is well established that subjective memory deficits are present during pregnancy and in the postpartum period (Cuttler et al. 2011; Casey et al. 1999; Glynn 2010; Janes et al. 1999; Logan et al. 2014; Woods et al. 2011; Henry and Rendell 2007). However, these subjective memory deficits are present in the absence of any objective prospective or episodic memory changes (Cuttler et al. 2011; Orchard et al. 2022). Both of these studies found that poorer wellbeing (increased scores on depressed mood and sleep/fatigue) during pregnancy and one-year postpartum was associated with worse subjective memory ratings in mothers compared to non-mothers (Orchard et al. 2022; Cuttler et al. 2011). Orchard et al. (2022) speculated that early motherhood represents a cognitive challenge with the new daily cognitive load required to care for an infant, and with this challenge comes higher consequences. Therefore, mothers are constantly evaluating and re-evaluating their subjective memory, better tuning themselves to awareness of memory performance, and providing potential explanations for increased memory lapses.

The peripartum period is noteworthy for other factors that could play a role in cognitive performance, including mood disturbances (see Chap. 10), sleep disturbances, physical symptoms and busyness (Cuttler et al. 2011; Orchard et al. 2022). Indeed, Ouellette and Hampson (2019) suggest that objective memory deficits found in pregnant people may be more prevalent in those experiencing anxiety and depression. Other studies have found that pregnant people exhibiting depressive and/or anxiety symptoms showed deficits in working memory compared to nulliparous controls (Hampson et al. 2015; Kataja et al. 2017) and these symptoms are correlated with subjective memory (Cuttler et al. 2011). Sleep disturbances and/or fatigue during the peripartum period are likely to contribute to cognitive performance, although most studies still find cognitive differences even after accounting for sleep disturbances (Casey 2000; Casey et al. 1999; Janes et al. 1999; Logan et al. 2014). Overall, pregnancy and the postpartum can influence certain forms of cognitive ability in humans, but in general, subjective memory is taxed to a greater degree than objective memory during this time, which is linked to poorer wellbeing.

As noted previously, pregnancy is not uniformly associated with impaired memory, as pregnant people in their third trimester of pregnancy exhibit enhanced spatial associative memory in addition to better initial learning about parenting-relevant stimuli (Callaghan et al. 2022). The enhancements in memory can continue into the postpartum period as working memory is enhanced in people pregnant with a male fetus (compared to non-pregnant and those pregnant with a female fetus) until at least 6 months postpartum (Vanston and Watson 2005). Buckwalter et al. (2001) also noted an improvement in memory tasks 2 years postpartum when compared to late pregnancy or early postpartum periods. Overall, these findings suggest that pregnancy and the postpartum period are associated with changes in cognitive performance, the direction of which depends on the nature of the cognitive task related to fetal sex, psychiatric symptoms and physical symptoms of pregnancy.

In rodents, cognition is often evaluated through various spatial mazes that can interrogate spatial working or reference memory. Spatial memory is particular to your surroundings, spatial relations between objects and locations, and performance in part relies on a part of the brain called the hippocampus (see below). One method of testing spatial memory in rodents is through the Morris Water Maze (Box 9.1). Spatial performance varies during pregnancy and the postpartum in rodents (Roes and Galea 2016; Workman et al. 2013). Rodents with maternal experience (**primiparous**) show improved spatial working memory when compared to **naive** (virgin; **nulliparous**) females in either early- or mid-gestation (Bodensteiner et al. 2006; Galea et al. 2000; Macbeth et al. 2008a). However, female rats in late gestation have impaired **spatial working memory** compared to non-pregnant females (Galea et al. 2000). In the early postpartum period, there is impaired spatial ability (Bridges 2016; Darnaudéry et al. 2007; Leuner et al. 2007), but during the later postpartum period and after the pups have been weaned, spatial ability (both **working** and **reference memory**) improves in comparison to nulliparous females (Kinsley et al. 1999; Pawluski et al. 2006). Improvements in **spatial reference memory** in the late postpartum period are more likely to be seen in primiparous rats and only after both pregnancy and experience with pups, but not in pregnant only or foster rat mothers, indicating that it is both pregnancy and experience with pups that contributes to these changes (Pawluski et al. 2006; Pawluski and Galea 2005). These findings suggest that spatial ability is enhanced during early pregnancy and in the late postpartum, but is impaired during late pregnancy and the early postpartum, which may be related to fluctuating and high levels of estradiol and adrenal steroids.

During the mid- and late postpartum periods, rodents exhibit enhanced **executive function**, measured through their performance in attention set-shifting behavioral tasks. Furthermore, improvements in cognitive flexibility during the postpartum requires both oxytocin signaling in the prefrontal cortex and the presence of pups (Albin-Brooks et al. 2017), because when the oxytocin receptors in the prefrontal cortex are blocked with an antagonist and pups are removed, the improved executive function is no longer observed (Albin-Brooks et al. 2017). The improved **cognitive flexibility** is not seen after the pups have been weaned, as performance in set-shifting tasks was not affected in either primiparous or **biparous** rodents (Workman et al. 2013). This suggests that once the postpartum period ends, the benefits for cognitive flexibility following pregnancy are no longer present.

As might be expected, **social learning** is improved in early postpartum (shortly after parturition) when compared to controls (Fleming et al. 1994). In this same period, female rats have also been observed to have reduced anxiety-like and stress responsiveness (Wartella et al. 2003) (see Chap. 4), although some studies show increased anxiety (Ladyman et al. 2018; Lemaire et al. 2006; Wartella et al. 2003) in the postpartum. The differences in anxiety-like outcomes in the postpartum period have been hypothesized to be due to the time since lactation (Tu et al. 2005), but may also be due to time since parturition. After weaning, reduced anxiety-like behaviors were still observed in primiparous female rats (Byrnes and Bridges 2006; Love et al. 2005), suggesting long-term effects of pregnancy and the postpartum on anxiety-like behaviors.

When discussing the findings regarding pregnancy and the postpartum in rodents and humans, it should be acknowledged that the hormone types and temporal patterns of hormones during pregnancy differ between rodents and humans (see Pawluski et al. 2009). For both species, there is an increase in estradiol, progesterone and cortisol/corticosterone during gestation. The levels of these hormones drop significantly around the time of labor/**parturition** in both humans and rodents, and levels remain low during the postpartum period for a few months in humans (Brett 2001; Brunton and Russell 2010; Pawluski et al. 2009).

Box 9.1 Morris Water Maze in Rodents

Spatial learning and memory can be tested in rodents through mazes. The Morris Water Maze **(MWM)** will be discussed as an example; further examples of spatial assessments can be found in Vorhees and Williams (2014). The MWM was developed by Richard Morris and has been continually improved upon over the years (Morris et al. 1982). The MWM uses a tank that is usually 200 cm in diameter (for rats) or 122 cm in diameter (for mice) and tall enough that the rodents cannot feel the bottom of the tank. The tank is filled with opaque water and a platform is placed at some part of the tank (location can affect difficulty). The task usually includes a hidden platform which is submerged 1–2 cm below the surface. Cues are placed around the room and remain unchanged during the duration of the assessment. These cues provide the opportunity for rodents to orient themselves, therefore allowing for the assessment of spatial memory. The basic premise is that rodents are placed in the pool and the instinct to find stable footing will lead them to swim around in search for the platform. The idea is that using the cues in the room, the rodents would be able to reliably learn where the platform is (as its location remains unchanged in the standard reference memory version of this task) and the latency or distance travelled to find the platform will decrease over time. It should be noted that latency measures will be affected by swim speed and therefore distance measurements are preferred to get a true estimate of memory performance. Variables such as temperature of the water and shape of the room will affect performance. There is some indication that male and female rats pay attention to different cues preferentially, and use different strategies to find the hidden platform which can vary with hormone levels (Spritzer et al. 2013; Williams et al. 1990). See Fig. 9.2 for diagrams of the MWM set-up. Refer to Vorhees and Williams (2006) for detailed procedures and analysis.

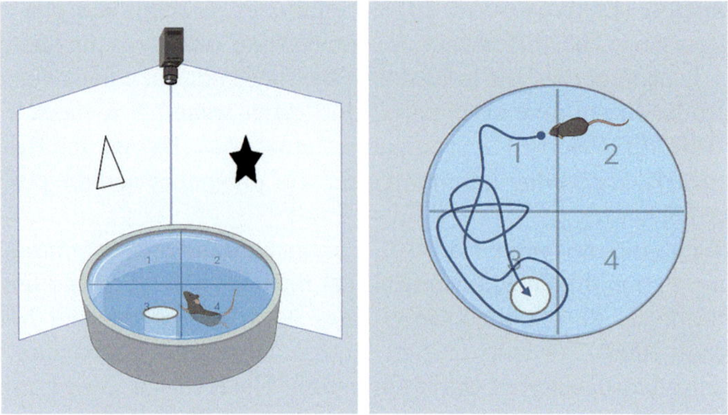

Fig. 9.2 Diagram of the Morris Water Maze, with examples of quadrants and cues (left), as well as a bird's eye view as to what is seen and recorded by the video camera (right). Created with BioRender.com

9.3 Neuroplasticity and Neurogenesis During Pregnancy and Postpartum

9.3.1 Neuroplasticity

Neuroplasticity can be used as an umbrella term regarding the ability of the nervous system to modify, change and adapt, both in structure and function, as a response to experience. Plasticity involves changes at the level of the neuron, such as altered **neurogenesis** or **synaptogenesis**.

9.4 Why the Hippocampus?

The hippocampus is a highly plastic area of the brain in adulthood and is, in fact, one of the only brain regions that retains the ability to produce a significant number of new neurons throughout life, in a process called neurogenesis. Neurogenesis is a form of neuroplasticity that is altered during pregnancy and the postpartum in the maternal brain. The hippocampus is located in the **medial temporal lobe** and is long recognized for its crucial role in certain forms of memory. Lesions to this region result in impairments to spatial, contextual and relational memory (Abrahams et al. 1999; Frankland et al. 1998; Good and Honey 1991; Maguire et al. 1998; Morris et al. 1982; Vargha-Khadem et al. 1997).

It is perhaps not surprising that the hippocampus is altered with pregnancy and during the postpartum period, as the hippocampus contains many receptors for the hormones that fluctuate during pregnancy and postpartum. For example, steroid hormones such as estrogens, progesterone and glucocorticoids are all increased during pregnancy and there is an abundance of these receptors in the hippocampus. Steroid receptors such as androgen receptors (AR), estrogen receptors (ER-alpha, ER-beta and G-protein coupled-ER), progesterone receptors (PR) and glucocorticoid (GR) and mineralocorticoid receptors (MR) are present at high levels in the hippocampus relative to other brain regions (Brailoiu et al. 2007; Cui et al. 2014; Hazell et al. 2009; Mitra et al. 2003). GR and MR are expressed at particularly high levels in the hippocampus compared to other brain regions, which can partially explain why the hippocampus may be more susceptible to stress (Chao et al. 1989; Mifsud and Reul 2016; Pawluski et al. 2015). Therefore, it is natural to draw a connection between the hippocampus, cognition and the dynamics of hormone and stress levels during pregnancy and postpartum.

9.4.1 Neurogenesis

Neurogenesis is the process in which new neurons are generated from neural progenitor cells and functionally integrated into the nervous system. Neurogenesis persists throughout adulthood in all mammals studied, including humans (Kempermann et al. 2018). The two primary sites of neurogenesis are the **subgranular zone (SGZ)** in the **dentate gyrus** of the hippocampus and the **subventricular zone (SVZ)** of the lateral ventricle, with cells migrating into the olfactory bulb. Broadly, the phases of neurogenesis include **cell proliferation**, **cell migration**, **cell differentiation** and **maturation** (Brandt et al. 2003; Encinas et al. 2006; Filippov et al. 2003; Fukuda et al. 2003; Steiner et al. 2006). Although two studies have suggested that neurogenesis in adult humans is absent (Cipriani et al. 2018; Sorrells et al. 2018), many other studies demonstrate that hippocampal neurogenesis exists in adult humans (Boldrini et al. 2018; Eriksson et al. 1998; Knoth et al. 2010; Mathews et al. 2017; Moreno-Jiménez et al. 2019; Spalding et al. 2013; Tobin et al. 2019). However, it is important to acknowledge that neurogenesis in the hippocampus declines dramatically with age (Moreno-Jiménez et al. 2019; Boldrini et al. 2018).

The different phases of **adult neurogenesis** are identified through stage-specific cell markers. These markers are selective for the presence of specific proteins (see Fig. 9.3 for an outline of commonly used markers of the different stages of neurogenesis). To label newly produced cells and their progeny, radioactivity or synthetic analogues of thymidine are typically used in vivo in animals. One such analogue is **bromodeoxyuridine (BrdU)**. BrdU is taken up by cells undergoing DNA synthesis, specifically during the S-phase of mitosis for approximately 2 h after injection. Depending on the timing between BrdU administration and brain collection, BrdU can be used to label cell proliferation or the survival of these newly produced cells (Taupin 2007). A BrdU-labelled cell can be co-labelled with mature neuronal protein markers, such as **neuronal nuclei (NeuN)**, to determine the presence of a newly formed mature neuron. Endogenous markers such as **Ki67** and **doublecortin (DCX)** are commonly used to label proliferating or immature neurons, respectively. Ki67 is a nuclear protein detected during all phases of the cell cycle except G0, thus it is exclusively a cell proliferation marker (Kee et al. 2002). DCX is a protein associated with microtubules and is expressed in migrating and immature neurons during the first 21 days post-proliferation in the dentate gyrus in a rat (Brown et al. 2003). The timeline of maturation of a newly formed cell into one that expresses mature neuronal protein is slower in mice compared to rats (Snyder et al., 2009) and in female compared to male rats (Yagi et al. 2020); thus, care must be taken when

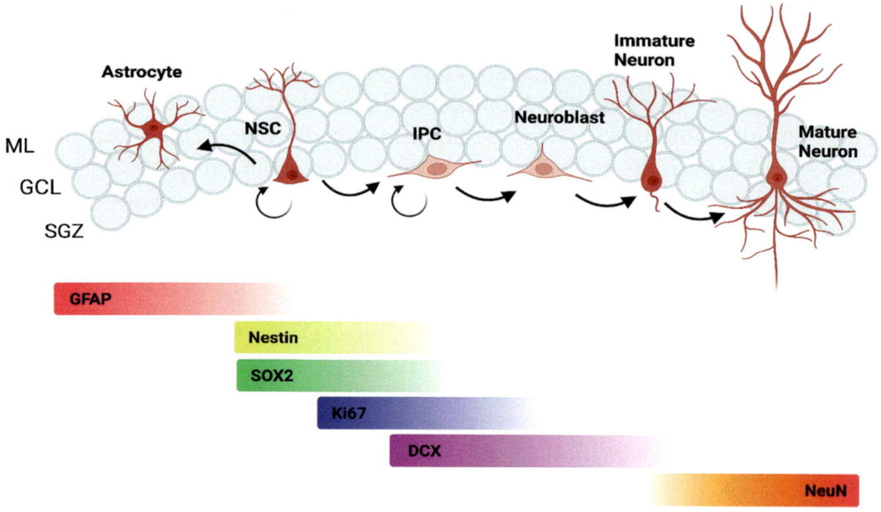

Fig. 9.3 Cell stage-specific neurogenesis markers. *DCX* doublecortin *GCL* granular cell layer *GFAP* glial fibrillary acidic protein *IPC* intermediate progenitor cell *Ki-67* antigen Kiel-67 *ML* molecular layer *NeuN* neuronal nuclear protein *NSC* neural stem cells *SGZ* subgranular zone *SOX2* sex determining region Y (SRY)-box 2

comparing timelines across studies. It is possible to visualize and quantify the dynamics of neurogenesis using various stage-specific cell markers.

Box 9.2 Utilizing BrdU to Visualize and Quantify Newly Formed Cells

BrdU is usually administered to animals through intraperitoneal injection and incorporates into the DNA of dividing cells within 2 h after injection. BrdU is a synthetic analogue of the nucleotide thymidine, thus cells undergoing mitosis during the 2-h period after the time of injection will incorporate this synthetic thymidine into their DNA instead of the regular nucleotide. This is valuable because BrdU can be visualized post-mortem using **immunohistochemistry** and microscopy to quantify newborn cells. Brain slices can be stained with antibodies that selectively bind to BrdU, and a secondary fluorescent antibody specific for the primary antibody allows BrdU to be localized using fluorescent microscopy. Furthermore, any of the stage-specific markers mentioned above can be located at the same time as BrdU using antibodies selective for those proteins (DCX, NeuN, etc.). Double-staining for BrdU and a stage-specific marker tells us how many newly formed cells express proteins consistent with that stage of neurogenesis, thus allowing us to classify new cells as immature neurons, mature neurons, etc. This is interesting because it allows us to compare the dynamics of neurogenesis between different treatment groups and in animal models of specific diseases that are related to altered neurogenesis. The timing of BrdU injection relative to treatment or experience, as well as dose of BrdU and any effects on the blood-brain barrier, can influence the results and should be noted (see Taupin 2007).

9.4.2 Neuroplasticity During Pregnancy and Postpartum in Humans

During pregnancy and the postpartum, the brain experiences a variety of dynamic changes. Total brain volume decreases during pregnancy, reaching a low point at parturition and recovering to a pre-pregnancy volume by 6 months postpartum (Oatridge et al. 2002), although this study did not include a nulliparous control group and other studies have failed to replicate the complete rebound at different time points postpartum (Hoekzema et al. 2017; Martínez-García et al. 2021a, b). However, the general finding of reductions across pregnancy followed by a potential rebound during the first 6 months postpartum is a useful framework to keep in mind when considering the literature, as different studies compare different time points. So, for example, if a study compares across pregnancy to the early postpartum and another from early postpartum to later in the postpartum, they may come to different conclusions, based on the potential rebound.

In a landmark study of first-time mothers, reduction in gray matter volume was observed in brain regions responsible for cognition following pregnancy compared to nulliparous controls. The brain areas showing a reduction in gray matter included the frontal and temporal regions, precuneus, parahippocampal regions, anterior cingulate cortex, the left insula and the hippocampus (Hoekzema et al. 2017). These gray matter volume reductions were still present 2 years (Hoekzema et al. 2017) and even 6 years after delivery (Martínez-García et al. 2021a, b). Gray matter volume is also reduced in the ventral striatum at 2 months postpartum in primiparous compared to nulliparous humans (Lisofsky et al. 2016; Hoekzema et al. 2020). Along with gray matter reductions, there were reductions in cortical thickness and cortical surface area but no changes in white matter volume (Carmona et al. 2019). Intriguingly, the reductions in gray matter across pregnancy are not associated with memory deficits but are associated with maternal cues, responsiveness and attachment. Reduced volume of gray matter in the ventral striatum during the postpartum is associated with higher activation of the ventral striatum in response to cues from a mother's own infant (Hoekzema et al. 2020). Indeed, prefrontal and temporal cortical gray matter reductions were not associated with memory scores but were associated with maternal sensitivity and attachment to children (Hoekzema et al. 2017). These are important findings as it must be recognized that a reduction in gray matter is not necessarily associated with "worse" outcomes or poorer performance. Thus, it is important to recognize that the reductions in brain region volume across pregnancy in primiparous females compared to age-matched nulliparous females are natural neural changes that likely support maternal transitions.

Gray matter increases are seen in various brain regions from the early postpartum to 3–5 months postpartum, underscoring the need to pay close attention to the timeline of image acquisition. These regions include the anterior cingulate cortex, middle frontal cortex, medial prefrontal cortex, cerebellum, nucleus accumbens, amygdala and hypothalamus, areas involved in maternal behavior and motivation (Kim et al. 2010; Lisofsky 2019) (see Chap. 8). These findings suggest an initial decrease in gray matter volume after childbirth that recovers over time in certain brain regions (for review Martínez-García et al. 2021a, b). Further analysis of mothers at various postpartum time points was conducted by Luders et al. (2018) using the BrainAGE index, which uses neuroimaging data to estimate the age of a brain, applied to mothers at various postpartum time points. Participants were analyzed for hallmarks of an aging brain including reduced overall size, reduction in gray matter tissue and increased size of the ventricles. Lower brain ages (i.e., younger brains) were found in mothers 6 weeks postpartum compared to 1–2 days postpartum, further demonstrating that the changes occurring throughout the postpartum period seem to recover after a low point reached around the time of childbirth [reviewed in Luders et al. (2018)]. Thus, taken together, the data suggest that there are a number of changes in gray matter volume across various brain regions, that occur from preconception to the early postpartum and from the early postpartum to the later postpartum.

Functional brain changes are also observed with pregnancy, as activation of the striatal region in the postpartum has been associated with enhanced maternal sensitivity (Kim et al. 2011), suggesting that the postpartum reduction in gray matter (and increase in activation) is related to optimization of behavior in preparation for motherhood. Resting state functional connectivity is reduced between the posterior cingulate cortex and the left medial prefrontal cortex in postpartum mothers compared to non-mothers (Zheng et al. 2020). These brain regions are part of the **default mode network**, which is most active at rest and less active during tasks involving attention or goal-directed behavior (Mantini et al. 2007). Furthermore, pregnancy-induced changes to neural network organization and neural architecture have been identified in brain regions associated with the default mode network, such as the posterior cingulate cortex (Hoekzema et al. 2022) which is critical for processing emotions and distressing information (Fransson and Marrelec 2008). These changes were associated with elevated estradiol levels characteristic of the third trimester (Hoekzema et al. 2022). Thus, pregnancy results in changes to gray matter volume and functional connectivity across a number of neural networks that are related to important processes in the transition to parenthood, including maternal sensitivity and attachment.

9.4.3 Neuroplasticity and Neurogenesis During Pregnancy and Postpartum in Rodents

Rodent studies indicate changes in neuroplasticity during pregnancy and the postpartum period. As seen in humans, hippocampal volume is decreased in pregnant and early postpartum rats when compared to nulliparous controls (Galea et al. 2000; Hillerer et al. 2014). The first indication that pregnancy may influence hippocampal neurogenesis was shown in wild meadow voles, where pregnant females captured during the breeding season displayed decreased cell proliferation compared to non-pregnant females captured in the non-breeding season (Galea and McEwen 1999). Neurogenesis in the hippocampus is reduced during pregnancy and postpartum (Eid et al. 2019; Leuner et al. 2010; Leuner and Sabihi 2016; Pawluski et al. 2020; Pawluski and Galea 2007; Rolls et al. 2008). During gestation, specifically gestation day 13 (GD13), and on postpartum days 8 and 30 (PD8; PD30), lower levels of neurogenesis (specifically cell proliferation and immature neurons) are observed in primiparous female rats compared to nulliparous counterparts (Eid et al. 2019; Rolls et al. 2008). New neurons produced in the hippocampus during mid-gestation have a reduced survival rate by the end of pregnancy in mice (Rolls et al. 2008), but there are no significant changes in cell proliferation observed in early (GD1 and GD7; Furuta and Bridges 2005; Pawluski et al. 2010) or in later pregnancy (GD18 and GD21; Furuta and Bridges 2005; Pawluski et al. 2010). In contrast, elevated neurogenesis (specifically cell proliferation and survival of new neurons) is observed in the subventricular zone of mice during early gestation (GD7) and early in the postpartum period (PD7) compared to age-matched virgin females (Shingo et al. 2003). Thus, reductions in hippocampal neurogenesis are observed during mid-gestation, with increased neurogenesis in the subventricular zone during early gestation in rodents.

Other aspects of neuroplasticity are also influenced by pregnancy, such as remodeling of synapses and dendrites in hippocampal neurons (Banasr et al. 2001; Pawluski et al. 2012, 2020). Analyses in rats revealed a reduction in dendritic complexity of CA3 pyramidal cells, but increased CA1 spine density, in late pregnancy (day 21) compared to non-pregnant females (Pawluski et al. 2012). Rats in late pregnancy display an increase in synaptophysin, a presynaptic membrane-associated protein involved in vesicle docking (Calakos and Scheller 1994), in the CA3 region and the dentate gyrus of the hippocampus (Pawluski et al. 2020). Furthermore, an adhesion molecule involved in cell migration and **synaptic reorganization** (polysialylated form of the neural cell adhesion molecule; **PSA-NCAM**) is also increased in late-pregnancy (GD18; Banasr et al. 2001) further implying that structural neuroplasticity is altered at this stage. In late pregnancy, female rats undergo a sharp decline in previously elevated progesterone levels, whereas estradiol levels are approaching their peak before birth. Thus, different characteristics of the dynamics of neuroplasticity throughout pregnancy may be reflective of the hormonal fluctuations that occur throughout pregnancy, birth, and postpartum.

In the postpartum period, changes in neuroplasticity are also observed as a reduction in cell proliferation is seen in the hippocampus of primiparous rats in the early postpartum (PD2 and 8) and mid-postpartum (PD13), but not in the late postpartum (PD28 and 30; Darnaudéry et al. 2007; Eid et al. 2019; Leuner et al. 2007; Pawluski and Galea 2007). Similar findings are seen in sheep, as cell proliferation in the dentate gyrus is down-regulated at birth and in the early postpartum (Brus et al. 2010). Therefore, it appears that a conserved mechanism is present across species regarding the reduction of cell proliferation in the early postpartum period.

Neurogenesis (specifically the survival of new neurons) is reduced across the postpartum in primiparous mice and rats compared to nulliparous controls (Eid et al. 2019; Glasper et al. 2011; Pawluski and Galea 2007), but may be increased in biparous female rats, given the lower levels of cell proliferation during the early postpartum seen in biparous rats (Pawluski and Galea 2007). Lastly, in the dorsal raphe, fewer neuronal precursor cells and decreased survival of young neurons was observed in late postpartum rats compared to rats in the early postpartum period (Holschbach and Lonstein 2017).

Other forms of hippocampal neuroplasticity also change across the postpartum including electrophysiology, dendritic spines and branching. Electrophysiological changes have been observed in mice as early as postpartum (PD3), where multiparous mothers exhibit more long-lasting **long-term potentiation (LTP)** along Schaffer collaterals (projections from CA3 to CA1 in the hippocampus) compared to nulliparous controls (Tomizawa et al. 2003). Primiparous rats, however, have exhibited reductions in dendritic branching and length in comparison to their biparous and nulliparous counterparts representing significant dendritic remodeling (Pawluski and Galea 2005). Additionally, spine density is increased in the postpartum (PD5-6) in the basal CA1 region, with similar increases mimicked by pregnancy-related hormones (Kinsley et al. 2006). This increase may be specific to multiple pregnancies and influenced by the sex of the offspring, as there was a positive correlation between spine density and number of male pups birthed (Pawluski and Galea 2005). These findings indicate that previous pregnancy, and the sex of the offspring, influence multiple facets of neuroplasticity in the postpartum period (Pawluski and Galea 2005).

As previously mentioned, altered neuroplasticity and neurogenesis observed in pregnancy is regulated through the characteristic fluctuations of steroid hormones observed across pregnancy and birth. When hormones are used to simulate pregnancy and the postpartum period in female rats, cell proliferation is down-regulated in the early postpartum as seen in natural pregnancy (Green and Galea 2008). The reduction in cell proliferation in the early postpartum is mimicked by a withdrawal from a hormone-simulated pregnancy (Green and Galea 2008), and is not a result of experience with pups, as foster mothers demonstrated increased cell proliferation at PD2 (Pawluski and Galea 2007), indicating that decreased cell proliferation in the postpartum is due to reproductive experience, but not pup exposure. The early postpartum reduction in cell proliferation is regulated by adrenal steroids, as adrenalectomy eliminates the expected decrease in cell proliferation (Leuner et al. 2007). In addition to steroid hormones, other pregnancy hormones have been associated with hippocampal neuroplasticity. In mice, high prolactin levels in early pregnancy increased the generation of neural progenitor cells in the SVZ and has been associated with reduced anxiety and maternal behavior observed in the postpartum (Larsen and Grattan 2010).

Several studies have observed changes in both learning and neurogenesis during pregnancy and postpartum. Impaired spatial ability is commensurate with decreased cell proliferation in the hippocampus in the early postpartum (Darnaudéry et al. 2007). However, at other time points (mid-postpartum period, late postpartum) there are reduced levels of neurogenesis coupled with improved spatial memory when compared to nulliparous controls (Eid et al. 2019; Hillerer et al. 2014; Pawluski and Galea 2007; Pawluski and Galea 2005). The observed decrease in neurogenesis during pregnancy and the postpartum is believed to contribute to "fine-tuning" of the maternal brain that results in optimized behavioral outcomes (Lambert et al. 2019). Although it may seem paradoxical that reduced neurogenesis in the late postpartum is also commensurate with enhancements in learning and memory, chronic restraint stress also reduces dendritic branching in the CA3 region in female rats while at the same time there is enhanced spatial memory (Luine et al. 2007). Therefore, these changes in neurogenesis in rodents, and gray matter reductions in humans, during pregnancy and the postpartum are natural, and the "reduced" neuroplasticity at this time is similar to other important developmental stages such as adolescence where a decrease in neurogenesis is important for normal cognitive function (Lambert et al. 2019).

9.5 Long-Term Effects of Pregnancy on the Hippocampus and Cognition

Motherhood has long-lasting impacts on brain aging and cognition in humans. Adult mothers, later in life, display a positive correlation between their number of children and the thickness of the parahippocampal gyrus (Orchard et al. 2020), a region that facilitates communication between the hippocampus and cortex (Burgmans et al. 2011), and is involved in memory consolidation (Köhler et al. 1998). Motherhood is protective against cellular aging, as Barha et al. (2016) found that increased telomere length correlated with an increased number of children. Furthermore, there are fewer signs of brain aging, including the hippocampus, in mothers compared to non-mothers as displayed in studies estimating brain age in mid-life (de Lange et al. 2019; Lange et al. 2020). Therefore, long-lasting alterations to brain structure and cellular aging persist with previous reproductive experience.

In terms of cognition, older mothers demonstrated better cognitive flexibility and working memory performance compared to teenage mothers (Chico et al. 2014). Although executive functioning generally improves from teenage to adult years (Geier et al. 2010; Luna et al. 2004), it was found that teenage mothers had the worst executive functioning compared to teen non-mothers, adult mothers and non-mothers. Furthermore, giving birth at a younger age (20 years or younger) was associated with worse cognitive performance on the Mini-Mental State Exam (MMSE) and on measures of verbal fluency when examined in women 65 years and older (Ryan et al. 2009). A later age of last pregnancy (24 years and older) is associated with better executive function, and verbal and cognitive performance is greater in women with their last pregnancy at age 35 years or older when tested in mid- and late-life women (Karim et al. 2016). Additionally, nulliparity is associated with less cognitive decline demonstrated by performance on the MMSE (McLay et al. 2003), thus pregnancy and the age at which a mother is pregnant has implications for cognitive performance on a variety of measures.

The cognitive changes associated with parity, however, are not limited to human females. In observing data collected from the UK Biobank, an association between parity and improved cognitive function, as measured by visual memory and response time, was found in both females and males (Ning et al. 2020). In females, this relationship appeared linear where the cognitive improvement increased with the number of children, whereas in males this relationship was quadratic (Ning et al. 2020). When compared to age-related cognitive decline, the protective effect of parity was significantly greater in human males compared to females, providing a recovery of 2–4 years in response time and visual memory with parity in comparison to a one-year improvement in females. This is in contrast to a similar relative brain age as predicted by MRI in both sexes, with a reduction in cognitive age 0.5–0.7 with 2 or 3 children, respectively (Ning et al. 2020). Although cognitive alterations may exist, structural changes in human males during pregnancy remain in question. While significant alterations in gray matter and cortical thickness have been demonstrated in pregnant women following birth, these changes are less pronounced or non-existent in male partners showing significance only when evaluated by surface volumetric reduction (Hoekzema et al. 2017; Paternina-Die et al. 2020). However, when comparing first-time fathers 2 weeks and 12 weeks postpartum, significant changes in gray matter volume were identified in regions consistent with those for parenting behaviors in animals (Kim et al. 2014). The strength and direction of cognitive adaptation may vary dependent on socioeconomic status (Read and Grundy 2017). Thus, cognitive and structural changes that human males experience, during and following their partner's pregnancy, may be related to alterations in lifestyle following childbirth or to the hormonal changes that are also noted in human males following childbirth (Saltzman and Ziegler 2014).

Long-term effects of parity on cognition have also been identified in female rodents, where in middle age and older age rats and mice, previous parity is associated with improved spatial learning and memory (Barha et al. 2015; Cui et al. 2014; Galea et al. 2018b; Gatewood et al. 2005; Kinsley et al. 1999, 2008; Lemaire et al. 2006; Love et al. 2005; Macbeth et al. 2008b). When repeatedly tested across the lifespan, better re-acquisition of spatial performance and attenuated age-dependent cognitive decline is seen in biparous female rats compared to nulliparous females at 12, 18, and 24 months old (Gatewood et al. 2005). In middle age, an improvement in spatial reference memory was observed in primiparous female rats when compared to nulliparous controls (Gatewood et al. 2005). Greater accuracy in spatial working memory tasks was observed in maternal rats 4 months post-weaning when compared to age-matched virgin females (Kinsley et al. 2008). These improvements in spatial memory have been found to last the entire lifespan of 22-month-old maternal rats that were tested at 6 months and 22 months of age (Lemaire et al. 2006). Furthermore, increased hippocampal LTP is observed in rodent mothers 2 weeks post-weaning that extends into old age (22 months) compared to nulliparous controls (Lemaire et al. 2006). Previous parity increases neurogenesis in the hippocampus and synaptic proteins such as PSD-95 in middle-age compared to nulliparous controls (Barha et al. 2015; Galea et al. 2018a; Eid et al. 2019; Duarte-Guterman et al. 2023; Lee et al. 2023). Intriguingly, parity effects vary by genotype, as primiparity increased neurogenesis in wild-type rats, but reduced neurogenesis in rats with an hAPOEe4 genotype, a genotype known to confer a high risk for developing Alzheimer's disease (Lee et al. 2023). Previously parous middle-aged rats also show increased cell proliferation in response to estrogens compared to age-matched nulliparous rats (Barha and Galea 2011). These results suggest that reproductive experience alters the capacity of the hippocampus to respond to estrogens later in life and may have implications for hormone therapy effectiveness. Thus, there are long-lasting alterations in cognition in rodents with reproductive experience that persists late into life.

9.6 Perspectives

In summary, various aspects of cognition are altered during pregnancy and the postpartum period that coincide with changes in neuroplasticity. Working memory, executive functioning and verbal memory scores are reduced in pregnant women, although an increase in recognition memory is observed. Cognitive impairments during pregnancy are exacerbated by factors such as presence of a mood disorder, younger age of pregnancy and carrying a female fetus. Studies in rodents have revealed that spatial memory, working memory and reference memory are improved in the late postpartum but impaired during late gestation, which has been linked to fluctuations in hormones such as estradiol and oxytocin. Executive functions are improved in mid- to late postpartum rats, but these effects disappear after weaning. It is likely that cognitive changes are related to the extreme hormonal fluctuations that occur during pregnancy, birth and lactation. Additionally, in first-time parents, it can also be possible that these cognitive differences are related to one's wellbeing as there are several novel experiences, challenges and behavioral adaptations that accompany parenthood, which may result in sleep disturbance and increased anxiety. Measuring and controlling for differences in mood and sleep are important in studying potential objective cognitive changes. At the same time, it is clear that new parenthood is a time of adaptive plasticity that results in reorganization of the brain to adapt to new behaviors.

Cognitive changes in the peripartum period are accompanied by changes in the brain, as total brain volume decreases during pregnancy and recovers 6 months postpartum in humans. Furthermore, changes in gray matter volume are observed in the postpartum period and are associated with maternal attachment in humans, which may serve as an evolutionary advantage for new mothers caring for their children. Hippocampal volume is diminished in pregnant rodents and humans, and in rodents this is accompanied by reductions in neurogenesis across pregnancy and into the postpartum. Furthermore, decreased dendritic complexity and increased spine density is observed in the rat hippocampus during late pregnancy, illustrating that structural neuroplasticity across the hippocampus is seen during pregnancy in rats. Hippocampal neurogenesis is decreased in the early- to mid-postpartum period and LTP is increased in the hippocampus during the early postpartum. These findings point to the fact that multiple levels of plasticity are affected by pregnancy, indicating that the female brain is highly plastic during this time, probably in preparation for the variety of new challenges faced by parents to ensure the survival of their offspring.

Mothers are not the only ones affected by the birth of a child, and non-birthing parents also experience a lifestyle change. While the research on subjective and objective cognition is limited, evidence suggests some cognitive and structural adaptations in men experiencing fatherhood for the first time. However, when compared to pregnant people, these changes may not be as pronounced and may affect fewer parts of the brain, further emphasizing the impact of the lifestyle change that comes with parenthood on neural adaptations (Paternina-Die et al. 2020; Kim et al. 2014; Hoekzema et al. 2017; Ning et al. 2020). Finally, it should be emphasized that pregnancy has life-long impacts on the maternal brain. Previous reproductive experience increases LTP and neurogenesis in the hippocampus of middle-aged and older females, along with slightly improved spatial learning ability in rodents. In humans, fewer signs of brain aging and increased thickness of the parahippocampal gyrus is observed in mothers compared to non-mothers. All of these data point to the notion that the maternal brain is highly plastic. It is important to understand the short- and long-term implications of this plasticity for maternal health. There is an elevated risk for development of a first-time mood disorder such as anxiety and/or depression in the early postpartum (see Chap. 10), which serves as an additional challenge faced by new parents (see review (Qiu et al. 2020)). Furthermore, risk for neurological and cardiovascular diseases later in life may be further exacerbated by reproductive experience (Galea et al. 2018a) and deserve more attention in the medical literature.

Key References (see main list for reference details)

Galea LAM, Ormerod BK, Sampath S, Kostaras X, Wilkie DM, Phelps MT (2000) Spatial working memory and hippocampal size across pregnancy in rats. Horm Behav 37:86–95. doi: https://doi.org/10.1006/hbeh.1999.1560

Evidence that memory can be affected during pregnancy. In one study, between the first and second trimester, pregnant rats were found to perform better than non-pregnant rats in the Morris Water maze (MWM). In another study, during the first and second trimester, pregnant rats performed better in a working memory variation of the MWM, but did worse during the third trimester.

Pawluski JL, Galea LAM (2007) Reproductive experience alters hippocampal neurogenesis during the postpartum period in the dam. Neuroscience 149(1):53–67. https://doi.org/10.1016/j.neuroscience.2007.07.031

Investigates the role of reproductive experience on neurogenesis in the hippocampus through cell proliferation or cell survival. Results show a significant decrease in cell proliferation in the dentate gyrus of both primiparous and multiparous rats in the early postpartum and a decrease in cell survival in the dentate gyrus during postpartum regardless of pup exposure. Overall, the results demonstrate that reproductive experiences affect hippocampal neurogenesis.

Orchard ER, Rutherford HJV, Holmes AJ, Jamadar SD (2023) Matrescence: lifetime impact of motherhood on cognition and the brain. Trends Cogn Sci 27(3):302–316. https://doi.org/10.1016/j.tics.2022.12.002

Reviews the impact of motherhood on the brain and cognition. This review discusses the changes that the maternal brain experiences in both structural and functional neuroplasticity during the peripartum period. It also addresses the cognitive adaptations that take place during this time that last throughout the lifespan. The review also examines the perspective that motherhood increases cognitive load and creates a complex environment that result in an increased cognitive reserve for mothers in late life.

de Lange A-MG, Kaufmann T, van der Meer D, Maglanoc LA, Alnæs D, Moberget T, Douaud G, Andreassen OA, Westlye LT (2019) Population-based neuroimaging reveals traces of childbirth in the maternal brain. Proc Natl Acad Sci U S A 116(44):22341–22346. https://doi.org/10.1073/pnas.1910666116

This study uses neuroimaging and machine learning to investigate structural brain characteristics in middle-aged women. The results showed that women with reproductive experience show less brain aging when compared to their nulliparous peers and the relationship between childbirth and a "younger-looking" brain are not a result of any common genetic variations or relevant confounders.

Hoekzema E, Barba-Müller E, Pozzobon C, Picado M, Lucco F, García-García D, Soliva JC, Tobeña A, Desco M, Crone EA, Ballesteros A, Carmona S, Vilarroya O (2017) Pregnancy leads to long-lasting changes in human brain structure. Nat Neurosci 20(2):287–296. https://doi.org/10.1038/nn.4458

Evidence showing that pregnancy creates substantial changes in brain structure in first-time mothers, primarily with reductions in gray matter volume in regions involved with social cognition. The changes observed were selective for the mothers (when compared to nulliparous peers and fathers) and highly consistent. Additionally, the changes in gray matter volume predicted measures of postpartum maternal attachment, suggesting that there is an adaptive process involved with the transition to motherhood. The study also includes a follow up of 2 years post-pregnancy and have found that the reductions persisted, demonstrating that pregnancy can result in long-lasting changes in a mother's brain.

Martínez-García M, Paternina-Die M, Desco M, Vilarroya O, Carmona S (2021) Characterizing the brain structural adaptations across the motherhood transition. Front Glob Women's Health 2:742775. https://doi.org/10.3389/fgwh.2021.742775

This review looks at the transition into motherhood and how the brain structurally adapts as a result. It discusses evidence regarding the trajectory, mediating factors and neural mechanisms behind the maternal brain and the relation that may have to postpartum mental health. The authors of the review also touch upon their own studies which contain preliminary evidence that changes in the maternal brain are persistent, predict mother-to-infant attachment and can be mediated by pregnancy-induced factors and parental experience with the infant.

Eid RS, Chaiton JA, Lieblich SE, Bodnar TS, Weinberg J, Galea L (2019) Early and late effects of maternal experience on hippocampal neurogenesis, microglia, and the circulating cytokine milieu. Neurobiol Aging 78:1–17. doi:https://doi.org/10.1016/j.neurobiolaging.2019.01.021

This study examines the short- and long-term effects of motherhood on neurogenesis in female rats. It was found that neurogenesis was suppressed during gestation and in the postpartum of primiparous rats as demonstrated by reduced DCX expression. Reduced cell proliferation was also observed during mid-gestation in the early postpartum of primiparous rats. Interestingly, long term trajectory of neurogenesis differed with reproductive experience, as nulliparous females showed a decrease in neurogenesis from 7 to 13 months while primiparous females showed a sli ght increase.

Puri TA, Richard JE, Galea LAM (2023) Beyond sex differences: short- and long-term effects of pregnancy on the brain. Trends Neurosci 46(6):459–471. doi:https://doi.org/10.1016/j.tins.2023.03.010

References

Abrahams S, Morris RG, Polkey CE, Jarosz JM, Cox TC, Graves M, Pickering A (1999) Hippocampal involvement in spatial and working memory: a structural MRI analysis of patients with unilateral mesial temporal lobe sclerosis. Brain Cogn 41(1):39–65. https://doi.org/10.1006/brcg.1999.1095

Adams Waldorf KM, Gammill HS, Lucas J, Aydelotte TM, Leisenring WM, Lambert NC, Nelson JL (2010) Dynamic changes in fetal microchimerism in maternal peripheral blood mononuclear cells, CD4+ and CD8+ cells in normal pregnancy. Placenta 31(7):589–594. https://doi.org/10.1016/j.placenta.2010.04.013

Albin-Brooks C, Nealer C, Sabihi S, Haim A, Leuner B (2017) The influence of offspring, parity, and oxytocin on cognitive flexibility during the postpartum period. Horm Behav 89:130–136. https://doi.org/10.1016/j.yhbeh.2016.12.015

Anderson MV, Rutherford MD (2012) Cognitive reorganization during pregnancy and the postpartum period: an evolutionary perspective. Evol Psychol 10(4):147470491201000. https://doi.org/10.1177/147470491201000402

Banasr M, Hery M, Brezun JM, Daszuta A (2001) Serotonin mediates oestrogen stimulation of cell proliferation in the adult dentate gyrus: serotonin-oestrogen relations and adult neurogenesis. Eur J Neurosci 14(9):1417–1424. https://doi.org/10.1046/j.0953-816x.2001.01763.x

Barda G, Mizrachi Y, Borokchovich I, Yair L, Kertesz DP, Dabby R (2021) The effect of pregnancy on maternal cognition. Sci Rep 11(1):12187. https://doi.org/10.1038/s41598-021-91504-9

Barha CK, Galea LAM (2011) Motherhood alters the cellular response to estrogens in the hippocampus later in life. Neurobiol Aging 32(11):2091–2095. https://doi.org/10.1016/j.neurobiolaging.2009.12.004

Barha CK, Lieblich SE, Chow C, Galea LAM (2015) Multiparity-induced enhancement of hippocampal neurogenesis and spatial memory depends on ovarian hormone status in middle age. Neurobiol Aging 36(8):2391–2405. https://doi.org/10.1016/j.neurobiolaging.2015.04.007

Barha CK, Hanna CW, Salvante KG, Wilson SL, Robinson WP, Altman RM, Nepomnaschy PA (2016) Number of children and telomere length in women: a prospective, longitudinal evaluation. PLoS One 11(1):e0146424. https://doi.org/10.1371/journal.pone.0146424

Bodensteiner KJ, Cain P, Ray AS, Hamula LA (2006) Effects of pregnancy on spatial cognition in female Hooded Long–Evans rats. Horm Behav 49(3):303–314. https://doi.org/10.1016/j.yhbeh.2005.08.002

Boldrini M, Fulmore CA, Tartt AN, Simeon LR, Pavlova I, Poposka V, Rosoklija GB, Stankov A, Arango V, Dwork AJ, Hen R, Mann JJ (2018) Human hippocampal neurogenesis persists throughout aging. Cell Stem Cell 22(4):589–599.e5. https://doi.org/10.1016/j.stem.2018.03.015

Brailoiu E, Dun SL, Brailoiu GC, Mizuo K, Sklar LA, Oprea TI, Prossnitz ER, Dun NJ (2007) Distribution and characterization of estrogen receptor G protein-coupled receptor 30 in the rat central nervous system. J Endocrinol 193(2):311–321. https://doi.org/10.1677/JOE-07-0017

Brandt MD, Jessberger S, Steiner B, Kronenberg G, Reuter K, Bick-Sander A, von der Behrens W, Kempermann G (2003) Transient calretinin expression defines early postmitotic step of neuronal differentiation in adult hippocampal neurogenesis of mice. Mol Cell Neurosci 24(3):603–613. https://doi.org/10.1016/S1044-7431(03)00207-0

Brett M (2001) Motherhood and memory: a review. Psychoneuroendocrinology 26(4):339–362. https://doi.org/10.1016/S0306-4530(01)00003-8

Bridges RS (2016) Long-term alterations in neural and endocrine processes induced by motherhood in mammals. Horm Behav 77:193–203. https://doi.org/10.1016/j.yhbeh.2015.09.001

Brown JP, Couillard-Després S, Cooper-Kuhn CM, Winkler J, Aigner L, Kuhn HG (2003) Transient expression of doublecortin during adult neurogenesis: doublecortin expression in adult neurogenesis. J Comp Neurol 467(1):1–10. https://doi.org/10.1002/cne.10874

Brunton PJ, Russell JA (2010) Endocrine induced changes in brain function during pregnancy. Brain Res 1364:198–215. https://doi.org/10.1016/j.brainres.2010.09.062

Brus M, Meurisse M, Franceschini I, Keller M, Lévy F (2010) Evidence for cell proliferation in the sheep brain and its down-regulation by parturition and interactions with the young. Horm Behav 58(5):737–746. https://doi.org/10.1016/j.yhbeh.2010.07.006

Buckwalter JG, Stanczyk FZ, McCleary CA, Bluestein BW, Buckwalter DK, Rankin KP, Chang L, Goodwin TM (1999) Pregnancy, the postpartum, and steroid hormones: effects on cognition and mood. Psychoneuroendocrinology 24(1):69–84. https://doi.org/10.1016/S0306-4530(98)00044-4

Buckwalter JG, Buckwalter DK, Bluestein BW, Stanczyk FZ (2001) Chapter 22 Pregnancy and postpartum: changes in cognition and mood. In: Progress in brain research, vol 133. Elsevier, pp 303–319. https://doi.org/10.1016/S0079-6123(01)33023-6

Burgmans S, van Boxtel MPJ, van den Berg KEM, Gronenschild EHBM, Jacobs HIL, Jolles J, Uylings HBM (2011) The posterior parahippocampal gyrus is preferentially affected in age-related memory decline. Neurobiol Aging 32(9):1572–1578. https://doi.org/10.1016/j.neurobiolaging.2009.09.008

Byrnes EM, Bridges RS (2006) Reproductive experience alters anxiety-like behavior in the female rat. Horm Behav 50(1):70–76. https://doi.org/10.1016/j.yhbeh.2006.01.006

Calakos N, Scheller RH (1994) Vesicle-associated membrane protein and synaptophysin are associated on the synaptic vesicle. J Biol Chem 269(40):24534–24537. https://doi.org/10.1016/S0021-9258(17)31422-9

Calcia MA, Bonsall DR, Bloomfield PS, Selvaraj S, Barichello T, Howes OD (2016) Stress and neuroinflammation: a systematic review of the effects of stress on microglia and the implications for mental illness. Psychopharmacology 233(9):1637–1650. https://doi.org/10.1007/s00213-016-4218-9

Callaghan B, McCormack C, Tottenham N, Monk C (2022) Evidence for cognitive plasticity during pregnancy via enhanced learning and memory. Memory (Hove, England) 30(5):519–536. https://doi.org/10.1080/09658211.2021.2019280

Carmona S, Martínez-García M, Paternina-Die M, Barba-Müller E, Wierenga LM, Alemán-Gómez Y, Pretus C, Marcos-Vidal L, Beumala L, Cortizo R, Pozzobon C, Picado M, Lucco E, García García D, Soliva JC, Tobeña A, Peper JS, Crone EA, Ballesteros A et al (2019) Pregnancy and adolescence entail similar neuroanatomical adaptations: a comparative analysis of cerebral morphometric changes. Hum Brain Mapp 40(7):2143–2152. https://doi.org/10.1002/hbm.24513

Casey P (2000) A longitudinal study of cognitive performance during pregnancy and new motherhood. Arch Womens Ment Health 3(2):65–76. https://doi.org/10.1007/s007370070008

Casey P, Huntsdale C, Angus G, Janes C (1999) Memory in pregnancy. II: implicit, incidental, explicit, semantic, short-term, working and prospective memory in primigravid, multigravid and postpartum women. J Psychosom Obstet Gynecol 20(3):158–164. https://doi.org/10.3109/01674829909075590

Chan WFN, Gurnot C, Montine TJ, Sonnen JA, Guthrie KA, Nelson JL (2012) Male microchimerism in the human female brain. PLoS One 7(9):e45592. https://doi.org/10.1371/journal.pone.0045592

Chao HM, Choo PH, McEwen BS (1989) Glucocorticoid and mineralocorticoid receptor mRNA expression in rat brain. Neuroendocrinology 50(4):365–371. https://doi.org/10.1159/000125250

Chico E, Gonzalez A, Ali N, Steiner M, Fleming AS (2014) Executive function and mothering: challenges faced by teenage mothers: executive Function and Parenting. Dev Psychobiol 56(5):1027–1035. https://doi.org/10.1002/dev.21185

Cipriani S, Ferrer I, Aronica E, Kovacs GG, Verney C, Nardelli J, Khung S, Delezoide A-L, Milenkovic I, Rasika S, Manivet P, Benifla J-L, Deriot N, Gressens P, Adle-Biassette H (2018) Hippocampal radial glial subtypes and their neurogenic potential in human fetuses and healthy and Alzheimer's disease adults. Cereb Cortex 28(7):2458–2478. https://doi.org/10.1093/cercor/bhy096

Cui J, Jothishankar B, He P, Staufenbiel M, Shen Y, Li R (2014) Amyloid precursor protein mutation disrupts reproductive experience-enhanced normal cognitive development in a mouse model of Alzheimer's disease. Mol Neurobiol 49(1):103–112. https://doi.org/10.1007/s12035-013-8503-x

Cuttler C, Graf P, Pawluski JL, Galea LAM (2011) Everyday life memory deficits in pregnant women. Can J Exp Psychol = Revue Canadienne De Psychologie Experimentale 65(1):27–37. https://doi.org/10.1037/a0022844

Darnaudéry M, Perez-Martin M, Del Favero F, Gomez-Roldan C, Garcia-Segura LM, Maccari S (2007) Early motherhood in rats is associated with a modification of hippocampal function. Psychoneuroendocrinology 32(7):803–812. https://doi.org/10.1016/j.psyneuen.2007.05.012

Davies SJ, Lum JA, Skouteris H, Byrne LK, Hayden MJ (2018) Cognitive impairment during pregnancy: a meta-analysis. Med J Aust 208(1):35–40. https://doi.org/10.5694/mja17.00131

de Lange A-MG, Kaufmann T, van der Meer D, Maglanoc LA, Alnæs D, Moberget T, Douaud G, Andreassen OA, Westlye LT (2019) Population-based neuroimaging reveals traces of childbirth in the maternal brain. Proc Natl Acad Sci U S A 116(44):22341–22346. https://doi.org/10.1073/pnas.1910666116

Duarte-Guterman P, Leuner B, Galea LAM (2019) The long and short term effects of motherhood on the brain. Front Neuroendocrinol 53:100740. https://doi.org/10.1016/j.yfrne.2019.02.004

Duarte-Guterman P, Richard JE, Lieblich SE, Eid, RS, Lamers, Y, Galea LAM (2023) Long-term cellular and molecular signatures of pregnancy in the adult and ageing brain. bio Rxiv 2023.02.24.529879. doi:https://doi.org/10.1101/2023.02.24.529879

Eid RS, Chaiton JA, Lieblich SE, Bodnar TS, Weinberg J, Galea LAM (2019) Early and late effects of maternal experience on hippocampal neurogenesis, microglia, and the circulating cytokine milieu. Neurobiol Aging 78:1–17. https://doi.org/10.1016/j.neurobiolaging.2019.01.021

Encinas JM, Vaahtokari A, Enikolopov G (2006) Fluoxetine targets early progenitor cells in the adult brain. Proc Natl Acad Sci U S A 103(21):8233–8238. https://doi.org/10.1073/pnas.0601992103

Enninga EAL, Nevala WK, Creedon DJ, Markovic SN, Holtan SG (2015) Fetal sex-based differences in maternal hormones, angiogenic factors, and immune mediators during pregnancy and the postpartum period. Am J Reprod Immunol 73(3):251–262. https://doi.org/10.1111/aji.12303

Eriksson PS, Perfilieva E, Björk-Eriksson T, Alborn A-M, Nordborg C, Peterson DA, Gage FH (1998) Neurogenesis in the adult human hippocampus. Nat Med 4(11):1313–1317. https://doi.org/10.1038/3305

Filippov V, Kronenberg G, Pivneva T, Reuter K, Steiner B, Wang L-P, Yamaguchi M, Kettenmann H, Kempermann G (2003) Subpopulation of nestin-expressing progenitor cells in the adult murine hippocampus shows electrophysiological and morphological characteristics of astrocytes. Mol Cell Neurosci 23(3):373–382. https://doi.org/10.1016/S1044-7431(03)00060-5

Fleming AS, Kuchera C, Lee A, Winocur G (1994) Olfactory-based social learning varies as a function of parity in female rats. Psychobiology 22(1):37–43. https://doi.org/10.3758/BF03327078

Frankland PW, Cestari V, Filipkowski RK, McDonald RJ, Silva AJ (1998) The dorsal hippocampus is essential for context discrimination but not for contextual conditioning. Behav Neurosci 112(4):863–874. https://doi.org/10.1037//0735-7044.112.4.863

Fransson P, Marrelec G (2008) The precuneus/posterior cingulate cortex plays a pivotal role in the default mode network: evidence from a partial correlation network analysis. NeuroImage 42(3):1178–1184. https://doi.org/10.1016/j.neuroimage.2008.05.059

Fukuda S, Kato F, Tozuka Y, Yamaguchi M, Miyamoto Y, Hisatsune T (2003) Two distinct subpopulations of nestin-positive cells in adult mouse dentate gyrus. J Neurosci 23(28):9357–9366. https://doi.org/10.1523/JNEUROSCI.23-28-09357.2003

Furuta M, Bridges RS (2005) Gestation-induced cell proliferation in the rat brain. Brain Res Dev Brain Res 156(1):61–66. https://doi.org/10.1016/j.devbrainres.2005.01.008

Galea LAM, McEwen BS (1999) Sex and seasonal changes in the rate of cell proliferation in the dentate gyrus of adult wild meadow voles. Neuroscience 89(3):955–964. https://doi.org/10.1016/S0306-4522(98)00345-5

Galea LAM, Ormerod BK, Sampath S, Kostaras X, Wilkie DM, Phelps MT (2000) Spatial working memory and hippocampal size across pregnancy in rats. Horm Behav 37(1):86–95. https://doi.org/10.1006/hbeh.1999.1560

Galea LAM, Frick KM, Hampson E, Sohrabji F, Choleris E (2017) Why estrogens matter for behavior and brain health. Neurosci Biobehav Rev 76:363–379. https://doi.org/10.1016/j.neubiorev.2016.03.024

Galea LAM, Qiu W, Duarte-Guterman P (2018a) Beyond sex differences: short and long-term implications of motherhood on Women's Health. Curr Opin Physio 6:82–88. https://doi.org/10.1016/j.cophys.2018.06.003

Galea LAM, Roes MM, Dimech CJ, Chow C, Mahmoud R, Lieblich SE, Duarte-Guterman P (2018b) Premarin has opposing effects on spatial learning, neural activation, and serum cytokine levels in middle-aged female rats depending on reproductive history. Neurobiol Aging 70:291–307. https://doi.org/10.1016/j.neurobiolaging.2018.06.030

Gammill HS, Harrington WE (2017) Microchimerism: defining and redefining the prepregnancy context – a review. Placenta 60:130–133. https://doi.org/10.1016/j.placenta.2017.08.071

Gatewood JD, Morgan MD, Eaton M, McNamara IM, Stevens LF, Macbeth AH, Meyer EAA, Lomas LM, Kozub FJ, Lambert KG, Kinsley CH (2005) Motherhood mitigates aging-related decrements in learning and memory and positively affects brain aging in the rat. Brain Res Bull 66(2):91–98. https://doi.org/10.1016/j.brainresbull.2005.03.016

Geier CF, Terwilliger R, Teslovich T, Velanova K, Luna B (2010) Immaturities in reward processing and its influence on inhibitory control in adolescence. Cereb Cortex 20(7):1613–1629. https://doi.org/10.1093/cercor/bhp225

Glasper ER, Kozorovitskiy Y, Pavlic A, Gould E (2011) Paternal experience suppresses adult neurogenesis without altering hippocampal function in Peromyscus californicus. J Comp Neurol 519(11):2271–2281. https://doi.org/10.1002/cne.22628

Glynn LM (2010) Giving birth to a new brain: hormone exposures of pregnancy influence human memory. Psychoneuroendocrinology 35(8):1148–1155. https://doi.org/10.1016/j.psyneuen.2010.01.015

Good M, Honey RC (1991) Conditioning and contextual retrieval in hippocampal rats. Behav Neurosci 105(4):499–509. https://doi.org/10.1037/0735-7044.105.4.499

Green AD, Galea LAM (2008) Adult hippocampal cell proliferation is suppressed with estrogen withdrawal after a hormone-simulated pregnancy. Horm Behav 54(1):203–211. https://doi.org/10.1016/j.yhbeh.2008.02.023

Hampson E, Phillips S-D, Duff-Canning SJ, Evans KL, Merrill M, Pinsonneault JK, Sadée W, Soares CN, Steiner M (2015) Working memory in pregnant women: relation to estrogen and antepartum depression. Horm Behav 74:218–227. https://doi.org/10.1016/j.yhbeh.2015.07.006

Hazell GGJ, Yao ST, Roper JA, Prossnitz ER, O'Carroll A-M, Lolait SJ (2009) Localisation of GPR30, a novel G protein-coupled oestrogen receptor, suggests multiple functions in rodent brain and peripheral tissues. J Endocrinol 202(2):223–236. https://doi.org/10.1677/JOE-09-0066

Henry JD, Rendell PG (2007) A review of the impact of pregnancy on memory function. J Clin Exp Neuropsychol 29(8):793–803. https://doi.org/10.1080/13803390701612209

Henry JF, Sherwin BB (2012) Hormones and cognitive functioning during late pregnancy and postpartum: a longitudinal study. Behav Neurosci 126(1):73–85. https://doi.org/10.1037/a0025540

Hillerer KM, Neumann ID, Couillard-Despres S, Aigner L, Slattery DA (2014) Lactation-induced reduction in hippocampal neurogenesis is reversed by repeated stress exposure: stress effects on hippocampal neurogenesis in lactating rats. Hippocampus 24(6):673–683. https://doi.org/10.1002/hipo.22258

Hoekzema E, Barba-Müller E, Pozzobon C, Picado M, Lucco F, García-García D, Soliva JC, Tobeña A, Desco M, Crone EA, Ballesteros A, Carmona S, Vilarroya O (2017) Pregnancy leads to long-lasting changes in human brain structure. Nat Neurosci 20(2):287–296. https://doi.org/10.1038/nn.4458

Hoekzema E, Tamnes CK, Berns P, Barba-Müller E, Pozzobon C, Picado M, Lucco F, Martínez-García M, Desco M, Ballesteros A, Crone EA, Vilarroya O, Carmona S (2020) Becoming a mother entails anatomical changes in the ventral striatum of the human brain that facilitate its responsiveness to offspring cues. Psychoneuroendocrinology 112:104507. https://doi.org/10.1016/j.psyneuen.2019.104507

Hoekzema E, Van Steenbergen H, Straathof M, Beekmans A, Freund IM, Pouwels PJW, Crone EA (2022) Mapping the effects of pregnancy on resting state brain activity, white matter microstructure, neural metabolite concentrations and grey matter architecture. Nat Commun 13(1):6931. https://doi.org/10.1038/s41467-022-33884-8

Holschbach MA, Lonstein JS (2017) Motherhood and infant contact regulate neuroplasticity in the serotonergic midbrain dorsal raphe. Psychoneuroendocrinology 76:97–106. https://doi.org/10.1016/j.psyneuen.2016.10.023

Janes C, Casey P, Huntsdale C, Angus G (1999) Memory in pregnancy. I: subjective experiences and objective assessment of implicit, explicit and working memory in primigravid and primiparous women. J Psychosom Obstet Gynaecol 20(2):80–87. https://doi.org/10.3109/01674829909075580

Karim R, Dang H, Henderson VW, Hodis HN, St. John, J., Brinton, R. D., & Mack, W. J. (2016) Effect of reproductive history and exogenous hormone use on cognitive function in mid- and late life. J Am Geriatr Soc 64(12):2448–2456. https://doi.org/10.1111/jgs.14658

Kataja E-L, Karlsson L, Huizink AC, Tolvanen M, Parsons C, Nolvi S, Karlsson H (2017) Pregnancy-related anxiety and depressive symptoms are associated with visuospatial working memory errors during pregnancy. J Affect Disord 218:66–74. https://doi.org/10.1016/j.jad.2017.04.033

Kee N, Sivalingam S, Boonstra R, Wojtowicz JM (2002) The utility of Ki-67 and BrdU as proliferative markers of adult neurogenesis. J Neurosci Methods 115(1):97–105. https://doi.org/10.1016/s0165-0270(02)00007-9

Keenan PA, Yaldoo DT, Stress ME, Fuerst DR, Ginsburg KA (1998) Explicit memory in pregnant women. Am J Obstet Gynecol 179(3 Pt 1):731–737. https://doi.org/10.1016/s0002-9378(98)70073-0

Kempermann G, Gage FH, Aigner L, Song H, Curtis MA, Thuret S, Kuhn HG, Jessberger S, Frankland PW, Cameron HA, Gould E, Hen R, Abrous DN, Toni N, Schinder AF, Zhao X, Lucassen PJ, Frisén J (2018) Human adult neurogenesis: evidence and remaining questions. Cell Stem Cell 23(1):25–30. https://doi.org/10.1016/j.stem.2018.04.004

Kim P, Leckman JF, Mayes LC, Feldman R, Wang X, Swain JE (2010) The plasticity of human maternal brain: Longitudinal changes in brain anatomy during the early postpartum period. Behav Neurosci 124(5):695–700. https://doi.org/10.1037/a0020884

Kim P, Ruth F, Linda CM, Virginia E, Nancy T, James FL, James ES (2011) Breastfeeding brain activation to own infant cry and maternal sensitivity. J Child Psychol Psychiatry 52(8) 907–915. https://doi.org/10.1111/j.1469-7610.2011.02406.x

Kim P, Rigo P, Mayes LC, Feldman R, Leckman JF, Swain JE (2014) Neural plasticity in fathers of human infants. Soc Neurosci 9(5):522–535. https://doi.org/10.1080/17470919.2014.933713

Kinsley CH, Madonia L, Gifford GW, Tureski K, Griffin GR, Lowry C, Williams J, Collins J, McLearie H, Lambert KG (1999) Motherhood improves learning and memory. Nature 402(6758):137–138. https://doi.org/10.1038/45957

Kinsley CH, Trainer R, Stafisso-Sandoz G, Quadros P, Marcus LK, Hearon C, Meyer EA, Hester N, Morgan M, Kozub FJ, Lambert KG (2006 Feb) Motherhood and the hormones of pregnancy modify concentrations of hippocampal neuronal dendritic spines. Horm Behav 49(2):131–142. https://doi.org/10.1016/j.yhbeh.2005.05.017

Kinsley CH, Bardi M, Karelina K, Rima B, Christon L, Friedenberg J, Griffin G (2008) Motherhood induces and maintains behavioral and neural plasticity across the lifespan in the rat. Arch Sex Behav 37(1):43–56. https://doi.org/10.1007/s10508-007-9277-x

Klein SL, Flanagan KL (2016) Sex differences in immune responses. Nat Rev Immunol 16(10):626–638. https://doi.org/10.1038/nri.2016.90

Knoth R, Singec I, Ditter M, Pantazis G, Capetian P, Meyer RP, Horvat V, Volk B, Kempermann G (2010) Murine Features of Neurogenesis in the Human Hippocampus across the Lifespan from 0 to 100 Years. PLoS One 5(1):e8809. https://doi.org/10.1371/journal.pone.0008809

Köhler S, Black SE, Sinden M, Szekely C, Kidron D, Parker JL, Foster JK, Moscovitch M, Wincour G, Szalai JP, Bronskill MJ (1998) Memory impairments associated with hippocampal versus parahippocampal-gyrus atrophy: An MR volumetry study in alzheimer's disease. Neuropsychologia 36(9):901–914. https://doi.org/10.1016/s0028-3932(98)00017-7

Ladyman SR, Khant Aung Z, Grattan DR (2018) Impact of pregnancy and lactation on the long-term regulation of energy balance in female mice. Endocrinology 159(6):2324–2336. https://doi.org/10.1210/en.2018-00057

Lambert K, Eisch AJ, Galea LAM, Kempermann G, Merzenich M (2019) Optimizing brain performance: identifying mechanisms of adaptive neurobiological plasticity. Neurosci Biobehav Rev 105:60–71. https://doi.org/10.1016/j.neubiorev.2019.06.033

Lange AG, Barth C, Kaufmann T, Anatürk M, Suri S, Ebmeier KP, Westlye LT (2020) The maternal brain: region-specific patterns of brain aging are traceable decades after childbirth. Hum Brain Mapp 41(16):4718–4729. https://doi.org/10.1002/hbm.25152

Larsen CM, Grattan DR (2010) Prolactin-induced mitogenesis in the subventricular zone of the maternal brain during early pregnancy is essential for normal postpartum behavioral responses in the mother. Endocrinology 151(8):3805–3814. https://doi.org/10.1210/en.2009-1385

Lee BH, Cevizci M, Lieblich, SE, Ibrahim M, Wen Y, Eid RS, Lamers Y, Duarte-Guterman P, Galea LAM (2023) Exploring the parity paradox: differential effects on neuroplasticity and neuroinflammation by APOEe4 genotype at middle-age bioRxiv 2023.07.12.548731. doi:https://doi.org/10.1101/2023.07.12.548731

Lemaire V, Billard J-M, Dutar P, George O, Piazza PV, Epelbaum J, Le Moal M, Mayo W (2006) Motherhood-induced memory improvement persists across lifespan in rats but is abolished by a gestational stress. Eur J Neurosci 23(12):3368–3374. https://doi.org/10.1111/j.1460-9568.2006.04870.x

Leuner B, Sabihi S (2016) The birth of new neurons in the maternal brain: hormonal regulation and functional implications. Front Neuroendocrinol 41:99–113. https://doi.org/10.1016/j.yfrne.2016.02.004

Leuner B, Mirescu C, Noiman L, Gould E (2007) Maternal experience inhibits the production of immature neurons in the hippocampus during the postpartum period through elevations in adrenal steroids. Hippocampus 17(6):434–442. https://doi.org/10.1002/hipo.20278

Leuner B, Glasper ER, Gould E (2010) Parenting and plasticity. Trends Neurosci 33(10):465–473. https://doi.org/10.1016/j.tins.2010.07.003

Lisofsky N, Wiener J, de Condappa O, Gallinat J, Lindenberger U, Kühn S (2016) Differences in navigation performance and postpartal striatal volume associated with pregnancy in humans. Neurobiol Learn Mem 134 Pt B:400–407. https://doi.org/10.1016/j.nlm.2016.08.022

Lisofsky (2019) Postpartal neural plasticity of the maternal brain: early renormalization of pregnancy-related decreases? Neurosignals 27(1):12–24. https://doi.org/10.33594/000000105

Logan DM, Hill KR, Jones R, Holt-Lunstad J, Larson MJ (2014) How do memory and attention change with pregnancy and childbirth? A controlled longitudinal examination of neuropsychological functioning in pregnant and postpartum women. J Clin Exp Neuropsychol 36(5):528–539. https://doi.org/10.1080/13803395.2014.912614

Love G, Torrey N, McNamara I, Morgan M, Banks M, Hester NW, Glasper ER, DeVries AC, Kinsley CH, Lambert KG (2005) Maternal experience produces long-lasting behavioral modifications in the rat. Behav Neurosci 119(4):1084–1096. https://doi.org/10.1037/0735-7044.119.4.1084

Lucassen PJ, Oomen CA, Naninck EFG, Fitzsimons CP, van Dam A-M, Czeh B, Korosi A (2015) Regulation of adult neurogenesis and plasticity by (early) stress, glucocorticoids, and

inflammation. Cold Spring Harb Perspect Biol 7(9):a021303. https://doi.org/10.1101/cshperspect.a021303
Luders E, Gingnell M, Poromaa IS, Engman J, Kurth F, Gaser C (2018) Potential brain age reversal after pregnancy: younger brains at 4–6 weeks postpartum. Neuroscience 386:309–314. https://doi.org/10.1016/j.neuroscience.2018.07.006
Luine VN, Beck KD, Bowman RE, Frankfurt M, Maclusky NJ (2007) Chronic stress and neural function: accounting for sex and age. J Neuroendocrinol 19(10):743–751. https://doi.org/10.1111/j.1365-2826.2007.01594.x. PMID: 17850456
Luna B, Garver KE, Urban TA, Lazar NA, Sweeney JA (2004) Maturation of cognitive processes from late childhood to adulthood. Child Dev 75(5):1357–1372. https://doi.org/10.1111/j.1467-8624.2004.00745.x
Macbeth AH, Gautreaux C, Luine VN (2008a) Pregnant rats show enhanced spatial memory, decreased anxiety, and altered levels of monoaminergic neurotransmitters. Brain Res 1241:136–147. https://doi.org/10.1016/j.brainres.2008.09.006
Macbeth AH, Scharfman HE, MacLusky NJ, Gautreaux C, Luine VN (2008b) Effects of multiparity on recognition memory, monoaminergic neurotransmitters, and brain-derived neurotrophic factor (BDNF). Horm Behav 54(1):7–17. https://doi.org/10.1016/j.yhbeh.2007.08.011
Maguire EA, Burgess N, Donnett JG, Frackowiak RS, Frith CD, O'Keefe J (1998) Knowing where and getting there: a human navigation network. Science (New York, N.Y.) 280(5365):921–924. https://doi.org/10.1126/science.280.5365.921
Mantini D, Perrucci MG, Del Gratta C, Romani GL, Corbetta M (2007) Electrophysiological signatures of resting state networks in the human brain. Proc Natl Acad Sci 104(32):13170–13175. https://doi.org/10.1073/pnas.0700668104
Martínez-García M, Paternina-Die M, Barba-Müller E, Martín De Blas D, Beumala L, Cortizo R, Pozzobon C, Marcos-Vidal L, Fernández-Pena A, Picado M, Belmonte-Padilla E, Massó-Rodriguez A, Ballesteros A, Desco M, Vilarroya Ó, Hoekzema E, Carmona S (2021a) Do pregnancy-induced brain changes reverse? The brain of a mother six years after parturition. Brain Sci 11(2):168. https://doi.org/10.3390/brainsci11020168
Martínez-García M, Paternina-Die M, Desco M, Vilarroya O, Carmona S (2021b) Characterizing the brain structural adaptations across the motherhood transition. Front Glob Women's Health 2:742775. https://doi.org/10.3389/fgwh.2021.742775
Mathews KJ, Allen KM, Boerrigter D, Ball H, Shannon Weickert C, Double KL (2017) Evidence for reduced neurogenesis in the aging human hippocampus despite stable stem cell markers. Aging Cell 16(5):1195–1199. https://doi.org/10.1111/acel.12641
McCormack C, Callaghan BL, Pawluski JL (2023) It's time to rebrand "mommy brain". JAMA Neurol. https://doi.org/10.1001/jamaneurol.2022.5180
McLay RN, Maki PM, Lyketsos CG (2003) Nulliparity and late menopause are associated with decreased cognitive decline. J Neuropsychiatry Clin Neurosci 15(2):161–167. https://doi.org/10.1176/jnp.15.2.161
Roes M, Liisa AMG (2016) Sex Differences in the Central Nervous System, The Maternal Brain: Short- and Long-Term Effects of Reproductive Experience on Hippocampus Structure and Function in Adulthood. Elsevier 197–220
Mifsud KR, Reul JMHM (2016) Acute stress enhances heterodimerization and binding of corticosteroid receptors at glucocorticoid target genes in the hippocampus. Proc Natl Acad Sci U S A 113(40):11336–11341. https://doi.org/10.1073/pnas.1605246113
Mitra SW, Hoskin E, Yudkovitz J, Pear L, Wilkinson HA, Hayashi S, Pfaff DW, Ogawa S, Rohrer SP, Schaeffer JM, McEwen BS, Alves SE (2003) Immunolocalization of estrogen receptor beta in the mouse brain: Comparison with estrogen receptor alpha. Endocrinology 144(5):2055–2067. https://doi.org/10.1210/en.2002-221069
Moreno-Jiménez EP, Flor-García M, Terreros-Roncal J, Rábano A, Cafini F, Pallas-Bazarra N, Ávila J, Llorens-Martín M (2019) Adult hippocampal neurogenesis is abundant in neurologically healthy subjects and drops sharply in patients with Alzheimer's disease. Nat Med 25(4):554–560. https://doi.org/10.1038/s41591-019-0375-9

Morris RGM, Garrud P, Rawlins JNP, O'Keefe J (1982) Place navigation impaired in rats with hippocampal lesions. Nature 297(5868):681–683. https://doi.org/10.1038/297681a0

Ning K, Zhao L, Franklin M, Matloff W, Batta I, Arzouni N, Sun F, Toga AW (2020) Parity is associated with cognitive function and brain age in both females and males. Sci Rep 10(6100). https://doi.org/10.1038/s41598-020-63014-7

Oatridge A, Holdcroft A, Saeed N, Hajnal JV, Puri BK, Fusi L, Bydder GM (2002) Change in brain size during and after pregnancy: study in healthy women and women with preeclampsia. AJNR Am J Neuroradiol 23(1):19–26

Orchard ER, Ward PGD, Egan GF, Jamadar SD (2022) Evidence of subjective, but not objective, cognitive deficit in new mothers at 1-year postpartum. J Women's Health 31(8):1087–1096. https://doi.org/10.1089/jwh.2021.0441

Orchard ER, Rutherford HJV, Holmes AJ, Jamadar SD (2023) Matrescence: lifetime impact of motherhood on cognition and the brain. Trends Cogn Sci 27(3):302–316. https://doi.org/10.1016/j.tics.2022.12.002

Orchard ER, Ward PG, Sforazzini F, Storey E, Egan GF, Jamadar SD (2020) Relationship between parenthood and cortical thickness in late adulthood. PLOS ONE 15(7). https://doi.org/10.1371/journal.pone.0236031

Ouellette SJ, Hampson E (2019) Memory and affective changes during the antepartum: a narrative review and integrative hypothesis. J Clin Exp Neuropsychol 41(1):87–107. https://doi.org/10.1080/13803395.2018.1485881

Paternina-Die M, Martinez-Garcia M, Pretus C, Hoekzema E, Barba-Muller E, Martin de Blas D, Pozzobon C, Ballesteros A, Vilarroya O, Desco M, Carmona S (2020) The paternal transition entails neuroanatomic adaptations that are associated with the fathers brain response to his infant cues. Cereb Cortex Commun 1:1–11. https://doi.org/10.1093/texcom/tgaa082

Pawluski JL, Galea LAM (2005) Hippocampal morphology is differentially affected by reproductive experience in the mother. J Neurobiol 66(1):71–81. https://doi.org/10.1002/neu.20194

Pawluski JL, Galea LAM (2007) Reproductive experience alters hippocampal neurogenesis during the postpartum period in the dam. Neuroscience 149(1):53–67. https://doi.org/10.1016/j.neuroscience.2007.07.031

Pawluski JL, Vanderbyl BL, Ragan K, Galea LAM (2006) First reproductive experience persistently affects spatial reference and working memory in the mother and these effects are not due to pregnancy or 'mothering' alone. Behav Brain Res 175(1):157–165. https://doi.org/10.1016/j.bbr.2006.08.017

Pawluski JL, Charlier TD, Lieblich SE, Hammond GL, Galea LAM (2009) Reproductive experience alters corticosterone and CBG levels in the rat dam. Physiol Behav 96(1):108–114. https://doi.org/10.1016/j.physbeh.2008.09.004

Pawluski JL, Barakauskas VE, Galea LAM (2010) Pregnancy decreases oestrogen receptor α expression and pyknosis, but not cell proliferation or survival, in the hippocampus. J Neuroendocrinol 22(4):248–257. https://doi.org/10.1111/j.1365-2826.2010.01960.x

Pawluski JL, Valença A, Santos AIM, Costa-Nunes JP, Steinbusch HWM, Strekalova T (2012) Pregnancy or stress decrease complexity of CA3 pyramidal neurons in the hippocampus of adult female rats. Neuroscience 227:201–210. https://doi.org/10.1016/j.neuroscience.2012.09.059

Pawluski JL, Császár E, Savage E, Martinez-Claros M, Steinbusch HWM, van den Hove D (2015) Effects of stress early in gestation on hippocampal neurogenesis and glucocorticoid receptor density in pregnant rats. Neuroscience 290:379–388. https://doi.org/10.1016/j.neuroscience.2015.01.048

Pawluski JL, Paravatou R, Even A, Cobraiville G, Fillet M, Kokras N, Dalla C, Charlier TD (2020) Effect of sertraline on central serotonin and hippocampal plasticity in pregnant and non-pregnant rats. Neuropharmacology 166:107950. https://doi.org/10.1016/j.neuropharm.2020.107950

Puri TA, Richard JE, Galea LAM (2023 Jun) Beyond sex differences: short- and long-term effects of pregnancy on the brain. Trends Neurosci 46(6):459–471. https://doi.org/10.1016/j.tins.2023.03.010

Qiu W, Hodges TE, Clark EL, Blankers SA, Galea LAM (2020) Perinatal depression: heterogeneity of disease and in animal models. Front Neuroendocrinol 59:100854. https://doi.org/10.1016/j.yfrne.2020.100854

Read SL, Grundy EMD (2017) Fertility history and cognition in later life. J Gerontol B Psychol Sci Soc Sci 72(6):1021–1031. https://doi.org/10.1093/geronb/gbw013

Rolls A, Schori H, London A, Schwartz M (2008) Decrease in hippocampal neurogenesis during pregnancy: a link to immunity. Mol Psychiatry 13(5):468–469. https://doi.org/10.1038/sj.mp.4002126

Ryan J, Carrière I, Scali J, Ritchie K, Ancelin M-L (2009) Life-time estrogen exposure and cognitive functioning in later life. Psychoneuroendocrinology 34(2):287–298. https://doi.org/10.1016/j.psyneuen.2008.09.008

Saltzman W, Ziegler TE (2014) Functional significance of hormonal changes in mammalian fathers. J Neuroendocrinol 26(10):685–696

Shingo T, Gregg C, Enwere E, Fujikawa H, Hassam R, Geary C, Cross JC, Weiss S (2003) Pregnancy-stimulated neurogenesis in the adult female forebrain mediated by prolactin. Science 299(5603):117–120. https://doi.org/10.1126/science.1076647

Sorrells SF, Paredes MF, Cebrian-Silla A, Sandoval K, Qi D, Kelley KW, James D, Mayer S, Chang J, Auguste KI, Chang EF, Gutierrez AJ, Kriegstein AR, Mathern GW, Oldham MC, Huang EJ, Garcia-Verdugo JM, Yang Z, Alvarez-Buylla A (2018) Human hippocampal neurogenesis drops sharply in children to undetectable levels in adults. Nature 555(7696):377–381. https://doi.org/10.1038/nature25975

Spalding KL, Bergmann O, Alkass K, Bernard S, Salehpour M, Huttner HB, Boström E, Westerlund I, Vial C, Buchholz BA, Possnert G, Mash DC, Druid H, Frisén J (2013) Dynamics of hippocampal neurogenesis in adult humans. Cell 153(6):1219–1227. https://doi.org/10.1016/j.cell.2013.05.002

Spritzer MD, Fox EC, Larsen GD, Batson CG, Wagner BA, Maher J (2013) Testosterone influences spatial strategy preferences among adult male rats. Horm Behav 63(5):800–812. https://doi.org/10.1016/j.yhbeh.2013.03.018

Steiner B, Klempin F, Wang L, Kott M, Kettenmann H, Kempermann G (2006) Type-2 cells as link between glial and neuronal lineage in adult hippocampal neurogenesis. Glia 54(8):805–814. https://doi.org/10.1002/glia.20407

Taupin P (2007) BrdU immunohistochemistry for studying adult neurogenesis: paradigms, pitfalls, limitations, and validation. Brain Res Rev 53(1):198–214. https://doi.org/10.1016/j.brainresrev.2006.08.002

Tobin MK, Musaraca K, Disouky A, Shetti A, Bheri A, Honer WG, Kim N, Dawe RJ, Bennett DA, Arfanakis K, Lazarov O (2019) Human hippocampal neurogenesis persists in aged adults and Alzheimer's disease patients. Cell Stem Cell 24(6):974–982.e3. https://doi.org/10.1016/j.stem.2019.05.003

Tomizawa K, Iga N, Lu Y-F, Moriwaki A, Matsushita M, Li S-T, Miyamoto O, Itano T, Matsui H (2003) Oxytocin improves long-lasting spatial memory during motherhood through MAP kinase cascade. Nat Neurosci 6(4):384–390. https://doi.org/10.1038/nn1023

Tronson NC, Collette KM (2017) (Putative) sex differences in neuroimmune modulation of memory. J Neurosci Res 95(1–2):472–486. https://doi.org/10.1002/jnr.23921

Tu MT, Lupien SJ, Walker C-D (2005) Measuring stress responses in postpartum mothers: perspectives from studies in human and animal populations. Stress 8(1):19–34. https://doi.org/10.1080/10253890500103806

Vanston CM, Watson NV (2005) Selective and persistent effect of foetal sex on cognition in pregnant women. Neuroreport 16(7):779–782. https://doi.org/10.1097/00001756-200505120-00024

Vargha-Khadem F, Gadian DG, Watkins KE, Connelly A, Van Paesschen W, Mishkin M (1997) Differential effects of early hippocampal pathology on episodic and semantic memory. Science (New York, N.Y.) 277(5324):376–380. https://doi.org/10.1126/science.277.5324.376

Vorhees CV, Williams MT (2006) Morris water maze: procedures for assessing spatial and related forms of learning and memory. Nat Protoc 1(2):848–858. https://doi.org/10.1038/nprot.2006.116

Vorhees CV, Williams MT (2014) Assessing spatial learning and memory in rodents. ILAR J 55(2):310–332. https://doi.org/10.1093/ilar/ilu013

Wang C, Wang L, Gu Y (2021) Microglia, synaptic dynamics and forgetting. Brain Res Bull 174:173–183. https://doi.org/10.1016/j.brainresbull.2021.06.005

Wartella J, Amory E, Macbeth A, McNamara I, Stevens L, Lambert KG, Kinsley CH (2003) Single or multiple reproductive experiences attenuate neurobehavioral stress and fear responses in the female rat. Physiol Behav 79(3):373–381. https://doi.org/10.1016/S0031-9384(03)00150-1

Williams CL, Barnett AM, Meck WH (1990) Organizational effects of early gonadal secretions on sexual differentiation in spatial memory. Behav Neurosci 104(1):84–97. https://doi.org/10.1037/0735-7044.104.1.84

Woods DL, Kishiyama MM, Yund EW, Herron TJ, Edwards B, Poliva O, Hink RF, Reed B (2011) Improving digit span assessment of short-term verbal memory. J Clin Exp Neuropsychol 33(1):101–111. https://doi.org/10.1080/13803395.2010.493149

Workman JL, Crozier T, Lieblich SE, Galea LAM (2013) Reproductive experience does not persistently alter prefrontal cortical-dependent learning but does alter strategy use dependent on estrous phase. Horm Behav 64(3):439–447. https://doi.org/10.1016/j.yhbeh.2013.06.005

Yagi S, Splinter JEJ, Tai D, Wong S, Wen Y, Galea LAM (2020) Sex differences in maturation and attrition of adult neurogenesis in the hippocampus. Eneuro 7(4):ENEURO.0468-19.2020. https://doi.org/10.1523/ENEURO.0468-19.2020

Zeng XX, Tan KH, Yeo A, Sasajala P, Tan X, Xiao ZC, Dawe G, Udolph G (2010) Pregnancy-associated progenitor cells differentiate and mature into neurons in the maternal brain. Stem Cells Dev 19(12):1819–1830. https://doi.org/10.1089/scd.2010.0046

Zheng J-X, Ge L, Chen H, Yin X, Chen Y-C, Tang W-W (2020) Disruption within brain default mode network in postpartum women without depression. Medicine 99(18):e20045. https://doi.org/10.1097/MD.0000000000020045

10 Neuroendocrinology of Perinatal Mental Illness

Jodi Pawluski, Molly Dickens, and Jamie Maguire

Abstract

Mental illness during pregnancy and the postpartum period is a growing health crisis. We know that at least 1 in 7 women during the perinatal period will experience a mood or anxiety disorder such as perinatal depression (PND) or general anxiety disorder. These mental illnesses significantly impact the well-being of the mother, as well as how she interacts with her infant and family. Over the past 10 years there has been a growing need to understand the neuroendocrine mechanisms mediating perinatal mental illness. Much of this research has focused on aspects of the hypothalamic-pituitary-adrenal (HPA) and -gonadal (HPG) axes and thus this chapter will focus on these two systems in perinatal mental illness. This chapter will review our current understanding of the neuroendocrinology of perinatal mental illness from research in women and animal models, with a focus on PND and anxiety, but will also consider the limited data we have for other perinatal mental illnesses such as postpartum psychosis and childbirth-related post-traumatic stress disorder. We have much research to do to fully understand the mechanisms behind these disorders and will discuss options for future research. Increased understanding of the unique neuroendocrine profiles associated with perinatal mental illness will allow us to predict risk, develop effective

J. Pawluski (✉)
Univ Rennes, INSERM, EHESP, IRSET (Institut de Recherche en Santé, Environnement et Travail), UMR_S 1085, Rennes, France

M. Dickens
College of Biological Science, University of California at Davis, Davis, CA, USA

J. Maguire
Neuroscience Department, Tufts University School of Medicine, Boston, MA, USA
e-mail: Jamie.Maguire@tufts.edu

P. J. Brunton, D. R. Grattan (eds.), *Neuroendocrine Regulation of Mammalian Pregnancy and Lactation*, Masterclass in Neuroendocrinology 15,
https://doi.org/10.1007/978-3-031-51138-7_10

treatments and, ultimately, prevent suffering for millions of mothers during this critical time in life.

Keywords
Perinatal depression · Anxiety · Mental illness · Stress · Postpartum · Steroid hormones · Hypothalamus

10.1 Introduction

Today it is estimated that 1 in every 7 women will experience a mental illness during pregnancy or the postpartum period (Pawluski et al. 2017; Riecher-Rössler and Steiner 2005). Perinatal depression (**PND**) is probably the most talked about and well-studied of the perinatal mental illnesses, but others include anxiety, childbirth-related post-traumatic stress disorder (CB-PTSD) and postpartum psychosis. The first descriptions of postpartum mental illnesses were by Hippocrates in 400 B.C., where he proposed that the brain is a key source of mental illness in new mothers. Hippocrates predicted that lochial discharge—the fluid that comes from the uterus after birth—if suppressed, could flow to the head and result in agitation, delirium and attacks of mania. With our advances in neuroscience, we know this scenario does not occur, but we are beginning to understand the specific neuroendocrine factors important for pregnancy, birth and lactation are involved in perinatal mental illness. Furthermore, a growing body of literature shows that mental illness during the perinatal period has unique neurophysiological and neurobiological profiles that overlap with the brain systems involved in maternal caregiving behaviors (Pawluski et al. 2017; Riecher-Rössler and Steiner 2005; also see Chap. 8).

Perinatal mental illness has enduring effects on maternal health, mother-infant interactions, family relationships and child development (Pawluski et al. 2017; Meltzer-Brody et al. 2018). Indeed, suicide is the leading cause of death in pregnancy and postpartum women in many countries studied to date (Grigoriadis et al. 2017). Additionally, the financial burden to society that perinatal mental illness causes is staggering: in the United Kingdom the known cost of perinatal mental health is £8.1 billion per year (Bauer et al. 2015). With the significant impact of these mental illnesses on the mother, child and society, assessing risk, implementing preventative measures and improving treatment options need to be the focus of current and future research programs.

The aim of this chapter is to review what we know about the neuroendocrine changes associated with perinatal mood and anxiety disorders (Box 10.1), focusing on the hypothalamic-pituitary-gonadal (**HPG**) and -adrenal (**HPA**) axes (for more on the HPA axis in healthy pregnancy, see Chap. 4). We will discuss how these axes interact and the role of other neuroendocrine factors, such as neuropeptides, on perinatal mental health. Research in clinical settings and animal models will be reviewed. We will also discuss key directions for future research in order to understand the unique nature of these mental illnesses of motherhood. The focus of this chapter will be on mothers: see Box 10.2 for information on perinatal mental illness in fathers.

Box 10.1 Perinatal Mental Illnesses

Neuroendocrine factors are likely to play a role in all of these illnesses, but research has yet to fully understand the mechanisms behind these disorders.

Perinatal Depression (PND)

PND is defined in the 5th Edition of the *Diagnostic and Statistical Manual of Mental Disorders* (DSM-V) as major depression with a perinatal onset. It is estimated to affect at least 13% of perinatal women. Major depression includes at least five of the following symptoms lasting for 2 or more weeks: persistent depressed mood and sadness, diminished pleasure in nearly all activities, changes in sleep patterns, changes in weight, fatigue, restlessness, feelings of worthlessness, poor concentration and reoccurring thoughts of death and suicide. Persistent depressed mood or loss of pleasure must be evident for a major depressive episode diagnosis. It should be noted that the DSM-V does not classify PND as a separate diagnosis and instead requires an episode of major (but not minor) depression with a perinatal onset occurring anywhere from the start of pregnancy through the first 4 weeks following delivery. Four weeks post-delivery is arguably an inappropriate cut off for a diagnosis of PND.

Perinatal Anxiety

At least 10% of women experience elevated rates of anxiety during pregnancy or the postpartum period. Generalized anxiety disorder (GAD) and obsessive-compulsive disorder (OCD) are the most common type of perinatal anxiety disorders. The DSM-V criteria for GAD at any time in life includes the presence of excessive anxiety and worry about a variety of topics, events or activities. This anxiety and worry are difficult to control and occur, more often than not, for at least 6 months and are clearly excessive. OCD at any time in life typically includes the persistent presence of obsessions and compulsions that significantly impact daily life. The DSM-V does not include a separate diagnosis of *perinatal* anxiety disorders and now places OCD under Obsessive-Compulsive and Related Disorders. However, the *International Classification of Diseases*, Tenth Revision (ICD-10) defines OCD as an anxiety disorder.

Childbirth-Related Post-Traumatic Stress Disorder (CB-PTSD)

CB-PTSD affects 6% of all women after birth, 15–19% of women in high-risk situations such as those who have pregnancy complications or preterm birth, and up to 39% of women whose babies die. CB-PTSD is not a separate diagnosis in the DSM-V and shares the same symptomatology as PTSD but with a postpartum onset. The criteria for PTSD, in brief, include intrusive symptoms, avoidance of stimuli related to the trauma, negative effects on cognition and mood and marked arousal or reactivity to the traumatic event which lasts for 1 month, causes clinically significant distress and is not attributable to a substance or other medical condition. CB-PTSD is also highly comorbid with depression. Despite this, PTSD still remains largely unrecognized in maternity services and, unlike depression, is not routinely screened for (Ayers et al. 2018).

(continued)

Box 10.1 (continued)

Postpartum Psychosis

In DSM-V, postpartum psychosis is classified in the "short psychotic disorder" section of the schizophrenia spectrum and other psychotic disorders. In most cases, the onset is rapid and occurs within the first two postpartum weeks. This disorder affects 1 in 500–1000 mothers after giving birth. It is sometimes called puerperal psychosis or postnatal psychosis. Symptoms can include hallucinations, delusions, manic mood, low mood and extreme confusion, to name a few. It is critical that urgent medical care is received for this disorder (Meltzer-Brody et al. 2018).

Box 10.2 Perinatal Mental Illness in Fathers

Fathers and partners are vulnerable to perinatal mental illness, particularly depression and anxiety. A recent meta-analysis, a review of the literature that combines data from many studies on the prevalence of perinatal depression in fathers, reports that 10% of fathers struggle with prenatal depression and 9% of fathers have postnatal depression (or depression in the first year after birth) (Rao et al. 2020). The rate of anxiety in fathers is slightly higher, with a recent meta-analysis using data from over 40,000 participants in 23 studies reporting that 11% of fathers will have an anxiety disorder during the perinatal period. That means that nearly 1 in every 10 fathers struggles with depression or anxiety and many are likely to suffer with both (Leiferman et al. 2021).

Research also indicates that fathers' depressive symptoms spike between 3 and 6 months postpartum and rather than feelings of sadness men may be more irritable, aggressive and hostile. They may use more drugs and alcohol, spend more time alone and be distracted more easily.

Not surprisingly, a couple's relationship suffers with perinatal depression, and if one parent has depression there is a 24–50% chance that the other parent will develop depression as well, which can lead to difficulties in caring for the newborn (Kim and Swain 2007).

Very little research has focused on the neurobiology of perinatal mental illness in human fathers or non-birthing parents. One study reports that lower levels of depressive symptoms, particularly those related to physical symptoms such as sleep and fatigue, are related to a greater increase in gray matter volume at 12–16 weeks postpartum in the father's brain—specifically in neural regions important for parental motivation such as the striatum, amygdala and subgenual cortex (Kim et al. 2014). Other studies have suggested steroid hormones as a potential factor contributing to paternal mental illness, with one recent study showing that the elevated concentrations of testosterone during pregnancy in fathers may be protective against paternal depression (Saxbe et al. 2017). Much more research on fathers, partners and LGBTQ+ parents is needed to understand the neuroendocrine mechanisms related to the emotional well-being of the partner [for review see Glasper et al. (2019)].

10.2 Perinatal Depression (PND)

Perinatal depression (PND) is a serious psychiatric disorder affecting 13–25% of women (Aktar et al. 2019), with the potential to have devastating effects on both the mother and child, increasing suicide risk (Lindahl et al. 2005) and adversely impacting infant behavioral, emotional and cognitive development (Murray and Cooper 1997). Gaining an understanding of the neurobiological mechanisms contributing to PND is a scientific and medical imperative. Several biological mechanisms have been proposed to contribute to PND, including genetic and **epigenetic** mechanisms, neuroendocrine changes, altered neurotransmission, neuroinflammatory changes and network-level dysfunction [for review see Pawluski et al. (2017), Riecher-Rössler and Steiner (2005), Payne and Maguire (2019)]. It is likely that many of these mechanisms are intertwined with environmental factors and potentially act synergistically (Payne and Maguire 2019).

10.2.1 The Hypothalamic-Pituitary-Gonadal (HPG) Axis and Perinatal Depression

The HPG axis is a neuroendocrine system involving hormone-mediated communication between the hypothalamus, pituitary gland and gonads (i.e., ovaries in females) (Fig. 10.1). This axis plays a critical role in the development and regulation of a number of the body's systems, such as the reproductive system. The reproductive steroids, estrogens, progesterone and testosterone, regulate **neurotransmitters** and **neurosteroids** implicated in depression, as well as **neuroplasticity** and a number of other neural functions, but how these reproductive steroids are related to PND in females is not known. The withdrawal of reproductive steroids with the removal of the placenta after birth has long been implicated in *postpartum* depression and this is where the majority of the research in this area has focused.

Although there has been speculation for decades that the decline in circulating levels of estradiol and progesterone after birth contributes to postpartum depression (Fig. 10.2), the clinical literature shows no clear role for these hormones in depression postpartum. For example, there is no reliable evidence that women who are diagnosed with postpartum depression have blunted circulating or salivary concentrations of reproductive hormones or greater or more rapid reductions in these hormones in the blood after birth, compared to postpartum women who are not diagnosed with depression (Schiller et al. 2016). This may be due to the fact that these hormones have many brain region-specific actions and their central actions are not able to be determined by measuring circulating concentrations in blood.

Treatment studies have been carried out to determine whether administration of exogenous estradiol, such as Premarin, during the postpartum period in women with depression or who are at risk for developing depression is beneficial in preventing or decreasing depressive symptoms. These studies have provided some evidence that treatment with exogenous estradiol can protect against the development of postpartum depression or act as a treatment. However, these studies often did not include

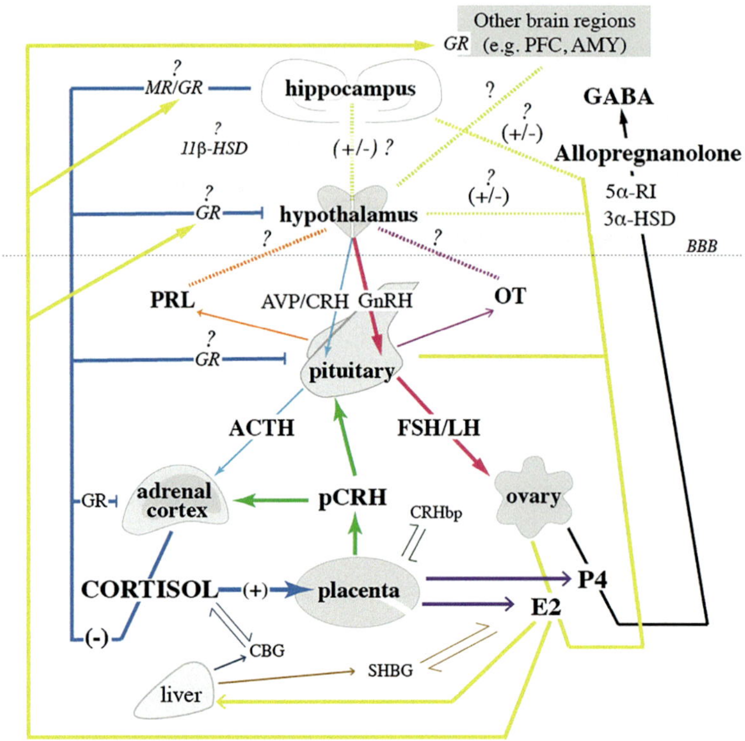

Fig. 10.1 Schematic representation summarizing the key components of the HPA and HPG axes that may play a role in mental health in pregnant humans. Points of potential dysregulation of the HPA/HPG axes related to perinatal mental illnesses that remain to be investigated are represented by question marks and/or dotted lines. How the different components of the HPA and HPG axes within the body interact with each other and with peptide hormones is indicated by arrows. Colors were chosen to match the hormones depicted in Fig. 10.2. *3α-HSD* 3α-hydroxysteroid dehydrogenase, *5α-RI* 5α-reductase, *11β-HSD* 11β-Hydroxysteroid dehydrogenase, *ACTH* adrenocorticotropic hormone, *AMY* amygdala, *AVP* arginine vasopressin, *BBB* blood-brain barrier, *CBG* corticosteroid binding globulin, *CRH* corticotropin-releasing hormone, *CRHbp* CRH-binding protein, *E2* estradiol, *FSH* follicle-stimulating hormone, *GnRH* gonadotropin-releasing hormone, *GR* glucocorticoid receptors, *LH* luteinising hormone, *MR* mineralocorticoid receptors, *OT* oxytocin, *pCRH* placental corticotropin-releasing hormone, *P4* progesterone, *pCRH* placental CRH, *PFC* prefrontal cortex, *PRL* prolactin, *SHBG* sex hormone-binding globulin

a control group, used a heterogeneous sample of women who were sometimes receiving other treatments, and did not control for confounding factors such as stress and sleep deprivation (for review see (Schiller et al. 2016)).

Others have taken a different approach to understanding how estradiol and progesterone may be related to postpartum depression by using a pseudo-pregnancy hormone administration regime in women. In this study, women with a history of postpartum depression and those without a history of postpartum depression were administered exogenous high estradiol and progesterone for 8 weeks, to mimic

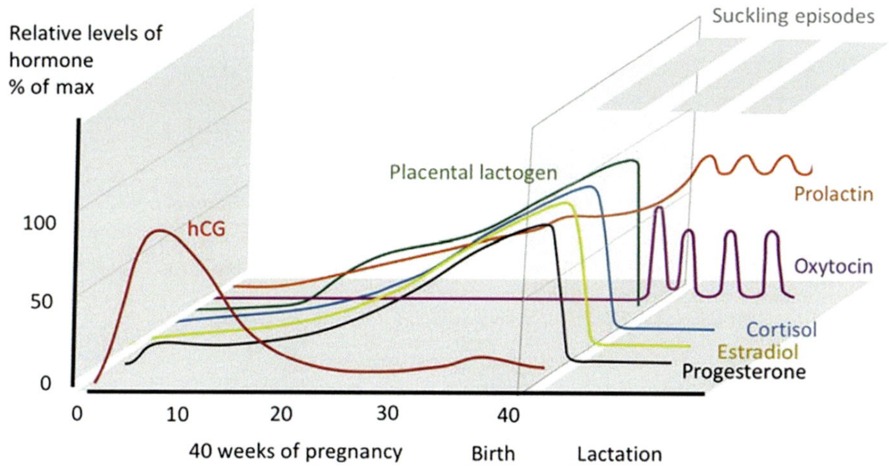

Fig. 10.2 Representation of the hormonal changes during human pregnancy and lactation. There is an abrupt reduction in hormones derived from the placenta [i.e., placental lactogen, estradiol and progesterone; also CRH (not shown)] at birth, in contrast to the activation and suckling-induced stimulation of prolactin and oxytocin. Reprinted from Neurology and Pregnancy Pathophysiology and Patient Care, Vol 171, D.R. Grattan and S.R. Ladyman, Chapter 2—Neurophysiological and cognitive changes in pregnancy, Pages 25-55, Copyright (2020), with permission from Elsevier

hormone levels during pregnancy, and mood measures were taken during and after hormone administration (Bloch et al. 2000). Women with a history of postpartum depression showed increased depressive symptoms during experimental estradiol and progesterone withdrawal, but no effects of hormone withdrawal were evident in the women who never experienced postpartum depression. This study points to a role for these reproductive steroids in postpartum depression, but the small sample size ($n = 8$) used in this study suggests that replication of this research is needed.

More recently, research has turned to a role for allopregnanolone, a neurosteroid derived from progesterone, in postpartum depression. Allopregnanolone is a positive allosteric modulator at $GABA_A$ receptors, meaning it increases the activity of this receptor and by doing so enhances inhibitory GABA signaling (Fig. 10.3). Elegant studies in both humans and animal models have shown that GABA and allopregnanolone are key players in postpartum depression. Lower plasma concentrations of GABA have been reported in women at risk of developing PPD (Deligiannidis et al. 2016). Experimental rodent models have demonstrated that deficits in the expression of neurosteroid-sensitive $GABA_A$ receptors is sufficient to induce abnormal postpartum behaviors and deficits in maternal care (Maguire and Mody 2008). These preclinical studies were the first to demonstrate the therapeutic potential of using GABA-positive allosteric modulators for the treatment of postpartum depression. Based on this work, the Food and Drug Administration (FDA) in the USA approved the drug brexanolone (commercially available as Zulresso™) (Box 10.3), a synthetic allopregnanolone analogue, for the treatment of postpartum depression [for review see Walton and Maguire (2019), Maguire (2019), Meltzer-Brody and Kanes (2020)].

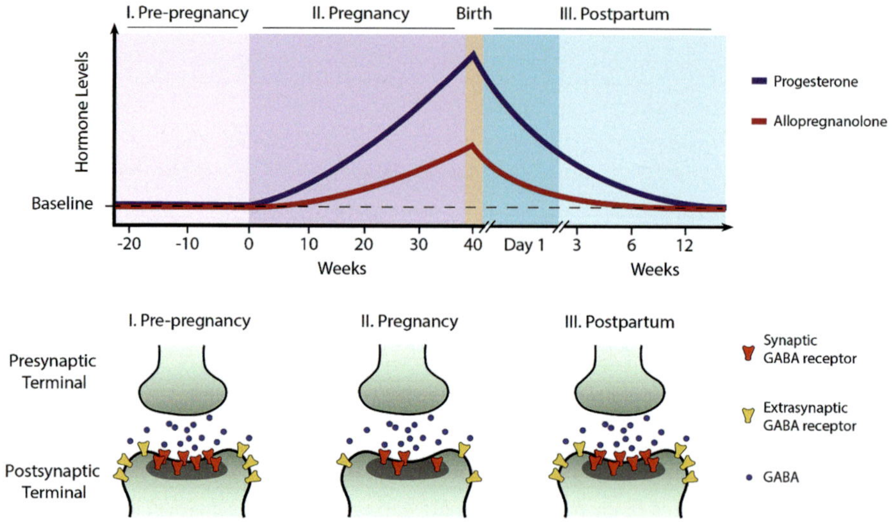

Fig. 10.3 GABAergic signaling in the perinatal period may be altered by fluctuations in allopregnanolone levels and changes in the expression of $GABA_AR$ subunits. Altered expression of $GABA_AR$ subunits could alter both the number and localization of $GABA_ARs$. (Reprinted from Neurobiology of Stress, 3/12, Meltzer-Brody S, Kanes SJ., Allopregnanolone in postpartum depression: Role in pathophysiology and treatment, 100212, Copyright (2020), with permission from Elsevier)

Box 10.3 Treating Perinatal Depression (PND)

Over the years, both psychotherapeutic and pharmacotherapeutic interventions have been recommended for a number of perinatal mental illnesses. Often psychotherapy in combination with pharmacotherapy results in the best outcomes for the parent and child. Pharmacotherapy for perinatal depression and anxiety is common with selective serotonin reuptake inhibitor medications (SSRIs) being prescribed to 3.5% of pregnant women globally (Gemmel et al. 2018; Oberlander et al. 2009). Questions arise as to the safety of using these and other medications during pregnancy and lactation, but recent research in clinical populations and animal models shows that if the medication is effective in alleviating depressed or anxious symptoms during pregnancy and/or breastfeeding this improves the outcomes for the mother and the developing child (Oberlander and Zwaigenbaum 2017; Pawluski et al. 2019).

(continued)

Box 10.3 (continued)
Thus, research is showing that the benefits of perinatal SSRI treatment outweigh the risks in most cases.

Most treatments used to treat perinatal mental illness are those developed to treat these illnesses when they occur outside the perinatal period, and thus are not specific to this reproductive period, which is characterized by a number of different physiological, social and psychological demands. In 2019 the FDA approved the first and only medicine specifically for postpartum depression, Zulresso™, a synthetic allopregnanolone, a naturally occurring metabolite of progesterone that through studies in animal models and clinical trials has shown significant efficacy in the treatment of postpartum depression (Walton and Maguire 2019; Maguire 2019; Meltzer-Brody and Kanes 2020). This medication is not yet readily available but is a first step in the development of a treatment specific for a perinatal mental illness. With continued research the hope is that further treatment options will be made available to alleviate mental illness in both mothers and fathers during the perinatal period.

Research using rodent models to investigate the role of the reproductive hormones estrogen and progesterone in postpartum depression has largely focused on hormone-simulated pregnancy (**HSP**) models of postpartum depression. HSP models essentially induce an artificial pregnancy in a female rodent. To achieve this, the ovaries (the primary source of estradiol and progesterone) are removed by ovariectomy. One week later a low dose of estradiol and a high dose of progesterone is administered for 16 days to simulate the pregnancy hormone profiles of the rat. On days 17–24 a high dose of estradiol is administered alone, to stmulate the increase in estradiol during late pregnancy, after which time hormone treatment is withdrawn, mimicking the decline in sex steroid secretion observed at birth (Galea et al. 2001). This model increases depressive-like behaviors during the simulated 'postpartum' period, as measured using the **forced swim test** (Box 10.4), in some, but not all studies (Galea et al. 2001; Suda et al. 2008).

Box 10.4 Common Tests of Depressive- and Anxiety-like Behaviors in Rodents

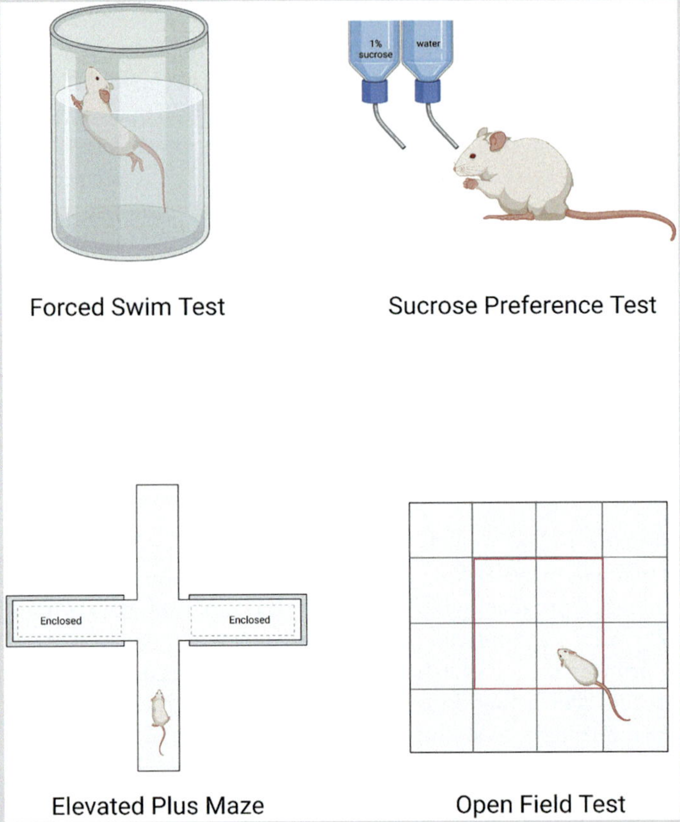

The **forced swim test** (FST) is one of the most commonly used animal models for assessing depressive-like behavior. The test involves the scoring of active (swimming and climbing) or passive (immobility, floating) behaviors when rodents are forced to swim in a cylinder from which they cannot escape. Antidepressant medications reduce the time spent immobile, hence a reduction in passive behavior is interpreted as an antidepressant-like effect, provided it does not increase general locomotor activity. For more see (Slattery and Cryan 2012).

The **sucrose preference test** (SPT) is used to measure anhedonia—a core symptom of depression in humans. The SPT is based on a two-bottle choice paradigm between water and a sucrose solution. A reduced preference for sucrose is indicative of anhedonia. For more see (Liu et al. 2018).

The **elevated plus maze** (EPM) is a widely used testing paradigm for anxiety-like behavior. It is the shape of a cross and is elevated above the

(continued)

Box 10.4 (continued)
ground. Two of the arms are enclosed with high walls and two of the arms are open, without walls. More time spent in the open arms reflects lower anxiety-like behavior, whereas more time in the enclosed arms reflects higher anxiety-like behavior. This test is validated to assess the anti-anxiety effects of pharmacological agents and steroid hormones in order to understand the mechanisms underlying anxiety. For more see (Walf and Frye 2007).

The **open field test** (OFT) is a common test of exploration, activity and anxiety-like behavior. It consists of an enclosed open field and is based on the idea that decreased levels of anxiety lead to increased exploration in an open area and increased anxiety leads to a preference to stay close to the walls of the open field. For more see (Seibenhener and Wooten 2015).

Figure created with BioRender.com

When investigating the impact of HSP models on the brain, the focus has been on the hippocampus. This brain structure plays a central role in stress regulation via the HPA axis, is implicated in the neurobiology of depression and is highly sensitive to changes in steroid hormones of the HPG and HPA axes (see Fig. 10.1). The first studies demonstrated that ovarian hormone withdrawal after HSP results in a significant decrease in cell proliferation in the dentate gyrus of the hippocampus (Zhang et al. 2016; Green and Galea 2008). However, this decrease in cell proliferation and neurogenesis in the HSP model is similar to what is reported during the early postpartum in healthy dams (Pawluski et al. 2016) (see Chap. 9) and thus, it cannot be clearly linked to a postpartum depressive phenotype.

Other studies have shown an effect of the HSP model on hippocampal gene expression (Suda et al. 2008). Real-time quantitative polymerase chain reaction techniques (PCR) show that the expression of several genes in the hippocampus are regulated by hormone withdrawal, including genes for the serotonin transporter (SERT), brain-derived neurotrophic factor (BDNF) and GABA type A receptor α4 ($GABA_AR\alpha4$), to name a few (Suda et al. 2008). These genes have been linked with postpartum depression in women, indicating altered expression in the hippocampus may be involved in mediating postpartum depression.

A recent study using a mouse HSP model reported that the withdrawal of exogenous estradiol and progesterone increased depressive-like behavior and reduced glucocorticoid receptor (GR) density in the hippocampus. Furthermore, central infusion of RU486, a GR and progesterone receptor (PR) antagonist, into the hippocampus after hormone withdrawal increased "postpartum" depressive- and anxiety-like behaviors in female mice (Wang et al. 2020), suggesting a possible role for hippocampal GR (and PR) in the regulation of postpartum depression-like behaviors following sex steroid hormone withdrawal.

There is much cross-talk between the HPG and HPA axes. For example, estrogens have an essential function in HPA axis regulation through direct effects on the hypothalamus by modulating estrogen response elements within the corticotropin-releasing hormone (CRH) gene, as well as through modulating GRs in the hypothalamus, hippocampus and other brain areas (Swaab et al. 2005). Thus, multiple mechanisms related to estradiol and progesterone withdrawal may be at play in postpartum depression but clearly a greater understanding of the impact of these reproductive hormones on neuroendocrine systems throughout the pregnant and postpartum brain is needed.

10.2.2 The Hypothalamic-Pituitary-Adrenal (HPA) Axis and Perinatal Depression

The HPA axis is a neuroendocrine system involving hormone-mediated communication between the hypothalamus, pituitary gland and adrenal glands. It serves to control homeostatic functions through diurnal secretions as well as mediating the body's physiological response to stress (Fig. 10.1). During healthy pregnancy, basal cortisol concentrations are elevated, serving critical functions throughout the perinatal period [Fig. 10.2; for review see Dickens and Pawluski (2018)]. However, during pregnancy and the postpartum period, stress-induced activation of the HPA axis is blunted, which is thought to be protective for both the mother and the neonate (see Chap. 4). Because of the impact of stress on mood and the dysregulation of the HPA axis in depression outside of the perinatal period, HPA axis dysfunction has long been suggested to play a role in the pathophysiology of PND and other perinatal mental illnesses. In addition, chronic stress and adverse childhood experiences are major risk factors for PND (Mahenge et al. 2018), again pointing to neuroendocrine disruptions as a significant factor in PND.

10.2.2.1 The HPA Axis and Perinatal Depression in Humans

To understand the mechanisms mediating the role of stress and HPA axis dysregulation in the development of PND, clinical and preclinical studies have focused on the stress hormones, including CRH and cortisol (or corticosterone in rodents), as potential neuroendocrine mediators.

A number of studies report higher plasma and salivary cortisol concentrations in pregnant women with depressive-symptoms. This elevation in cortisol with depression in pregnancy appears to be related to a lower concentration of cortisol at waking and a less sharp decline in cortisol over the course of the day [for review see Szpunar and Parry (2018)]. Others have shown that both morning serum cortisol concentrations and corticosteroid-binding globulin (CBG) concentrations are negatively correlated with depressive symptoms in late pregnancy in women, such that lower scores of depressive symptoms are associated with higher circulating concentrations of cortisol and CBG (Pawluski et al. 2012). However, a number of studies have found no association between cortisol concentrations and depressive-symptoms during pregnancy in women, probably due to a number of methodological differences, such as the timing of sample collection and the types of measurements used to assess depressive symptoms [for review see Seth et al. (2016)]. Taken together, when examining studies that found correlations between peripheral (blood or saliva) 'stress' hormones and depressive symptoms during pregnancy, higher cortisol concentrations are evident with increased depressive-symptoms, however further work is needed in order to fully determine the relationship between HPA axis function and depression in pregnancy (Seth et al. 2016; Dickens and Pawluski 2018).

The relationship between HPA axis function and depression during the *postpartum* period also remains to be fully determined. Perhaps the first research suggesting a link between the changes in gonadal steroid levels with pregnancy and HPA axis dysfunction in postpartum depression was carried out by Bloch et al. (2005). They administered supraphysiological doses of gonadal steroids for 8 weeks to women with or without a history of postpartum depression (as mentioned in Sect. 10.2.1) and found that women with a history of postpartum depression had increased urinary concentrations of cortisol during the hormone administration, suggesting enhanced sensitivity of the HPA axis in women at risk for postpartum depression (Bloch et al. 2005).

Recent research reports an increase in salivary cortisol concentrations with self-reported depressed mood in postpartum women (Laurent et al. 2018). Other clinical studies suggest a time course in the relationship between circulating cortisol concentrations and postpartum depressed mood, with higher cortisol concentrations associated with more depressed mood in women during the first week postpartum, but lower cortisol concentrations associated with depressed mood weeks or months later (Seth et al. 2016). Postpartum depressive-symptoms are also consistently associated with low concentrations of salivary morning cortisol [for review see Szpunar and Parry (2018)].

In addition to circulating cortisol concentrations, CRH may be a valuable biomarker of postpartum depression. CRH is produced by neurons in the paraventricular nucleus (PVN) of the hypothalamus, but during pregnancy is also synthesized and secreted by the placenta. High concentrations of CRH in the blood at mid-pregnancy (week 17) are associated with postpartum depressive symptoms (Iliadis et al. 2016). Others have found that elevated concentrations of CRH in the blood in the early postpartum period are positively correlated with depressive-symptoms in women and propose blood CRH concentrations as a potential diagnostic criteria for postpartum depression (Yim et al. 2009; Rich-Edwards et al. 2009; Magiakou et al. 1996).

It is clear that the relationship between HPA axis dysfunction and perinatal mental illness has not been firmly established, and given the inconsistencies in the clinical findings, there is a need for additional studies to resolve these conflicting reports. However, a clear relationship between HPA axis dysfunction in major depression in men and women is established, so it is reasonable to hypothesize that HPA axis dysfunction may contribute to the underlying pathophysiology of depression during the perinatal period.

To better understand the relationship between the HPA axis and perinatal mental illness, future studies should investigate not only circulating concentrations of cortisol, but also circulating concentrations of ACTH, CRH and their binding proteins—CBG and CRH-binding protein—which modulate glucocorticoid and CRH actions by binding these ligands and thereby reducing their availability to bind to their receptors [for a review see (Hammond 2016)]. Studies also need to consider the absolute concentrations of these factors as well as the rate of change across the perinatal period and ratios of hormone concentrations to develop a more comprehensive understanding of HPA axis function throughout this period in relation to PND and other perinatal mental illnesses.

10.2.2.2 The HPA Axis and Animal Models of Perinatal Depression

As highlighted above, clinical studies suggest a relationship between neuroendocrine changes, stressful life events and PND; however, challenges remain in deciphering the precise role with respect to the underlying neurobiological mechanisms. Our current understanding of the relationship between stress and maternal mental health has benefited primarily from rodent models. These studies often use stress paradigms involving repeated restraint stress or repeated unpredictable stress during pregnancy to model aspects of perinatal depression. Rodent models have demonstrated that early-life stress, pre-gestational stress, postpartum social stress and mother-offspring separation can increase depressive-like behaviors in the dam postpartum (Gemmel et al. 2018; Li and Chou 2016; Qiu et al. 2020). Exogenous administration of extra-physiological doses of corticosterone to rats during pregnancy or postpartum is another animal model of postpartum depression (Qiu et al. 2020; Brummelte et al. 2006).

These rodent models show that stress exposure or glucocorticoid administration during pregnancy or in the postpartum period may be sufficient to induce depressive-like postpartum behaviors and deficits in maternal care (Murgatroyd et al. 2015; Qiu et al. 2020; Payne and Maguire 2019). This further supports the hypothesis that alterations in the HPA axis can directly contribute to the underlying pathophysiology of PND.

Animal models have also contributed to our understanding of the neurobiology of postpartum depression, revealing alterations in synaptic plasticity, neurogenesis, microglia density, neurotransmitters and neuroendocrine function in brain areas implicated in postpartum depression in women (see Chap. 9 and Pawluski et al. 2017). Affected brain areas include the PFC, nucleus accumbens (NAc), amygdala, hippocampus and hypothalamus (Pawluski et al. 2019; Post and Leuner 2019; Melón et al. 2018). For example, Haim et al. (2017) show that gestational stress decreases **dendritic spine** density in the mPFC and NAc, and increases dendritic spine density in the basolateral amygdala of the rat dam. Given 90% of the excitatory synapses in the brain occur on dendritic spines, this shows an impact of maternal stress on brain areas important in decision making (mPFC), reward and motivation (NAc), as well as emotional regulation (amygdala); all of which play a significant role in maternal care-giving behaviors (see Chap. 8). In the hippocampus, repeated stress during pregnancy decreases GR density in the CA3 region in late pregnancy, increases neurogenesis in the postpartum period and decreases dendritic complexity of CA3 pyramidal cells in the maternal brain postpartum [for review see Pawluski et al. (2016)], indicating that stress-related disorders such as PND may occur due to dysregulation of several neural processes in different areas of the maternal brain.

Perhaps the most compelling animal research supporting a role for the HPA axis in postpartum depression is work showing that a direct manipulation (genetic or **chemogenetic**) of the HPA axis, resulting in the inability to suppress HPA axis activity during the perinatal period (see Chap. 4), results in abnormal postpartum affective behaviors (both depression and anxiety) and deficits in maternal care (Melón et al. 2018). The $K+/Cl^{-}$-co-transporter KCC2 plays a key role in GABAergic regulation of CRH neurons in the PVN of the hypothalamus. Mice lacking KCC2, specifically in CRH neurons, display HPA axis dysregulation, increased depression- and anxiety-like behaviors and deficits in maternal care-giving behaviors (Melón et al. 2018).

Additional work in mice shows that these and other alterations in the PVN may be central to postpartum depression. Using a chronic psychosocial stress paradigm applied during pregnancy that results in altered postpartum maternal care-giving behaviors and increased depressive-like behaviors in the mouse postpartum, researchers found greater CRH and lower CRH receptor 1 (CRH-R1) gene expression in the PVN (Zoubovsky et al. 2020). Given PVN neurons project both to the median eminence (to regulate secretion of anterior pituitary gland hormones) and intracerebrally, these changes in CRH/CRH-R1 may contribute to both the HPA axis and behavioral changes in gestationally stressed dams. Furthermore, blocking CRH signaling with Antalarmin (a CRH-R1 antagonist) during the perinatal period prevents abnormal postpartum behaviors in $GABA_{A}R$ δ subunit-knockout mice, which typically exhibit postpartum depression-like behaviors (Maguire and Mody 2016).

In summary, rodent models demonstrate that HPA axis dysfunction is capable of inducing behavioral deficits relevant to postpartum depression, yet there is still much to be learned about the relationship between the HPA axis and depression during pregnancy. Studies such as these are critical for empirically testing the potential role for HPA axis dysfunction in maternal mental illness.

10.2.3 Other Neuroendocrine Factors Implicated in Perinatal Depression

Oxytocin is a key hypothalamic peptide hormone, vital for maternal caregiving behaviors. A large body of rodent research shows that reducing PVN production of oxytocin disrupts maternal care-giving behaviors (Numan 2020). As poor maternal feelings of attachment and struggles with breastfeeding are often evident in mothers with PND, the role of oxytocin in this disorder has been the subject of a growing body of research, with mixed results. In women, the majority of studies indicate that lower plasma oxytocin concentrations during pregnancy or postpartum are associated with an increased risk of postpartum depression (Skrundz et al. 2011), whereas others have found higher prenatal plasma oxytocin concentrations predicts greater postpartum depressive symptoms only in women with a history of depression and that prenatal plasma oxytocin concentrations are positively associated with depressive symptoms *during* pregnancy (Saxbe et al. 2019). These findings are difficult to reconcile, however, they suggest a complex role for oxytocin and PND, which may be influenced by social factors and prior experience.

Research has also investigated how intravenous synthetic oxytocin administration may affect PND risk, as this is commonly used for labor induction/augmentation and for postpartum hemorrhage prevention. However, despite one study reporting a higher risk of a depressive disorder diagnosis during the first year postpartum in women given synthetic oxytocin compared with those not exposed (Kroll-Desrosiers et al. 2017), others report the opposite or no association [for systematic review see Thul et al. 2020)]. Hence, the existing evidence does not allow any firm conclusions to be drawn on the relationship between exogenously administered oxytocin and postpartum depression.

Others have administered intranasal oxytocin after birth as a preventative measure or in an attempt to treat PND, but a clear link between oxytocin administration and PND has not been found (Thul et al. 2020; Peled-Avron et al. 2020).

Many factors may contribute to the seemingly conflicting effects of oxytocin on PND reported in the clinical literature, including how peripheral measures of oxytocin may relate to central levels of this neuropeptide, the timing of sample collection, immunoassay procedures (e.g., whether samples are extracted or not, prior to the assay), assessment of depressive symptoms, the duration, dose and route of oxytocin administration, and the rate and duration of breastfeeding.

Research using a rodent model of postpartum depression provides evidence linking central oxytocin and PND via its actions on the HPA axis. Using a rat gestational stress model, Wang et al. (2018) report that postpartum oxytocin microinjections into the PVN reverse the effects of gestational stress on postpartum depressive-like behaviors in the dam and prevent the prenatal stress-induced increase in postpartum plasma corticosterone concentrations. This suggests that the central actions of oxytocin on the maternal HPA axis may be a significant factor in modulating PND.

Prolactin is another peptide hormone recently implicated in PND. As well as milk production during lactation (Chap. 7), prolactin is also important for the regulation of maternal behavior (Chap. 8) and the function of the HPA axis during the perinatal period (Chap. 4). Recent studies show that in the first week postpartum, women with depression have elevated serum concentrations of prolactin, but that weeks later no relationship between prolactin and depressive symptoms is evident [for review see Szpunar and Parry (2018)]. These data suggest an early postpartum relationship between prolactin and PND, however, more research is needed to understand the role of prolactin in the development of PND in women and animal models.

10.3 Perinatal Anxiety

Perinatal anxiety is a serious psychiatric disorder affecting 13–21% of pregnant women and 11–17% of postpartum women (Thorsness et al. 2018). Risk factors for perinatal anxiety mirror those associated with PND, including poor family functioning, low social support, challenging partner relationships, unintended pregnancy, lower income and occupational stress, with many of these risk factors falling under the umbrella of 'stress'. Perinatal anxiety and PND often co-occur and potentially have overlapping pathophysiological mechanisms. As with PND, it is generally considered that the large or rapid fluctuations in steroid or peptide hormones across the perinatal period act on the brain to trigger anxiety. This may be true for some women who are susceptible to significant changes in circulating hormones (Schiller et al. 2016). Unfortunately, studies (both clinical and preclinical) investigating the neuroendocrine mechanisms contributing to perinatal anxiety are lacking, as perinatal anxiety is often included as part of PND. Below we will discuss the available neuroendocrine data related specifically to perinatal anxiety.

10.3.1 The HPG Axis and Perinatal Anxiety

The significant increase in the steroid hormones, estrogens and progesterone during pregnancy and the rapid decrease in these hormones at birth, with the expulsion of the placenta (Fig. 10.2), have been proposed as contributing to mental illness in pregnancy and postpartum, as mentioned in Sect. 10.2.1. Indeed, a recent study specifically investigating anxiety levels in women within 12 h of delivery showed that a decrease in estradiol, but not in progesterone or cortisol, was associated with increased anxiety in the immediately postpartum period (de Rezende et al. 2019).

In the perinatal period, anxiety levels in women have been positively correlated with plasma concentrations of allopregnanolone and pregnanolone measured across the third trimester of pregnancy and the first 2 months postpartum (Deligiannidis et al. 2016). Others have found that low serum levels of allopregnanolone during the second trimester of pregnancy predict increased symptoms of anxiety in mothers 6 weeks postpartum (Osborne et al. 2019). Interestingly, intravenous infusion of synthetic allopregnanolone in the first 6 months postpartum has not consistently resulted in improved anxiety in women (Meltzer-Brody et al. 2018), but further research is underway in this area.

Despite the lack of a clear relationship between progesterone and its metabolite, allopregnanolone, on perinatal anxiety in women, preclinical research has focused on how allopregnanolone may affect anxiety-like behavior (in addition to depressive-like behavior discussed above) via its effects on $GABA_A$ receptors (Maguire and Mody 2008; Payne and Maguire 2019). The GABAergic system has long been implicated in the pathophysiology of various psychiatric disorders, including depression and anxiety. Brain areas where reproduction-induced $GABA_A$ receptor subunit modulation and expression occur and overlap with the traditional brain network mediating anxiety include the PFC, hippocampus, hypothalamus, basolateral amygdala and midbrain periaqueductal gray (PAG) (Payne and Maguire 2019). For example, ovarian hormones alter $GABA_A$ receptor subunit expression in the hippocampus and PAG across the estrous cycle (Maguire et al. 2005) and in the hippocampus, striatum, cortex and ventral respiratory column neurons during late pregnancy (Maguire and Mody 2008; Licheri et al. 2015). These changes in $GABA_A$ receptor subunit expression associated with altered ovarian hormone levels are thought to be a homeostatic mechanism to maintain an optimum level of inhibition in the face of dramatically elevated neurosteroid levels.

The rodent HSP model described above can also be used to model postpartum anxiety. Adult female rats, mice or Syrian hamsters given exogenous estradiol and progesterone to mimic a pregnancy, followed by rapid hormone withdrawal, show an increase in anxiety-like behaviors (Doornbos et al. 2009). Further supporting this relationship between estradiol and progesterone withdrawal with increased postpartum anxiety-like behavior are studies in female rats showing that a 10- to 11-day **pseudopregnancy** followed by a rapid decline in ovarian hormones via **ovariectomy** increases anxiety-like behaviors (Box 10.4) (Bitran and Solano 2005).

Increased anxiety behavior in a HSP model is also associated with neural changes. Following hormone withdrawal in a mouse HSP model, expression of the estrogen receptors GPR30, ERα and ERβ in the basolateral amygdala are lower compared to controls. Moreover, estradiol withdrawal after HSP causes dysfunction of $GABA_AR$-mediated inhibition in the basolateral amygdala through reduced GPR30 expression, which impairs long-term depression induction and causes anxiety-like behaviors (Yang et al. 2017). In view of the amygdala's central role in mediating anxiety, depression and maternal care-giving behaviors, this is a key brain area where future work related to perinatal anxiety is needed.

10.3.2 The HPA Axis and Perinatal Anxiety

Studies of the relationship between peripheral components of the HPA axis and anxiety in pregnant or early postpartum women are limited and often are carried out in women with PND. As with anxiety disorders at other times in life, it is likely that HPA axis dysregulation plays an important role (de Kloet et al. 2005) but further clinical research is needed to delineate the exact direction of this HPA dysfunction with perinatal anxiety.

Recent research using animal models suggests that CRH plays an important role in perinatal anxiety. As mentioned above, psychosocial stress during pregnancy in rats, which increases both anxiety-like and depressive-like behaviors postpartum, is associated with elevated CRH and CRH-R1 mRNA expression in the PVN (Zoubovsky et al. 2020). Directly potentiating central CRH signaling by infusing CRH into the cerebral ventricles of the brain generally increases postpartum anxiety in rats. Similarly, CRH-R1 agonists injected into the medial preoptic area or bed nucleus of the stria terminalis also increase postpartum anxiety-related behavior (Klampfl and Bosch 2019).

10.3.3 Other Neuroendocrine Factors Implicated in Perinatal Anxiety

As with PND, changes in oxytocin concentrations have been proposed to play a role in perinatal anxiety. For example, the administration of synthetic oxytocin at birth is associated with increased risk of being diagnosed with an anxiety disorder the first year postpartum (Kroll-Desrosiers et al. 2017). However, as mentioned above, the relationship between peripheral oxytocin concentrations, which are often measured in serum or plasma of women, and intracerebral oxytocin concentrations that more directly control anxiety and social affiliation, remain to be determined [see Lefevre et al. (2017)].

It is likely that cross-talk between the oxytocin system and the HPA and HPG axes plays a role in perinatal anxiety. Recent research shows that estradiol withdrawal after HSP in female Syrian hamsters results in increased anxiety, in the absence of depressive-like behavior. This effect is associated with increased oxytocin-immunoreactive cells, oxytocin messenger RNA in the PVN and increased oxytocin receptor density in the dorsal raphe nucleus. Moreover, blocking oxytocin receptor activity in the dorsal raphe nucleus in these animals prevents the increased anxiety-like behavior after estrogen withdrawal (Hedges et al. 2021), suggesting that estrogen withdrawal induces changes in the oxytocin system in the PVN and dorsal raphe nucleus to increase anxiety-like behavior during the postpartum period.

Prolactin may also play an important role in perinatal anxiety. Clinical research shows plasma prolactin concentrations measured 3 days postpartum are negatively associated with anxiety symptoms in nursing mothers (Asher et al. 1995). Additional research has reported a negative correlation between serum prolactin concentrations and anxiety symptoms in women with preterm infants, but not in women with full-term infants (Weigl et al. 2020), suggesting that the timing of birth and intensity of postpartum stressors may be important to consider.

In mice, low prolactin concentrations (achieved by administration of bromocriptine, a dopamine agonist used to treat hyperprolactinemia) early in pregnancy are associated with increased anxiety-like behaviors in the postpartum period. This effect of increasing anxiety-like behavior in the postpartum period seems to be a result of a failure to induce an increase in neurogenesis in the subventricular zone of the maternal brain that is normally mediated via prolactin actions in early pregnancy. Hence factors that influence prolactin in pregnancy may also impact on postpartum mood, however further work is needed to determine how prolactin may interact with the HPG and HPA axes to modulate perinatal anxiety (Larsen and Grattan 2010).

Taken together, a number of neuroendocrine factors likely play a role in perinatal anxiety disorders, but more research is needed to understand the central mechanisms involved in anxiety during pregnancy and postpartum.

10.4 Other Perinatal Mental Illnesses

10.4.1 Childbirth-Related Post-Traumatic Stress Disorder (CB-PTSD)

CB-PTSD occurs in a significant number of women (Box 10.1) but it is unclear how CB-PTSD is related to changes in neuroendocrine profiles in mothers. Brain imaging studies show that mothers with interpersonal violence-related PTSD have increased insula and dorsolateral PFC activity as well as decreased cingulate, hippocampal and mPFC activity when viewing videos of child separation (Moses-Kolko et al. 2014). In addition, patterns of brain activity were associated with the mother's self-reported feelings of stress or dissociative symptoms. This suggests there is likely a modification of brain circuitry related to the functioning of the HPA axis that may also be contributing to CB-PTSD.

We also know that PTSD in women outside the perinatal period is associated with HPA axis dysregulation (Bangasser and Valentino 2014). Some studies report that women with PTSD have lower cortisol concentrations than their healthy counterparts, an effect not observed in men. These results suggest that, in some cases, low cortisol levels may be a hallmark of PTSD in women.

Recent research on mothers with a previous diagnosis of PTSD shows that ongoing self-reported trauma symptoms are associated with blunted effects of life events on salivary cortisol reactivity, suggesting that CB-PTSD may also be associated with lower cortisol concentrations (Ramdas et al. 2021). How cortisol concentrations reflect the experience of CB-PTSD remains to be determined, but it is likely that dysregulation of the HPA axis, as well as the oxytocinergic system, important for uterine contractions at birth (Chaps. 5 and 6), may be key neural systems implicated in CB-PTSD [for review see Witteveen et al. (2020)].

It should also be noted that, similarly to anxiety disorders during the perinatal period, CB-PTSD is often comorbid with PND or misdiagnosed as PND. With better screening tools and improved understanding of the psychological and biological symptom profile of these perinatal mental illnesses, mothers will be able to access more effective care.

10.4.2 Postpartum Psychosis

Postpartum psychosis is the most severe psychiatric disorder associated with childbirth (Box 10.1) and affects 1–2 in every 1000 women after birth. The risk of postpartum psychosis is higher for women with a history of bipolar disorder, schizoaffective disorder or those who have suffered a previous episode of postpartum psychosis (Hazelgrove et al. 2021). There is evidence linking stress to psychosis unrelated to childbirth, however the role of stress and changes in the HPA axis in the onset of postpartum psychosis has not been fully investigated. Recent research shows that severe childhood maltreatment and higher daily cortisol in the third trimester of pregnancy predict psychiatric relapse in the first 4 weeks postpartum in women at risk of postpartum psychosis (Hazelgrove et al. 2021). Interestingly, brain imaging research is beginning to point to unique neural profiles in this disorder (Sambataro et al. 2021) and it has been hypothesized that altered development of many brain regions, including the hypothalamus, rapid changes in steroid hormones with parturition, genetic background and environmental factors all act to increase risk of postpartum psychosis (Hazelgrove et al. 2021). Ongoing research is needed to determine the neuroendocrine mechanisms behind this and other perinatal mental illnesses.

10.5 Future Directions

As this chapter has illustrated, PND and anxiety disorders are very likely to be related to the significant changes in neuroendocrine function that occur during pregnancy and postpartum. A full understanding of the normal neurobiological changes in a healthy pregnancy still has a long way to go and these gaps widen when considering the potential links between brain changes in pregnancy and perinatal mental illness, maternal care-giving behaviors and mother-infant attachment. Future research must continue to explore and describe these links in both human research and animal models.

Human research and animal studies have their own limitations. For example, invasive investigations into neuroendocrine profiles with perinatal mental illness in women are very limited. Brain imaging techniques permit better understanding of changes in neural activity or brain area volumes, but not neuroendocrinology. Animal models can allow a 'deep dive' into the brain, but the gestational and postpartum periods in rodents are quite different from those in humans, both in circulating hormone profiles, e.g., humans have much greater changes in estradiol than rodents, but also in the number of offspring being produced and the different demands of lactation. In addition, animal models are limited in that they allow for understanding specific symptoms of depressive-like behaviors or anxiety, but they are not precise models of perinatal mood disorders, which are uniquely human.

Furthermore, much research focuses on small aspects of perinatal mental illness, e.g., CRH in the PVN, and a broader understanding of the complex and interconnected mechanisms behind these disorders is needed. One way to better conceptualize future directions in research is to employ a recent comprehensive conceptual model to determine mechanisms to further investigate and/or target for treatment. The Maternal Reactive Scope Model, recently published as an adaptable, heuristic framework merging known pathways and physiological interactions, considers the natural fluctuations in physiological pathways that support a healthy pregnancy (Dickens et al. 2020). The critical, yet "extreme" physiological changes are then discussed in the context of vulnerability to pathology due to genetic susceptibility, perinatal stress, or early life stress. By conceptualizing multiple, interacting physiological pathways, the Maternal Reactive Scope Model removes the need to focus on a single system or compartmentalize specific physiological contributions when considering risks and pathologies associated with pregnancy. Connecting back to the neuroendocrinology of perinatal mental illness, use of this model may be helpful for integrating known and unknown changes along the HPG and HPA axes within pregnancy with mental health outcomes linked to genetic disposition for mental illness or exposure to life stress.

10.6 Perspectives

The maternal mental health crisis is not improving: if anything, it is worsening, with a related increase in perinatal mental illness during the Covid-19 pandemic and growing racial disparities that are likely to be tied to the life-long and daily stress of racism and inequities in healthcare provision (Masters et al. 2021). One key take-away from this chapter on the neuroendocrinology of perinatal mental illness is that there are many knowledge gaps in the literature that remain to be investigated. The repercussions of our lack of understanding across maternal mental health extends into drug development, health technology, healthcare delivery and closing the equity gaps for marginalized and at-risk communities. Those developing treatments and innovations for perinatal mental illness cannot do so without understanding the neuroendocrine mechanisms underlying maternal mental illness; and those making clinical decisions and determining new care pathways for maternal mental health need a depth and breadth of data and evidence to affect long-standing diagnostic and treatment options.

In order to improve our understanding of the neuroendocrine mechanisms mediating perinatal mental illness we need to continue creatively to find ways to investigate the mechanisms behind these disorders, accelerate the pace of discovery and disseminate knowledge. By doing this we will be able to alleviate the burden of suffering that many mothers face.

Key References

1. Meltzer-Brody and Kanes 2020 is a review of allopregnanolone in postpartum mental illness.
2. Pawluski et al. 2017 is a current review of how postpartum anxiety and depression affect the human brain.
3. Payne and Maguire 2019 is a review of mechanisms implicated in postpartum mental illness.
4. Schiller et al. 2016 is an excellent review of the role of reproductive steroids in mood and mental illness.
5. Seth et al. 2016 is an excellent review of cortisol and PND in women.

References

Aktar E, Jin Q, Lawrence PJ, Tollenaar MS, Elzinga BM, Bögels SM (2019) Fetal and infant outcomes in the offspring of parents with perinatal mental disorders: earliest influences. Front Psychiatry/Front Res Found 10(June):391

Asher I, Kaplan B, Modai I, Neri A, Valevski A, Weizman A (1995) Mood and hormonal changes during late pregnancy and puerperium. Clin Exp Obstet Gynecol 22(4):321–325

Ayers S, Wright DB, Thornton A (2018) Development of a measure of postpartum PTSD: the city birth trauma scale. Front Psychiatry/Front Res Found 9(September):409

Bangasser DA, Valentino RJ (2014) Sex differences in stress-related psychiatric disorders: neurobiological perspectives. Front Neuroendocrinol 35:303–319

Bauer A, Pawlby S, Plant DT, King D, Pariante CM, Knapp M (2015) Perinatal depression and child development: exploring the economic consequences from a South London Cohort. Psychol Med 45(1):51–61

Bitran D, Solano SM (2005) Termination of pseudopregnancy in the rat alters the response to progesterone, chlordiazepoxide, and MK-801 in the elevated plus-Maze. Psychopharmacology 180(3):447–454

Bloch M, Schmidt PJ, Danaceau M, Murphy J, Nieman L, Rubinow DR (2000) Effects of gonadal steroids in women with a history of postpartum depression. Am J Psychiatry 157(6). https://doi.org/10.1176/appi.ajp.157.6.924

Bloch M, Rubinow DR, Schmidt PJ, Lotsikas A, Chrousos GP, Cizza G (2005) Cortisol response to ovine corticotropin-releasing hormone in a model of pregnancy and parturition in euthymic women with and without a history of postpartum depression. J Clin Endocrinol Metab 90(2):695–699

Brummelte S, Pawluski JL, Galea LAM (2006) High post-partum levels of corticosterone given to dams influence postnatal hippocampal cell proliferation and behavior of offspring: a model of post-partum stress and possible depression. Horm Behav 50(3):370–382

de Kloet ER, Joels M, Holsboer F (2005) Stress and the brain: from adaptation to disease. Nat Rev Neurosci 6:463–475

Deligiannidis KM, Kroll-Desrosiers AR, Mo S, Nguyen HP, Svenson A, Jaitly N, Hall JE, Barton BA, Rothschild AJ, Shaffer SA (2016) Peripartum neuroactive steroid and γ-aminobutyric acid profiles in women at-risk for postpartum depression. Psychoneuroendocrinology 70(August):98–107

Dickens MJ, Pawluski JL (2018) The HPA axis during the perinatal period: implications for perinatal depression. Endocrinology 159(11):3737–3746

Dickens MJ, Pawluski JL, Michael Romero L (2020) Moving forward from COVID-19: bridging knowledge gaps in maternal health with a new conceptual model. Front Glob Women's Health 1(November). https://doi.org/10.3389/fgwh.2020.586697

Doornbos B, Fokkema DS, Molhoek M, Tanke MAC, Postema F, Korf J (2009) Abrupt rather than gradual hormonal changes induce postpartum blues-like behavior in rats. Life Sci 84(3-4):69–74

Galea LA, Wide JK, Paine TA, Holmes MM, Ormerod BK, Floresco SB (2001) High levels of estradiol disrupt conditioned place preference learning, stimulus response learning and reference memory but have limited effects on working memory. Behav Brain Res 126(1–2):115–126

Gemmel M, Harmeyer D, Bögi E, Fillet M, Hill LA, Hammond GL, Charlier TD, Pawluski JL (2018) Perinatal fluoxetine increases hippocampal neurogenesis and reverses the lasting effects of pre-gestational stress on serum corticosterone, but not on maternal behavior, in the rat dam. Behav Brain Res 339(February):222–231

Glasper ER, Kenkel WM, Bick J, Rilling JK (2019) More than just mothers: the neurobiological and neuroendocrine underpinnings of allomaternal caregiving. Front Neuroendocrinol 53:100741

Green AD, Galea LAM (2008) Adult hippocampal cell proliferation is suppressed with estrogen withdrawal after a hormone-simulated pregnancy. Horm Behav 54(1):203–211

Grigoriadis S, Wilton AS, Kurdyak PA, Rhodes AE, VonderPorten EH, Levitt A, Cheung A, Vigod SN (2017) Perinatal suicide in Ontario, Canada: a 15-year population-based study. Can Med Assoc J = Journal de l'Association Medicale Canadienne 189(34):E1085–E1092

Haim A, Julian D, Albin-Brooks C, Brothers HM, Lenz KM, Leuner B (2017) A survey of neuroimmune changes in pregnant and postpartum female rats. Brain Behav Immun 59(January):67–78

Hammond GL (2016) Plasma steroid-binding proteins: primary gatekeepers of steroid hormone action. J Endocrinol 230:R13–R25

Hazelgrove K, Biaggi A, Waites F, Fuste M, Osborne S, Conroy S, Howard LM et al (2021) Risk factors for postpartum relapse in women at risk of postpartum psychosis: the role of psychosocial stress and the biological stress system. Psychoneuroendocrinology 128(April):105218

Hedges VL, Heaton EC, Amaral C, Benedetto LE, Bodie CL, D'Antonio BI, Davila Portillo DR et al (2021) Estrogen withdrawal increases postpartum anxiety via oxytocin plasticity in the paraventricular hypothalamus and dorsal raphe nucleus. Biol Psychiatry 89(9):929–938
Iliadis SI, Sylven S, Hellgren C, Olivier JD, Schijven D, Comasco E, Chrousos GP, Sundstrom Poromaa I, Skalkidou A (2016) Mid-pregnancy corticotropin-releasing hormone levels in association with postpartum depressive symptoms. Depress Anxiety 33:1023–1030
Kim P, Swain JE (2007) Sad dads: paternal postpartum depression. Psychiatry (Edgmont) 4:35–47
Kim P, Rigo P, Mayes LC, Feldman R, Leckman JF, Swain JE (2014) Neural plasticity in fathers of human infants. Soc Neurosci 9(5):522–35. https://doi.org/10.1080/17470919.2014.933713
Klampfl SM, Bosch OJ (2019) Mom doesn't care: when increased brain CRF system activity leads to maternal neglect in rodents. Front Neuroendocrinol 53(April):100735
Kroll-Desrosiers AR, Nephew BC, Babb JA, Guilarte-Walker Y, Moore TA, Simas, and Kristina M. Deligiannidis. (2017) Association of peripartum synthetic oxytocin administration and depressive and anxiety disorders within the first postpartum year. Depress Anxiety 34(2):137–146
Larsen CM, Grattan DR (2010) Prolactin-induced mitogenesis in the subventricular zone of the maternal brain during early pregnancy is essential for normal postpartum behavioral responses in the mother. Endocrinology 151(8):3805–3814
Laurent H, Goodman SH, Stowe ZN, Halperin M, Khan F, Wright D, Nelson BW et al (2018) Course of ante- and postnatal depressive symptoms related to mothers' HPA axis regulation. J Abnormal Psychol. https://doi.org/10.1037/abn0000348
Lefevre A, Mottolese R, Dirheimer M, Mottolese C, Duhamel J-R, Sirigu A (2017) A comparison of methods to measure central and peripheral oxytocin concentrations in human and non-human primates. Sci Rep 7(1):17222
Leiferman JA, Farewell CV, Jewell J, Rachael L, Walls J, Harnke B, Paulson JF (2021) Anxiety among fathers during the prenatal and postpartum period: a meta-analysis. J Psychosom Obstet Gynaecol 42:152–161
Li M, Chou SY (2016) Modeling postpartum depression in rats: theoretic and methodological issues. Dongwuxue Yanjiu 37:229–236
Licheri V, Talani G, Gorule AA, Mostallino MC, Biggio G, Sanna E (2015) Plasticity of GABAA Receptors during Pregnancy and Postpartum Period: From Gene to Function. Neural Plast. 170435. https://doi.org/10.1155/2015/170435
Lindahl V, Pearson JL, Colpe L (2005) Prevalence of suicidality during pregnancy and the postpartum. Arch Womens Ment Health 8(2):77–87
Liu MY, Yin CY, Zhu LJ, Zhu XH, Xu C, Luo CX, Chen H, Zhu DY, Zhou QG (2018) Sucrose preference test for measurement of stress-induced anhedonia in mice. Nat Protoc 13:1686–1698
Magiakou MA, Mastorakos G, Rabin D, Dubbert B, Gold PW, Chrousos GP (1996) Hypothalamic corticotropin-releasing hormone suppression during the postpartum period: implications for the increase in psychiatric manifestations at this time. J Clin Endocrinol Metab 81(5):1912–1917
Maguire J (2019) Neuroactive steroids and GABAergic involvement in the neuroendocrine dysfunction associated with major depressive disorder and postpartum depression. Front Cell Neurosci 13(March):83
Maguire J, Mody I (2016) Behavioral Deficits in Juveniles Mediated by Maternal Stress Hormones in Mice. Neural Plast. 2762518. https://doi.org/10.1155/2016/2762518
Maguire J, Mody I (2008) GABA(A)R plasticity during pregnancy: relevance to postpartum depression. Neuron 59(2):207–213
Maguire JL, Stell BM, Rafizadeh M, Mody I (2005) Ovarian cycle-linked changes in GABA(A) receptors mediating tonicinhibition alter seizure susceptibility and anxiety. Nat Neurosci 8(6):797–804
Mahenge B, Stöckl H, Mizinduko M, Mazalale J, Jahn A (2018) Adverse childhood experiences and intimate partner violence during pregnancy and their association to postpartum depression. J Affect Disord. https://doi.org/10.1016/j.jad.2017.12.036

Masters GA, Asipenko E, Bergman AL, Person SD, Brenckle L, Moore TA, Simas JY, Ko CL, Robbins, and Nancy Byatt. (2021) Impact of the COVID-19 pandemic on mental health, access to care, and health disparities in the perinatal period. J Psychiatr Res 137(May):126–130

Melón LC, Hooper A, Yang X, Moss SJ, Maguire J (2018) Inability to suppress the stress-induced activation of the HPA Axis during the peripartum period engenders deficits in postpartum behaviors in mice. Psychoneuroendocrinology 90(April):182–193

Meltzer-Brody S, Kanes SJ (2020) Allopregnanolone in postpartum depression: role in pathophysiology and treatment. Neurobiol Stress 12(May):100212

Meltzer-Brody S, Colquhoun H, Robert Riesenberg C, Epperson N, Deligiannidis KM, Rubinow DR, Li H et al (2018) Brexanolone injection in post-partum depression: two multicentre, double-blind, randomised, placebo-controlled, phase 3 trials. Lancet 392(10152):1058–1070

Moses-Kolko EL, Horner MS, Phillips ML, Hipwell AE, Swain JE (2014) In search of neural endophenotypes of postpartum psychopathology and disrupted maternal caregiving. J Neuroendocrinol 26(10):665–684

Murgatroyd CA, Peña CJ, Podda G, Nestler EJ, Nephew BC (2015) Early life social stress induced changes in depression and anxiety associated neural pathways which are correlated with impaired maternal care. Neuropeptides 52(August):103–111

Murray L, Cooper PJ (1997) Effects of postnatal depression on infant development. Arch Dis Child. https://doi.org/10.1136/adc.77.2.99

Numan M (2020) The parental brain: mechanisms, development, and evolution. Oxford University Press

Oberlander TF, Zwaigenbaum L (2017) Disentangling maternal depression and antidepressant use during pregnancy as risks for autism in children. JAMA 317:1533–1534

Oberlander TF, Gingrich JA, Ansorge MS (2009) Sustained neurobehavioral effects of exposure to SSRI antidepressants during development: molecular to clinical evidence. Clin Pharmacol Ther 86:672–677

Osborne LM, Betz JF, Yenokyan G, Standeven LR, Payne JL (2019) The role of allopregnanolone in pregnancy in predicting postpartum anxiety symptoms. Front Psychol 10(July):1033

Pawluski JL, Brain UM, Underhill CM, Hammond GL, Oberlander TF (2012) Prenatal SSRI exposure alters neonatal corticosteroid binding globulin, infant cortisol levels, and emerging HPA function. Psychoneuroendocrinology 37(7):1019–1028

Pawluski JL, Lambert KG, Kinsley CH (2016) Neuroplasticity in the maternal hippocampus: Relation to cognition and effects ofrepeated stress. Horm Behav 77:86–97. https://doi.org/10.1016/j.yhbeh.2015.06.004

Pawluski JL, Lonstein JS, Fleming AS (2017) The neurobiology of postpartum anxiety and depression. Trends Neurosci 40(2):106–120

Pawluski JL, Li M, Lonstein JS (2019) Serotonin and motherhood: from molecules to mood. Front Neuroendocrinol 53:100742

Payne JL, Maguire J (2019) Pathophysiological mechanisms implicated in postpartum depression. Front Neuroendocrinol. https://doi.org/10.1016/j.yfrne.2018.12.001

Peled-Avron L, Abu-Akel A, Shamay-Tsoory S (2020) Exogenous effects of oxytocin in five psychiatric disorders: a systematic review, meta-analyses and a personalized approach through the lens of the social salience hypothesis. Neurosci Biobehav Rev 114(July):70–95

Post C, Leuner B (2019) The maternal reward system in postpartum depression. Arch Womens Ment Health 22(3):417–429

Qiu W, Hodges TE, Clark EL, Blankers SA, Galea LAM (2020) Perinatal depression: heterogeneity of disease and in animal models. Front Neuroendocrinol 59 (October):100854. Psychoneuroendocrinology 101 (March). https://doi.org/10.1016/j.psyneuen.2018.11.006

Ramdas DL, Sbrilli MD, Laurent HK (2021) Impact of maternal trauma-related psychopathology and life stress on HPA axis stress response. Arch Womens Ment Health

Rao WW, Zhu XM, Zong QQ, Zhang Q, Hall BJ, Ungvari GS, Xiang YT (2020) Prevalence of prenatal and postpartum depression in fathers: a comprehensive meta-analysis of observational surveys. J AffectDisord 263:491–499

Rich-Edwards J, Hacker M, Gillman M (2009) Premature recommendation of corticotropin-releasing hormone as screen for postpartum depression. Arch Gen Psychiatry
Riecher-Rössler A, Steiner M (2005) Perinatal stress, mood and anxiety disorders: from bench to bedside. Karger Medical and Scientific Publishers
Sambataro F, Cattarinussi G, Lawrence A, Biaggi A, Fusté M, Hazelgrove K, Mehta MA et al (2021) Altered dynamics of the prefrontal networks are associated with the risk for postpartum psychosis: a functional magnetic resonance imaging study. Transl Psychiatry 11(1):238
Saxbe DE, Schetter CD, Simon CD, Adam EK, Shalowitz MU (2017) High paternal testosterone may protect against postpartum depressive symptoms in fathers, but confer risk to mothers and children. Horm Behav 95(September):103–112
Saxbe D, Khaled M, Horton KT, Mendez AJ (2019) Maternal prenatal plasma oxytocin is positively associated with prenatal psychological symptoms, but method of immunoassay extraction may affect results. Biol Psychol 147(October):107718
Schiller CE, Johnson SL, Abate AC, Schmidt PJ, Rubinow DR (2016) Reproductive steroid regulation of mood and behavior. Compr Physiol. https://doi.org/10.1002/cphy.c150014
Seibenhener ML, Wooten MC (2015) Use of the Open Field Maze to measure locomotor and anxiety-like behavior in mice. J Vis Exp JoVE:e52434
Seth S, Lewis AJ, Galbally M (2016) Perinatal maternal depression and cortisol function in pregnancy and the postpartum period: a systematic literature review. BMC Pregnancy Childbirth 16(1):124
Skrundz, Marta, Margarete Bolten, Irina Nast, Dirk H. Hellhammer, and Gunther Meinlschmidt. 2011. "Plasma oxytocin concentration during pregnancy is associated with development of postpartum depression." Neuropsychopharmacology 36(9): 1886–1893.
Slattery DA, Cryan JF (2012) Using the rat forced swim test to assess antidepressant-like activity in rodents. Nat Protoc 7:1009–1014
Suda S, Segi-Nishida E, Newton SS, Duman RS (2008) A postpartum model in rat: behavioral and gene expression changes induced by ovarian steroid deprivation. Biol Psychiatry 64(4):311–319
Swaab DF, Bao A-M, Lucassen PJ (2005) The stress system in the human brain in depression and neurodegeneration. Ageing Res Rev 4(2):141–194
Szpunar MJ, Parry BL (2018) A systematic review of cortisol, thyroid-stimulating hormone, and prolactin in peripartum women with major depression. Arch Womens Ment Health 21(2):149–161. https://doi.org/10.1007/s00737-017-0787-9
Thorsness KR, Watson C, LaRusso EM (2018) Perinatal anxiety: approach to diagnosis and management in the obstetric setting. Am J Obstet Gynecol 219:326–345
Thul TA, Corwin EJ, Carlson NS, Brennan PA, Young LJ (2020) Oxytocin and postpartum depression: a systematic review. Psychoneuroendocrinology 120(October):104793
Walf AA, Frye CA (2007) The use of the elevated plus maze as an assay of anxiety-related behavior in rodents. Nat Protoc 2:322–328
Walton N, Maguire J (2019) Allopregnanolone-based treatments for postpartum depression: why/how do they work? Neurobiol Stress 11(November):100198
Wang T, Shi C, Li X, Zhang P, Liu B, Wang H, Wang Y et al (2018) Injection of oxytocin into paraventricular nucleus reverses depressive-like behaviors in the postpartum depression rat model. Behav Brain Res 336(January):236–243
Wang J, Yun Q, Ma SF, Song HR, Guo MN, Zhang WN (2020) Inhibition of expression of glucocorticoids receptors maycontribute to postpartum depression. Biochem Biophys Res Commun 26;523(1):159–164. https://doi.org/10.1016/j.bbrc.2019.12.040
Weigl T, Schneider N, Stein A, Felderhoff-Müser U, Schedlowski M, Engler H (2020) Postpartal affective and endocrine differences between parents of preterm and full-term infants. Front Psychiatry/Front Res Foundation 11. https://doi.org/10.3389/fpsyt.2020.00251
Witteveen AB, Stramrood CAI, Henrichs J, Flanagan JC, van Pampus MG, Olff M (2020) The oxytocinergic system in PTSD following traumatic childbirth: endogenous and exogenous oxytocin in the peripartum period. Arch Womens Ment Health 23(3):317–329

Yim IS, Glynn LM, Dunkel-Schetter C, Hobel CJ, Chicz-DeMet A, Sandman CA (2009) Risk of postpartum depressive symptoms with elevated corticotropin-releasing hormone in human pregnancy. Arch Gen Psychiatry 66(2):162–169

Zhang Z, Hong J, Zhang S, Zhang T, Sha S, Yang R, Qian Y, Chen L (2016) Postpartum estrogen withdrawal impairs hippocampal neurogenesis and causes depression- and anxiety-like behaviors in mice. Psychoneuroendocrinology 66(April):138–149

Zoubovsky SP, Hoseus S, Tumukuntala S, Schulkin JO, Williams MT, Vorhees CV, Muglia LJ (2020) Chronic psychosocial stress during pregnancy affects maternal behavior and neuroendocrine function and modulates hypothalamic CRH and nuclear steroid receptor expression. Transl Psychiatry 10(1):6

Glossary

Adipocytes Also known as fat cells; they store energy as fat and are the primary component of adipose tissue.

Adrenalectomy Surgical procedure to remove the adrenal glands.

Adrenergic Nerves that synthesize and release adrenaline (also called epinephrine).

Adult neurogenesis The process of generating functional neurons from adult neural precursors, occurring throughout life in restricted brain regions in mammals.

Afferents Nerve inputs to a specific area in the central nervous system.

Altricial Refers to young born in a relatively underdeveloped state physiologically and behaviorally compared to the young of other species.

Altricial young Young that are born underdeveloped and require care from parents (or other adults) to reach maturity.

Alveolar cells In the mammary gland, the alveoli are small cavities in which milk is produced.

Amnion The fetal membrane closest to the fetus.

Angiogenesis The formation of new blood vessels from existing blood vessels.

Anorectic Has appetite-suppressing properties.

Anovulation The absence of an egg being ovulated, or released from the ovary, which is a key step that is required for fertility.

Antidromic A key breakthrough in the identification of magnocellular neurons recorded *in vivo* was the introduction of "antidromic identification." When an electrical stimulus is applied to the posterior pituitary, an action potential is elicited at that site which travels "antidromically" to the cell body (i.e., in the opposite direction to spontaneous action potentials). This antidromic spike arrives at the cell body with a constant latency—unless it collides with a descending (orthodromically propagated) action potential, in which case both action potentials will be eliminated by the collision. These two tests ("constant latency" and "collision") are key to the antidromic identification of neurons as projecting to the posterior pituitary.

Anxiolytic To reduce anxiety.

Atosiban Modified peptide-based oxytocin receptor antagonist.

Autonomic Involuntary or unconscious; relating to the autonomic nervous system.

P. J. Brunton, D. R. Grattan (eds.), *Neuroendocrine Regulation of Mammalian Pregnancy and Lactation*, Masterclass in Neuroendocrinology 15,
https://doi.org/10.1007/978-3-031-51138-7

Axonal varicosity The axons of peptide-secreting neurons contain numerous swellings that contain clusters of large dense-cored peptide-containing vesicles. It is now clear that the vesicles can be released from these varicosities in an activity-dependent manner.

Baby Used to describe the neonate immediately after birth (see also: embryo and fetus).

Barusiban and Retosiban Both are non-peptide oxytocin receptor antagonists.

Biparous Having given birth to two offspring i.e. twins.

Blastocyst Hollow, fluid-filled cellular mass formed by dividing cells during early development (after a sperm fertilizes an oocyte), consisting of cells forming the trophectoderm and the inner cell mass.

Blood–brain barrier (BBB) Specialized tight junctions between vascular endothelial cells in the brain that form a highy selective semi-permeable barrier and regulate the transfer of substances between the circulatory system and the central nervous system.

Body mass index (BMI) An indicator of healthy body weight calculated by dividing a person's body weight in kgs by their height in meters squared.

Bouton-like structure Axonal swelling associated with a close apposition to another neuronal element, potentially representing a site of neurotransmitter release.

Brain plasticity The ability of the nervous system to change its activity in response to intrinsic or extrinsic stimuli by reorganizing its structure, functions, or connections; see neuroplasticity.

Brainbow A genetic cell-labeling technique by which individual neurons in the brain can be distinguished from neighboring neurons using fluorescent proteins.

Bromodeoxyuridine (BrdU) A thymidine analog that incorporates into newly synthesized DNA of dividing cells during the S-phase of the cell cycle; BrdU is used for monitoring cell proliferation.

Catabolic Used to describe metabolic pathways that break down molecules into smaller units.

Cell differentiation The process in which a stem cell changes from one type to another; usually, the cell changes to a more specialized type.

Cell maturation The cellular developmental process, independent of morphogenetic (shape) change, that is required for a specific cell to attain its fully functional state.

Cell migration A central process in the development and maintenance of multicellular organisms, wherein old or damaged cells are replaced by the migration of newly formed cells from the underlying tissue layers.

Cell proliferation An increase in the number of cells as a result of cell growth and cell division.

Cervical dilation Softening of the cervical connective tissue and shortening of the cervical length, which continues during labor to allow the passage of the fetus.

Chemogenetic The engineering of protein receptors to respond to previously unrecognized small molecules.

Chorioamnionitis The inflammation of the fetal membranes, usually by the presence of bacteria which may progress to infection of the amniotic fluid and then to fetal sepsis (infection).

Choriodecidual The fetal chorion membrane and maternal uterine lining (decidua).

Chorion Outermost fetal derived membrane that lies between the maternal tissue and the fetal amnion. The amnion and chorion are usually attached and adherent.

Chorion The outermost membrane around the embryo.

Chorionic villi The villus structures of the placenta that are coated with chorion and in the human placenta are bathed in maternal blood. The outer chorion layer forms the maternal–fetal interface.

Circadian An internal process that occurs in approximately 24-h cycles. From the Latin *circa*, meaning "approximately," and *diēm*, meaning "day."

Circadian rhythm A circadian rhythm is a physiological or behavioral change that follows a 24-h cycle. These natural processes are entrained to the light and dark phases of the light cycle, predominantly through a hypothalamic structure known as the suprachiasmatic nucleus, which receives direct input from the retina.

Cognitive flexibility The ability to adapt our behavior and thinking in response to the environment.

Connexins Gap-junction proteins between cells, particularly in muscle, that allow electrical conduction and transmission of contractile activity such that the muscle contraction is coordinated and forceful.

Corpus luteum A temporary hormone secreting structure found in the female ovaries that is formed from the remnants of the ovarian follicle that has released a mature oocyte during ovulation. It produces large amounts of progesterone until it is eliminated by luteolysis.

Corticosteroid-binding globulin (CBG) Synthesized by the liver and secreted into the bloodstream where it binds to glucocorticoids, modulating their tissue availability. Only free (unbound) glucocorticoids are biologically active.

Cortisol awakening response (CAR) The change in cortisol concentration that occurs in the first hour after waking from sleep. It is used to assess hypothalamic-pituitary-adrenal (HPA) axis function and mental health risk.

CpG islands Cytosine-phosphate-guanine (CpG)-rich regions of the DNA, typically in gene promotors that may be methylated.

Cytokines Inflammatory mediators including interleukin (IL)-1β, IL-6 and tumor necrosis factor-α (TNFα).

Cytotrophoblasts The placental trophoblast layer that is made up of single cells.

Decidua Modified endometrium of the uterus formed in preparation for pregnancy.

Decidualization Hormone-induced differentiation (both functional and morphological changes) of the endometrial lining of the uterus, in preparation for implantation. In humans, the initial phase of this process occurs during each menstrual cycle, and the decidualized lining of the uterus must be shed if pregnancy does not occur in a period of menstruation.

Deciduum See decidua.

Default mode network A set of brain regions, in the prefrontal cortex and medial temporal lobe, that are active during mind wandering.

Delayed free recall The ability to recollect information acquired earlier. Frequently used in laboratory studies of memory, delayed recall is also used in neuropsychological examinations to determine the rate of loss of information presented earlier, in comparison with established norms.

Dendritic spine A small membranous protrusion from a neuron's dendrite. Typically, a spine receives input from a single axon at the synapse.

Dentate gyrus A region in the mammalian brain critical for memory encoding with a neuronal architecture and function that deviates considerably from other cortical areas.

Depolarized Changed cellular membrane potential resulting in less negative charge inside the cell compared to the outside.

Diestrus See estrous cycle.

Dinoprostone A prostaglandin E2 (PGE_2) analogue.

Diurnal Pertaining to the daylight hours, i.e. opposite of nocturnal; repeating once every 24 h, e.g. diurnal rhythm.

DNA methyltransferases Enzymes that transfer methyl groups to certain nucleotides (cytosine residues) in DNA.

Doublecortin (DCX) A microtubule (MT)-stabilizing protein essential for neuronal migration during human brain development.

Dyad Represents a group of two interacting individuals.

Embryo Early stages of development, formed from the inner cell mass of the blastocyst. The trophectoderm will form the supporting membranes and the placenta. Embryonic development involves establishing the fundamental tissues (see also: fetus).

Endogenous Substances originating within an organism.

Endometrium The inner epithelial layer of the mammalian uterus together with its mucosal lining.

Energy expenditure Energy the body uses to function.

Energy homeostasis Biological systems that co-ordinate to maintain a relatively constant balance between the body's energy intake (food intake) and energy expenditure.

Enkephalin Endogenous opioid peptide naturally produced in the brain.

Epigenetics The study of how different factors (e.g., environment, early life experience) can alter the way genes function, without changing the DNA sequence.

Estrous cycle The reproductive cycle in rodents that lasts for 4-5 days and has four phases—proestrus, estrus, metestrus, and diestrus. It is characterized by a period ("estrus") during which the females will display reproductive behavior. This is tightly coordinated by the hormonal changes leading up to ovulation on the day preceding estrus (termed proestrus), thereby increasing the chance that mating with result in a pregnancy. After a brief luteal phase on the day of estrus, the days of metestrus then diestrus represent the "follicular phase" of the next ovulatory cycle. Note the term is spelled "estrus," when used as a noun (e.g., day of estrus); but "estrous," when used as an adjective (e.g., estrous cycle).

Estrus A recurring period of sexual receptivity and fertility in many female mammals. Also see 'estrous cycle'.

Executive function An umbrella term for the mental processes involved in higher cognitive thinking: decision-making, planning, attention, etc.

Ferguson reflex A reflex initiated by stretching of the cervix during birth that leads to increased neural stimulation of hypothalamic neurosecretory oxytocin neurons leading to increased secretion of oxytocin from the posterior pituitary gland. This further increases uterine contractions and is particularly important for delivery of the placenta.

Fertilized oocyte Single diploid cell formed by fusion of an oocyte with a sperm, creating the zygote, from which a new individual organism will develop.

Fetal microchimerism Low levels of fetal cells harboring in maternal blood and tissues during pregnancy and for years after.

Fetus Later stages of development (from about 8 weeks after fertilization in human pregnancy, or 10 weeks of pregnancy using the convention of counting from the last menstrual period) through to birth. Fetal development is generally considered as growth and development after attaining the basic structural plan (see also: embryo).

Forced swim test Rodent behavioral test used to measure susceptibility to negative mood.

Free recall A type of memory task in which participants attempt to remember previously studied information in any order.

GABA The primary inhibitory neurotransmitter in the nervous system.

GABAergic agent A chemical that modifies the effects of GABA in the body.

Galactophore Duct that carries milk from the mammary gland alveoli to the nipple, also known as milk ducts, mammary ducts and lactiferous ducts.

Genomic imprinting An epigenetic process exclusive to mammals (and some flowering plants) that results in parent-of-origin monoallelic expression of approximately 0.4% of all protein coding genes.

Gestation Equivalent term to pregnancy, describing the full period of development of the baby in the mother, from conception through to parturition (giving birth).

Gluconeogenesis The process of generating glucose from its own breakdown products or from the breakdown of lipids or proteins.

Hedonic Relating to a state or feeling of pleasure.

Homeostasis A state of stable equilibrium.

HPA axis See hypothalamic-pituitary-adrenal (HPA) axis.

Hypothalamic-pituitary-gonadal (HPG) axis A system involving the hypothalamus, anterior pituitary, and the gonads (testes or ovaries) that regulates reproduction.

Hormone-simulated pregnancy (HSP) A regime of hormones are administered (typically estradiol and progesterone) to mimic pregnancy. This can be used to model postpartum depression in rodents.

Human cytotrophoblast cells A layer of mononuclear cells in the human placenta that are proliferative and continually differentiate into syncytiotrophoblast during villous formation and development.

Human syncytiotrophoblast In the human placenta, the syncytiotrophoblast is *the* outer layer of the human placenta, in direct contact with maternal blood, with which exchange of metabolites, nutrients, waste products, and gases occurs. The syncytiotrophoblast is a key source of placental hormones in humans.

Hypercorticosteronemia A state associated with high circulating concentrations of corticosterone in the blood.

Hyperpolarized Change in cellular membrane potential resulting in more negative charge inside the cell compared to the outside.

Hypophysectomy/Hypophysectomized Surgical removal of the pituitary gland.

Hypophysiotropic Action of stimulating the pituitary gland.

Hypothalamic-pituitary-adrenal (HPA) axis A system involving the hypothalamus, anterior pituitary, and adrenal glands that regulates stress responses. Release of corticotropin-releasing hormone (CRH) from hypothalamic neurosecretory neurons stimulates ACTH secretion by the anterior pituitary corticotrophs which in turn stimulates glucocorticoid (e.g. cortisol) secretion from the adrenal glands.

Immediate early genes Genes that are rapidly activated (before any new proteins are synthesized) in response to a wide variety of cellular stimuli, including stress.

Immunohistochemistry A laboratory method that uses antibodies to check for certain antigens (markers) in a sample of tissue; the antibodies are usually linked to an enzyme or a fluorescent dye.

Implantation The attachment of the blastocyst to the lining of the maternal uterus. In humans, this occurs approximately 7–9 days after fertilization. There are significant species differences in the type of implantation and the type of placental development that occurs.

Ki67 A nuclear protein associated with cellular proliferation.

Kyphosis The upright, crouched (or arched-back) posture displayed by many small mammals while they stand over and nurse their young.

Long-term potentiation (LTP) A process involving persistent strengthening of synapses that leads to a long-lasting increase in signal transmission between neurons; it is an important process in the context of synaptic plasticity; LTP recording is widely recognized as a cellular model for the study of memory.

Lordosis A postural reflex expressed by female rodents during copulation, involving a pronounced inward arching of the spinal column (dorsiflexion) and raising of the tail. This reflex is stimulated by mounting behavior of the male, triggering mechanoreceptors along the flank and around the tail, and serves to facilitate successful mating.

Long-term depression (LTD) A reduction in synaptic activity that can last for hours or longer.

Luteal phase Period of the reproductive cycle dominated by progesterone secretion from the corpus luteum (formed from the post-ovulatory follicle). Important stage for early pregnancy supporting the early stages of embryonic development after fertilization. The corpus luteum is supported by luteotrophic factors and inhibited by luteolytic factors (see also: luteolysis).

Luteolysis The structural and functional degradation of the corpus luteum (also known as luteal regression), which occurs at the end of the luteal phase of both the estrous and menstrual cycles in the absence of pregnancy, resulting in a decrease in progesterone secretion. Luteolysis also occurs at the end of pregnancy in the lead up to parturition. The mechanisms of luteolysis differ between species, but in many species it is a consequence of the production of prostaglandin F2α by the ovaries (during the normal ovarian cycle) or by the uterus (at the end of pregnancy).

Luteotrophic Action of stimulating the corpus luteum.

Magnocellular neuron Large neuroendocrine cells located in the supraoptic and paraventricular nuclei of the hypothalamus and project to the posterior pituitary, where they secrete neuropeptides into the circulation.

Maternal sensitivity The ability to respond in a temporally and emotionally appropriate way to the offspring's needs.

Metestrus See estrous cycle.

Milk let-down Release of milk due to oxytocin-induced contraction of myoepithelial cells in mammary gland tissue.

Morris Water Maze A test of spatial learning for rodents that relies on distal cues to navigate from start locations around the perimeter of an open swimming arena to locate a submerged escape platform.

Mouse junctional zone Region of the rodent placenta where cells with an endocrine function are located.

Mouse spongiotrophoblast Major endocrine lineage of the mouse placenta located within the junctional zone. The spongiotrophoblast synthesizes most prolactin family members, including Prl3b1, which encodes placental lactogen.

Mouse syncytiotrophoblast In the mouse placenta there are two adjacent syncytiotrophoblast layers which separate and regulate material exchanges between maternal and fetal circulation. The syncytiotrophoblast is not a key source of placental hormones in mice.

Mouse trophoblast giant cells These cells are defined by their giant nuclei with a high degree of polyploidy thought to enable large-scale protein synthesis. In mice there are six subtypes: primary parietal trophoblast giant cells, secondary parietal trophoblast giant cells, spiral artery trophoblast giant cells, channel trophoblast giant cells, sinusoidal trophoblast giant cells, and canal trophoblast giant cells. These cells are a source of placental hormones in rodents.

Multiparity, multiparous Having given birth more than once; producing multiple offspring (more than one).

Myometrium The middle layer of the uterine wall, consisting mainly of uterine smooth muscle cells that control uterine contractile activity.

Naive In this context, naïve rodents are virgin and have no reproductive experience.

Negative energy balance When energy expenditure is greater than energy intake, it leads to body mass loss due to using body tissue to meet the energy deficit.

Neurotransmitter Chemical messengers used by the nervous system for neuron to neuron or neuron to muscle communication.

Neurobiotin A water-soluble fluorescent neuroanatomical tracer.

Neurogenesis The process by which new neurons are formed in the brain.

Neurohormone Hormone that is secreted by a neuroendocrine cell.

Neuronal nuclear protein (NeuN) A protein which is a homologue to the protein product of a sex-determining gene in *Caenorhabditis elegans*; a neuronal nuclear antigen that is commonly used as a biomarker for neurons.

Neuroplasticity The ability of the nervous system to change its activity in response to intrinsic or extrinsic stimuli by reorganizing its structure, functions, or connections; see brain plasticity.

Neuroactive steroids Steroids (synthesized in the brain or periphery) that modulate neuronal excitability by rapid non-genomic actions.

Neurosteroids Steroids synthesized within the brain that modulate neuronal excitability by rapid non-genomic actions.

Noradrenergic Nerves that synthesizes and releases noradrenaline (also called norepinephrine).

Nulliparous A female that has not given birth to offspring.

Objective memory A term used to describe memory that is measurable using standardized tests.

Orexigenic Having a stimulating effect on appetite.

Orthodromic See antidromic.

Ovariectomy Surgical removal of the ovaries.

Paced mating An experimental model in rodents where the sexually receptive female is able to control mating interactions with a male, typically by providing a separated chamber only accessible to the female. Under these conditions, the female can approach and withdraw from the male, controlling the timing of receipt of mounts, intromissions and ejaculations. This is thought to enable a full array of female reproductive behaviors, rather than allowing the male to dominate interactions, which can be stressful for the female.

Paraventricular nucleus Nucleus in the hypothalamus that contains large neurosecretory (magnocellular) neurons that project to the posterior pituitary gland and secrete oxytocin and vasopressin, as well as parvocellular neurons that project to the median eminence and other brain regions.

Parturition, parturient The action of giving birth to young; childbirth.

Parvocellular division of the hypothalamic paraventricular nucleus (pPVN) Parvocellular refers to the small size of the neurons compared to the larger magnocellular neurons, also located in the PVN. The axons of the parvocellular neurons in the PVN project to the median eminence, where they release their hormones into the hypophyseal portal system.

Perikaryon Neuronal cell body containing the cell nucleus.

Peripartum The last month of gestation and the first few weeks after delivery.

Pharmacological adrenalectomy A drug regime used to inhibit glucocorticoid synthesis and induce functional adrenalectomy (removal of the adrenal glands).

Pinnipeds A clade of semiaquatic, usually marine mammals commonly known as seals and sea lions.

Placenta A fetally derived and transient organ of pregnancy that develops from the trophectoderm of the blastocyst. The placenta is attached to the endometrial lining of the uterus and functions in the transport of nutrients and waste, as a barrier between the maternal and fetal circulation and to synthesize placental hormones.

Placental lactogen The key hormone of mammalian pregnancy, responsible for inducing maternal metabolic adaptations required to support fetal growth and development of the mammary gland for lactation. In humans, placental lactogen is encoded by *chorionic somatomammotropin hormone 1* and 2. In rodents, placental lactogen is encoded by *prolactin family 3, subfamily d, member 1* (*Prl3d1*; previously named *PL-I*) and *prolactin family 3, subfamily b, member 1* (*Prl3b1*; previously named *PL-II*). Infusion experiments identified this hormone as being critical for the induction of maternal behavior in rodents.

Placentation The formation of the placenta, the organ that allows for maternal–fetal transfer during pregnancy.

Pleiotropic hormone When one hormone influences two or more seemingly unrelated functions.

Positive energy balance When energy intake is greater than energy expenditure, leading to gain in body mass due to storage of excessive energy mostly in the form of adipose tissue.

Postpartum Period of time after birth.

Precocial Refers to young born in a relatively developed state physiologically and behaviorally compared to the young of other species.

Pregnancy See gestation.

Primiparity, primiparous Having given birth only once or borne only one litter of offspring.

Proestrus See estrous cycle.

Proinflammatory cytokines A type of signaling molecule (cytokine) that is secreted from immune cells such as helper T cells and macrophages, and certain other cell types that promote inflammation.

Prostaglandin Eicosanoids that stimulate inflammatory processes and increase uterine contractions during labor and increase the breakdown of cervical connective tissue leading to dilation.

Prostaglandin-endoperoxide synthase type-2 The rate-limiting enzyme in the formation of prostaglandins from arachidonic acid. The type-2 enzyme is highly inducible (with the expression of type-1 enzyme more constitutive).

Polysialylated neuronal cell adhesion molecule (PSA-NCAM) A marker of developing and migrating neurons and of synaptogenesis in the immature vertebrate nervous system.

Pseudopregnancy A condition in females that physiologically/behaviorally resembles pregnancy in the absence of an implanted embryo or fetus. An important phenomenon in rodents, as it can be induced by the mating stimulus in the absence of fertilization (for example using a vasectomized male), because the mating stimulus provides a surrogate "maternal recognition of pregnancy" in these species with a short luteal phase.

Psychogenic stress Negative stimulus involving mental or emotional components that is perceived as stressful (e.g., fear, threat). Also called psychological or neurogenic stress.

Reference memory A type of long-term memory; a spatial task, it mimics two aspects of episodic memory, namely the "what" (content) and "where" (place) dimensions of an event.

Salience Property of a stimulus that has relative importance related to the context in which it occurs.

Secretagogue A substance that promotes secretion of another substance.

Social learning A theory of learning how to process social behavior which proposes that new behaviors can be acquired by observing and imitating others.

Spatial reference memory A form of long-term memory representing the spatial, contextual, and factual aspects of a task that remains constant between trials.

Spatial working memory The ability to keep spatial information active in working memory over a short period of time.

Subgranular zone A brain region in the hippocampus where adult neurogenesis occurs.

Subjective memory A term used to indicate how individuals interpret, feel, or think about their own memory—in other words, one's perceptions regarding memory performance.

Synaptic reorganization Includes progressive and regressive changes on branches of dendritic and/or axonal processes that accompany the formation and elimination of synapses.

Synaptogenesis The formation of synapses, which are the points of contact where information is transmitted between neurons.

Syncytialization The formation of a syncytium.

Syncytiotrophoblast The outer trophoblast layer of the placenta after the membranes of many cells of the cytotrophoblast have fused to form a syncytium layer.

Syncytiotrophoblast The placental barrier between maternal and fetal blood that allows exchanges in nutrients and gases and also represents the endocrine tissue of the human placenta. Formed from the trophectoderm of the developing blastocyst.

Syncytium A tissue layer in which many of the cell membranes have fused, leading to very large multi-nucleated cells.

Systemic stress An immediate and real threat to an organism's homeostasis (e.g., dehydration, extreme temperature, hemorrhage, hypoxia, infection). Also called physiological or physical stressor.

Tonic A state of sustained or continuous activity.

Trier Social Stress Test (TSST) A procedure used to reliably induce stress in human research subjects that involves a presentation and mental arithmetic component in the presence of an unknown audience of judges.

Trimester An arbitrary division of the duration of pregnancy into three portions of approximately 3 months each in humans. The first trimester (from time of last menstrual period until 12 weeks of pregnancy) is predominantly focused on embryonic development, while the second and third trimester are associated with fetal growth and maturation.

Trophectoderm/trophoblast A specialized lineage that develops early in pregnancy to generate the trophoblast giant cells and the chorioallantoic placenta.

Tuberoinfundibular dopamine (TIDA) neurons A population of neuroendocrine dopamine neurons found in the arcuate nucleus of the hypothalamus, which project to the median eminence and release dopamine as a hormone into the pituitary portal blood. These have a hypophysiotrophic action (maintaining the structure and function of the hypophysis, or pituitary gland) to inhibit secretion of prolactin.

Ultradian A recurrent period or cycle repeated throughout a 24-h day with a period shorter than a day.

Ungulates A clade of mammals, most of which have hooves. Includes sheep, goats, pigs, horses, and deer, among other animals.

Verbal memory Broad concept that refers to memory for verbally presented information.

Working memory The small amount of information that can be held in mind and used in the execution of cognitive tasks.

Printed by Printforce, the Netherlan